SANITARY
& DRAINAGE

Peter Wenning

PLUMBING SKILLS Series

SANITARY & DRAINAGE

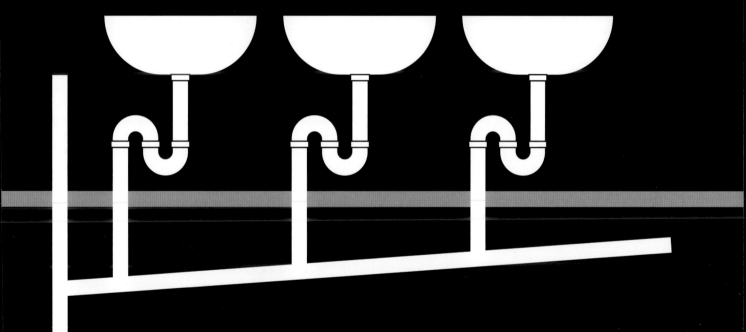

3e

Sanitary and Drainage
3rd Edition
Peter Wenning

Head of content management: Dorothy Chiu
Content manager: Chee Ng / Sandy Jayadev
Senior content developer: James Cole / Raphael Solarsh / Stephanie Davis
Senior project editor: Nathan Katz
Cover designer: Chris Starr (MakeWork)
Text designer: Linda Davidson (Original design by Norma Van Rees)
Permissions/Photo researcher: Liz McShane
Editor: Anne Mulvaney
Proofreader: James Anderson
Indexer: Max McMaster
Art direction: Linda Davidson
Typeset by KnowledgeWorks Global Ltd.

Any URLs contained in this publication were checked for currency during the production process. Note, however, that the publisher cannot vouch for the ongoing currency of URLs.

This third edition published in 2021

© 2021 Cengage Learning Australia Pty Limited

For product information and technology assistance,
in Australia call **1300 790 853**;
in New Zealand call **0800 449 725**

For permission to use material from this text or product, please email **aust.permissions@cengage.com**

National Library of Australia Cataloguing-in-Publication Data
ISBN: 9780170424899
A catalogue record for this book is available from the National Library of Australia.

Cengage Learning Australia
Level 7, 80 Dorcas Street
South Melbourne, Victoria Australia 3205

Cengage Learning New Zealand
Unit 4B Rosedale Office Park
331 Rosedale Road, Albany, North Shore 0632, NZ

For learning solutions, visit **cengage.com.au**

Printed in China by 1010 Printing International Limited.
6 7 25 24

BRIEF CONTENTS

CONTENTS

Guide to the text

As you read this text you will find a number of features in every chapter to enhance your study of plumbing and help you understand how the theory is applied in the real world.

PART- AND CHAPTER-OPENING FEATURES

Refer to the **Unit list** for an outline of the units covered in each part.

PART 1

PLANNING AND PREPARATION

- **Unit 1** Worksite preparation and maintenance
- **Unit 2** Excavation and trench support
- **Unit 3** Levelling
- **Unit 4** Planning the layout for a residential external sanitary drainage system
- **Unit 5** Planning the layout for a residential sanitary plumbing system
- **Unit 6** Weld polyethylene and polypropylene pipes using fusion method
- **Unit 7** Locating and clearing blockages

1

WORKSITE PREPARATION AND MAINTENANCE

Identify the key concepts you will engage with through the **Learning objectives** at the start of each chapter.

Learning objectives

This unit provides information needed to prepare, maintain and drain the worksite. Areas addressed in this unit include:
- safety and environmental considerations
- considerations when coordinating the work with other trades
- preparing the worksite before excavation
- locating stormwater drainage points
- removing water from a worksite by gravity through stormwater and sub-soil drainage systems
- removing water from a worksite by pumping
- installing submersible and non-submersible-type pumps

For a comprehensive study of stormwater drainage, read this unit in conjunction with Unit 11, 'Installing stormwater and sub-soil drainage systems'.

FEATURES WITHIN CHAPTERS

Actively engage with the learning by completing the practical activities in the **NEW Learning Task** boxes.

LEARNING TASK 1.1

Research submersible pumps and find the brand and model number of a single-phase submersible pump that can deliver dirty water and has a float switch.

Caution boxes highlight important advice on safe work practices for plumbers by identifying safety issues and providing urgent safety reminders.

From Experience boxes explain the responsibilities of employees, including skills they need to acquire and real-life challenges they may face at work, to enhance employability skills on the job site.

Green Tip boxes highlight the applications of sustainable technology, materials or products relevant to plumbers and the plumbing industry.

Example boxes highlight a theoretical or practical task with step-by-step walkthroughs.

The **Know your code** icons highlight where Plumbing Standards are addressed to strengthen knowledge and hone research skills.

 The primary purpose of worksite preparation and maintenance is to provide safe conditions for workers and the public, and to protect the environment from harmful discharges.

FROM EXPERIENCE

Before preparing the worksite, it is important to read the project plans and specifications to identify any areas of proposed building works. You will also need to communicate with your supervisor regarding the location of temporary site storage and facilities.

GREEN TIP

The main issues that you must consider are minimising soil erosion and preventing the escape of soil sediment from the worksite.

EXAMPLE 5.2

We now follow Example 5.1, and add ventilation to the graded discharge pipe. In the fully vented modified system of plumbing, group vents are provided for each 10 fixtures or part thereof. In our example, following the previous section, there are only four fixtures, and so only one group vent is required. Refer to the table in AS/NZS 3500.2, 'Size of group vents'.

 AS/NZS 3500.3 PLUMBING AND DRAINAGE. STORMWATER DRAINAGE

 NEW

GET IT RIGHT

PROVIDE FOR EXPANSION

FIGURE 5.12 PVC junction fitting with cracks

Failure to provide for thermal movement can result in pipe or fitting fracture and leaks. In Figure 5.12, a sweep junction to a vent in elevated pipework has fractured and leaked, leading to a building dispute and causing an accommodation building to be vacated.

GET IT RIGHT

5

As you work through the chapter, reinforce the practical component of your training with the **NEW Get It Right** section.

At the end of each chapter you will find several tools to help you to review, practise and extend your knowledge of the key learning objectives.

Review your understanding of the key chapter topics with the **NEW Summary**.

SUMMARY

- Worksite preparation and maintenance are important to protect buildings and other properties and to make working conditions on-site safe and efficient. Among other things, this includes providing for drainage of the worksite.
- When preparing for the work, you will need to have the plans and specifications at hand to plan out where to drain any excess water and where to lay drains and safely store materials and equipment.
- Preparing and excavating the site involves knowing the design building and site surface levels to lay drains with the correct size, gradient and depth of cover. Pits must be of adequate dimensions. Erosion and sediment controls will assist in controlling and diverting surface run-off with clean water to the legal points of discharge.

- When providing temporary drainage and pumping of stormwater, you will need to check site levels and position pumps at a location and depth to be able to effectively drain the worksite. The pump must be suitable and capable of discharging dirty water at adequate flow rates, and it will need to protect against silt-laden water entering any stormwater drains.
- Any drainage system for permanent pumping must be installed to comply with AS/NZS 3500.3 and be correctly commissioned and installed to provide for future maintenance.
- Always ensure that you clean up as you go and dispose of rubbish. Ensure tools are always serviceable, maintained and securely stored.

PART 1

1

Worksheets give you the opportunity to test your knowledge and consolidate your understanding of the chapter competencies.

WORKSHEET 1

To be completed by teachers
Student competent ☐
Student not yet competent ☐

Student name: _____

Enrolment year: _____

Class code: _____

Competency name/Number: _____

Task: Review the section 'Preparing for work', then complete the following.

1   Stormwater drainage plans are usually prepared for new building projects. What are five pieces of information that can be found on the plan?

a   _____

b   _____

c   _____

d   _____

e   _____

2   Complete the following: All workers installing plumbing and drainage must be aware of the contract requirements – in particular, the _____ _____ and _____ _____ .

WORKSHEETS

1

Worksheet icons indicate in the text when a student should complete an end-of-chapter worksheet.

COMPLETE WORKSHEET 1

Guide to the online resources

Cengage is pleased to provide you with a selection of resources that will help you prepare your lectures and assessments. These teaching tools are accessible via cengage.com.au/instructors for Australia or cengage.co.nz/instructors for New Zealand.

MINDTAP

Premium online teaching and learning tools are available on the *MindTap* platform – the personalised eLearning solution.

MindTap is a flexible and easy-to-use platform that helps build student confidence and gives you a clear picture of their progress. We partner with you to ease the transition to digital – we're with you every step of the way.

The *Cengage Mobile App* puts your course directly into students' hands with course materials available on their smartphone or tablet. Students can read on the go, complete practice quizzes or participate in interactive real-time activities. MindTap is full of innovative resources to support critical thinking, and help your students move from memorisation to mastery!

The *Series MindTap for Plumbing* is a premium purchasable eLearning tool. Contact your Cengage learning consultant to find out how *MindTap* can transform your course.

SOLUTIONS MANUAL

The **Solutions Manual** provides detailed answers to every question in the text.

MAPPING GRID

The **Mapping Grid** is a simple grid that shows how the content of this book relates to the units of competency needed to complete the Certificate III in Plumbing.

INSTRUCTORS' CHOICE RESOURCE PACK

This optional, purchasable pack of premium resources provides additional teaching support, saving time and adding more depth to your classes. These resources cover additional content with an exclusive selection of engaging features aligned with the text. Contact your Cengage learning consultant to find out more.

COGNERO® TEST BANK

A **bank of questions** has been developed in conjunction with the text for creating quizzes, tests and exams for your students. Create multiple test versions in an instant and deliver tests from your LMS, your classroom, or wherever you want using Cognero. Cognero test generator is a flexible online system that allows you to import, edit, and manipulate content from the text's test bank or elsewhere, including your own favourite test questions.

POWERPOINT™ PRESENTATIONS

Cengage **PowerPoint lecture slides** are a convenient way to add more depth to your lectures, covering additional content and offering an exclusive selection of engaging features aligned with the textbook, including teaching notes with mapping, activities, and tables, photos and artwork.

ARTWORK FROM THE TEXT

Add the **digital files** of graphs, pictures and flow charts into your learning management system, use them in student handouts, or copy them into your lecture presentations.

FOR THE STUDENT

MINDTAP

MindTap is the next-level online learning tool that helps you get better grades!

MindTap gives you the resources you need to study – all in one place and available when you need them.

In the MindTap Reader, you can make notes, highlight text and even find a definition directly from the page.

If your instructor has chosen MindTap for your subject this semester, log in to MindTap to:

- Get better grades
- Save time and get organised
- Connect with your instructor and peers
- Study when and where you want, online and mobile
- Complete assessment tasks as set by your instructor.

When your instructor creates a course using *MindTap*, they will let you know your course link so you can access the content.

Please purchase *MindTap* only when directed by your instructor.

Course length is set by your instructor.

PREFACE

Sanitary plumbing and drainage is inextricably linked to public health. In order to sustain health, we must have access to safe drinking water and sanitary facilities. In modern economies, our plumbing standards, work practices, materials and products have evolved considerably; however, the basic objectives remain timeless. The basic objectives and performance requirements, fundamental to the work of plumbing practitioners, can be found in the National Construction Code, Volume 3, Plumbing Code of Australia, which is produced and maintained by the Australian Building Codes Board (ABCB).

In general terms, the Plumbing Code of Australia refers to AS/NZS 3500 Plumbing and drainage as the standard for the installation and alterations to plumbing systems. This is one step towards achieving a uniform and efficient regulatory environment for plumbing across Australia and New Zealand. The author understands that not all jurisdictions have fully adopted AS/NZS 3500 as an acceptable standard for plumbing, and that there are some state/territory regulatory variations. It may be necessary for teachers/trainers to amend and/or append information to suit local regulatory variations. In this book, the reader will find many references to AS/NZS 3500. The reason is to prompt use of, and encourage familiarity with, what is effectively a regulatory reference book. Also, to develop employability skills, the learner must acquire a culture of using regulatory reference books as they are constantly updated. At the end of each chapter there are theoretical and practical tasks to enable the learner to demonstrate competency.

The author encourages teachers/trainers to apply exam conditions to learners completing the theoretical tasks. It is also best practice to engage with learners during practical tasks while testing underpinning knowledge. Also in this book, the reader will find regular references to work health and safety (WHS), environmental and quality assurance requirements. The author encourages teachers/trainers to integrate these into the context of the subject matter in class discussion.

The title of this book is *Sanitary and Drainage*. An objective in writing this book was to address relevant AQF Certificate II and III level competencies, and not to confuse the learner with unpredictable problems or advanced concepts. Another objective was to provide the learner with additional basic information and references to encourage further study and continuing professional development. It is the author's desire that these objectives will be achieved, and that this book is easy to use.

Peter Wenning
Wenning Technical Services
http://www.wenning.com.au

ACKNOWLEDGEMENTS

I wish to thank plumbing teachers who gave feedback in the preparation of this book. Their helpful suggestions concerning some aspects of the technical content were drawn from their broad experience in the plumbing trade. In particular, I would like to thank Gary Cook for providing technical content and advice.

Thanks to Cengage for the opportunity to be involved with this book and for the continued support.

Cengage would like to thank the following reviewers for their feedback:
Grant Hasite, South Regional TAFE
Dean Carter, TAFE NSW
Tony Pingnam, TAFE NSW
Tony Backhouse, TAFE NSW
Tom Colley, Victoria University.

UNIT CONVERSION TABLES

TABLE 1   Length units

Millimetres mm	Metres m	Inches in	Feet ft	Yards yd
1	0.001	0.03937	0.003281	0.001094
1000	1	39.37008	3.28084	1.093613
25.4	0.0254	1	0.083333	0.027778
304.8	0.3048	12	1	0.333333
914.4	0.9144	36	3	1

TABLE 2   Area units

Millimetre square mm^2	Metre square m^2	Inch square in^2	Yard square yd^2
1	0.000001	0.00155	0.000001
1 000 000	1	1550.003	1.19599
645.16	0.000645	1	0.000772
836 127	0.836127	1296	1

TABLE 3   Volume units

Metre cube m^3	Litre L	Inch cube in^3	Foot cube ft^3
1	1000	61024	35
0.001	1	61	0.035
0.000016	0.016387	1	0.000579
0.028317	28.31685	1728	1

TABLE 4   Mass units

Grams g	Kilograms kg	Pounds lb	Ounces oz
1	0.001	0.002205	0.035273
1000	1	2.204586	35.27337
453.6	0.4536	1	16
28	0.02835	0.0625	1

TABLE 5   Volumetric liquid flow units

Litre/second L/sec	Litre/minute L/min	Metre cube/hour m^3/hr	Foot cube/minute ft^3/min	Foot cube/hour ft^3/hr
1	60	3.6	2.119093	127.1197
0.016666	1	0.06	0.035317	2.118577
0.277778	16.6667	1	0.588637	35.31102
0.4719	28.31513	1.69884	1	60
0.007867	0.472015	0.02832	0.01667	1
0.06309	3.785551	0.227124	0.133694	8.019983

TABLE 6   High pressure units

Bar bar	Pound/square inch psi	Kilopascal kPa	Megapascal mPa	Kilogram force/ centimetre square kgf/cm^2	Millimetre of mercury mm Hg	Atmospheres atm
1	14.50326	100	0.1	1.01968	750.0188	0.987167
0.06895	1	6.895	0.006895	0.070307	51.71379	0.068065
0.01	0.1450	1	0.001	0.01020	7.5002	0.00987
10	145.03	1000	1	10.197	7500.2	9.8717
0.9807	14.22335	98.07	0.09807	1	735.5434	0.968115
0.001333	0.019337	0.13333	0.000133	0.00136	1	0.001316
1.013	14.69181	101.3	0.1013	1.032936	759.769	1

TABLE 7   Temperature conversion formulas

Degree Celsius (°C)	(°F − 32) × 0.56
Degree Fahrenheit (°F)	(°C × 1.8) + 32

TABLE 8   Low pressure units

Metre of water mH_2O	Foot of water ftH_2O	Centimetre of mercury cmHg	Inches of mercury inHg	Inches of water inH_2O	Pascal Pa
1	3.280696	7.356339	2.896043	39.36572	9806
0.304813	1	2.242311	0.882753	11.9992	2989
0.135937	0.445969	1	0.39368	5.351265	1333
0.345299	1.13282	2.540135	1	13.59293	3386
0.025403	0.083339	0.186872	0.073568	1	249.1
0.000102	0.000335	0.00075	0.000295	0.004014	1

PART 1

PLANNING AND PREPARATION

- **Unit 1** Worksite preparation and maintenance
- **Unit 2** Excavation and trench support
- **Unit 3** Levelling
- **Unit 4** Planning the layout for a residential external sanitary drainage system
- **Unit 5** Planning the layout for a residential sanitary plumbing system
- **Unit 6** Weld polyethylene and polypropylene pipes using fusion method
- **Unit 7** Locating and clearing blockages

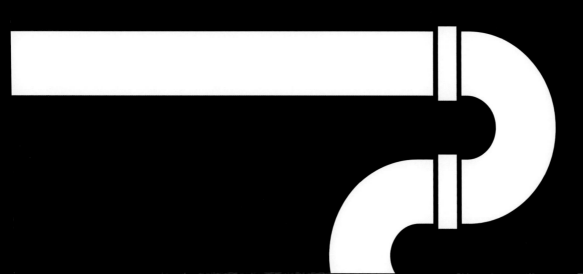

1

WORKSITE PREPARATION AND MAINTENANCE

Learning objectives

This unit provides information needed to prepare, maintain and drain the worksite. Areas addressed in this unit include:

- safety and environmental considerations
- considerations when coordinating the work with other trades
- preparing the worksite before excavation
- locating stormwater drainage points
- removing water from a worksite by gravity through stormwater and sub-soil drainage systems
- removing water from a worksite by pumping
- installing submersible and non-submersible-type pumps.

For a comprehensive study of stormwater drainage, read this unit in conjunction with Unit 11, 'Installing stormwater and sub-soil drainage systems'.

Introduction

Preparation and maintenance of the worksite is an essential part of safe and efficient work practice. You must be able to control access to the site and recognise and remove hazards. Also, you must be able to control and remove excess water from the site, through either permanent or temporary means such as stormwater and sub-soil drainage systems. It may be necessary to install pumps suitable for pumping unscreened roof water, sub-soil water and surface water.

Site drainage is used to move water from a catchment area to another location where it can be discharged in such a way as not to create a nuisance. The reasons for this may be to:

- increase the stability of the ground around a building or property
- avoid any run-off water entering a neighbouring property
- stop any catchment water ponding on the ground after rain periods
- prevent damage to buildings or roads caused by excess water making the ground waterlogged and unstable
- enable pipe laying and associated works in waterlogged or high water table areas.

This unit refers to both domestic and commercial worksites. The application may be a new worksite, or an existing structure being renovated, extended, restored or maintained.

At the end of this unit, you will be required to complete tasks by answering questions relating to worksite preparation and maintenance, and applying what you have learned in a practical task.

Before studying this subject, you should have completed training in work health and safety (WHS) as outlined in Chapter 3 of *Basic Plumbing Services Skills*.

Preparing for work

Plans and specifications

FROM EXPERIENCE

Before preparing the worksite, it is important to read the project plans and specifications to identify any areas of proposed building works. You will also need to communicate with your supervisor regarding the location of temporary site storage and facilities. This will ensure that your worksite preparation and maintenance does not interfere with proposed works.

Before attempting any work on-site, you should refer to the project plans and specifications. These can be obtained from the owner or project manager. If there are no plans and specifications, such as for restoration or maintenance works, or works not involving building construction, the minimum requirements will be that you comply with the relevant regulations and AS/NZS 3500 Plumbing and drainage.

AS/NZS 3500 PLUMBING AND DRAINAGE

A preliminary site inspection will assist you in identifying any differences between the plans and the site conditions, and in gathering information on levels and options for drainage.

During the progress of the work, you will need to refer to the plans and specifications. You may frequently need to query the owner/project manager about changes to the plans, and seek approval for any variations. Such variations from the original plans may need to be recorded. Some projects may require that you submit an as-constructed plan when the work has been completed.

Stormwater drainage plans are usually prepared for new building projects. The information typically required for stormwater drainage plans can be found in AS/NZS 3500.3 Plumbing and drainage. Stormwater drainage.

AS/NZS 3500.3 PLUMBING AND DRAINAGE. STORMWATER DRAINAGE

You should have a plan of the proposed building site showing details of the following:

- size, layout, location relevant to north, fall of the land (surface levels or contours), street number, suburb, etc.
- catchment area of the roof and paved areas
- position of downpipes and catchment pits
- layout of drainage and pipe sizing
- connection point at street, sump well, absorption pits (soak-away hole). (These are trenches or holes dug in permeable soil and filled with aggregate or gravel that allows the water to soak into the surrounding ground.)

Safety and environmental requirements

The primary purpose of worksite preparation and maintenance is to provide safe conditions for workers and the public, and to protect the environment from harmful discharges.

The safety of all workers and visitors to the site may depend on your awareness and recognition of, and adherence to, the relevant workplace and state/territory WHS requirements. *You need to be fully aware of these requirements and regulations.* Each state and territory has WHS legislation and regulations that require employers and self-employed people to identify hazards, and assess and control risks, in the workplace in consultation with their workers before undertaking any work in the construction industry. Prosecutions can result from breaches of WHS legislation or regulations.

The WHS requirements can include:

- obtaining and adhering to material safety data sheets (MSDS) for hazardous substances that may be required, including solvent cement and primer
- ensuring power tools and leads are tested and tagged as required
- the use of protective clothing and equipment, including safety goggles or glasses, boots, gloves, hard hats, overalls, masks and respirators
- the safe use of tools and equipment, powered excavation equipment, hand excavation equipment, measuring equipment, electrical power tools and temporary de-sludging pumps
- methods of manual handling in the workplace
- workplace induction
- the identification of site-specific hazards and the use of hazard controls (e.g. overhead electrical wires and in-ground electrical cables)
- the use of safe work method statements (SWMS) and job safety analysis (JSA)
- methods of handling and disposing of different types of waste
- the availability and use of fire-fighting equipment
- site fencing to prevent or control public access and pedestrian traffic.

In order to work safely, you must be able to recognise and control or remove hazards associated with: using tools, plant and equipment; uneven or unstable terrain; trees; pits; poles; trip hazards; dirt mounds; underground services; building structure and site facilities; dangerous materials; recently filled trenches; elevated work platforms; traffic control; working in proximity to others; and worksite visitors and the public.

Quality assurance

The quality assurance requirements for worksite preparation and maintenance include:

- the specific quality procedures of the employer and principal contractor
- individual state/territory regulations and standards
- compliance with the contract plans, specifications and the work program
- inspection and testing.

These are broad descriptions, and the specific details will depend upon the nature of the work, the type of project and the local authority's requirements.

Ensure that you are absolutely clear on the requirements of any contract, whether it is verbal or in writing. It is best practice to have every contract in writing, with both parties signing and receiving a copy. All workers installing plumbing and drainage must be aware of the contract requirements – in particular, the project plans and specifications. Whenever there is a change or variation to the plan or specifications, workers should return the old plans and specifications, and be issued with the new version.

Sequencing of tasks

For the smooth running of any project, it is important to communicate with the project supervisor and other trades working around you. Tasks at the worksite should be planned and completed in the correct sequence. Proper sequencing of tasks will ensure that the project is completed efficiently and progressively, with minimum delays. This is especially important when your work involves earthmoving machinery and other plant. Other people who can be affected by the sequencing of tasks are listed in Table 1.1.

TABLE 1.1 Working with other trades

Builder	The builder is the most important person in relation to planning the installation of a stormwater system. The program of construction will allow you to organise a work schedule to fit in with all the other contractors.
Bricklayers	Bricklayers can have packs of bricks positioned to suit their own needs, so the position of packs can create problems. Completing earthworks before bricks are delivered may save time.
Electricians	Electricians may need to put in underground cables. It can be an advantage to know where, and at what depth, these cables are to run.
Roof workers	It is normal for the stormwater drainage to be in position before the roof work is commenced, so as to catch any run-off water after the roof installation. The roof worker may need access around the building site to lift roofing materials or crane packs of roofing iron up to higher levels. It is important, therefore, that all stormwater drainage is laid and backfilled prior to there being any traffic around the site.
Public	Worksites are potentially hazardous for inexperienced persons, and so access should be controlled by the erection of fencing. The worksite must be fenced and locked when work has stopped for the day. If services may be affected during work, appropriate notification must be given to those concerned.

Tools and equipment

Tools of the plumbing trade are described in Chapter 6 of *Basic Plumbing Services Skills*.

Consideration should be given to whether to hire or purchase any necessary tools or equipment, including personal protective equipment (PPE), that is not already owned by the company or available for use on-site.

Tools and equipment that may be required for worksite preparation and maintenance must be free of defects and serviceable, and include those listed in Table 1.2.

TABLE 1.2 Tools and equipment

Hand and mechanical excavation equipment (see Unit 2)	Backhoe/excavator, bobcat, trenching machine, jackhammer, pick, shovel, crowbar
Hand and power tools	Angle grinder, electric jackhammer, shovels, picks, masonry drills
Trench shoring equipment (see Unit 2)	Closed timber shoring, soldier sets, metal trench shields
Measuring equipment	Rulers, tape measures
Levelling equipment (see Unit 3)	Laser levels, boning rods, spirit levels, string line and plumb-bob
Formwork	Different-sized formwork for the construction of pits, sumps or wells
Personal protective equipment (PPE)	Overalls, work boots, gloves, hats, raincoats, breathing apparatus for working in a well or for cutting concrete, eye and hearing protection
Lifting and handling equipment	Hand trolleys, rollers, forklifts, chain blocks, slings, hoists, jacks, cranes *Note*: All lifting equipment is tested to specific design limits. These load limits must not be exceeded.

Work area

Worksite preparation and maintenance will often involve excavation by machine. Before excavation, preparatory work should be done to avoid potentially costly delays in the operation of machinery. These preparations include determining the requirements for the task.

The types of stormwater site drainage systems you might use are:

- pipe system to the street gutter or main stormwater drain
- pipe system to a sump with pumps and rising main to street main drain

- pipe system to an absorption, or soakage, pit
- on-site stormwater detention (OSD) system
- temporary drainage to remove surface or groundwater, generally using pumps during site ground works.

The permanent systems mentioned above are discussed in more detail in Unit 11.

Before machinery arrives on the worksite, you should prepare by:

- determining the in-fall and outfall points
- determining the layout of pipework, including the width and depth of trenches
- marking out the trenches on the ground with lime or something similar
- determining the width of the machinery and ensuring there is enough room to access various areas of the site
- determining whether the excavation requires trench shoring, as this will need to be organised prior to the commencement of excavation
- ensuring that materials that are required have been ordered, received, checked and stored in a safe location
- ensuring that pipe bedding material has been delivered and positioned to allow easy access
- determining where excavated soil may be stockpiled, so as not to disrupt other work.

 COMPLETE WORKSHEET 1

Identifying installation requirements

The location of the stormwater connection points, including any sumps, wells and pumps, should be noted from the project drawings and specifications. You will then be able to determine the quantity and types of materials that are required.

It may be necessary to excavate within the property, and in some cases out in a public area such as a road or footpath.

Several points will need to be considered, including:

- the depth of the trench
- the location of underground services and obstructions
- the disruption to public footpaths and roadways (see Unit 2)
- permits required from relevant authorities (e.g. Road Opening Permits – see Unit 2)
- working with others within the building schedule.

Determining the location of work

It is now practical to determine whether the system can be installed from the inlet connection points to the suggested outfall point.

Inlet connections to the stormwater drains include:

- downpipe outlets
- pre-cast grating channels or pits
- sub-soil drains that collect groundwater for the protection of foundations or to stop water from ponding.

The types of sumps and wells include:

- stormwater pits
- inlet pits
- arrestors
- inspection chambers
- wet wells
- sump pits.

Stormwater outlets may connect to:

- the local council's reticulated stormwater drains
- street kerbs and channels
- absorption pits and soak-aways (soakage pits).

Note that stormwater must only be discharged to a location approved by the relevant local authority. This is often referred to as the legal point of discharge (LPD).

Materials and material quantities

Identifying materials

The materials for stormwater pipes and fittings must be approved for installation as listed in AS/NZS 3500.3, bearing the relevant standards marking. For example, PVC (polyvinyl chloride) fittings are labelled either AS/NZS 1254 or AS/NZS 1260. The full names of these standards are: AS/NZS 1254 PVC-U Pipes and fittings for stormwater and surface water applications and AS/NZS 1260 PVC-U pipes and fittings for drain, waste and vent application.

AS/NZS 3500.3 PLUMBING AND DRAINAGE. STORMWATER DRAINAGE

AS/NZS 1254 PVC-U PIPES AND FITTINGS FOR STORMWATER AND SURFACE WATER APPLICATIONS

AS/NZS 1260 PVC-U PIPES AND FITTINGS FOR DRAIN, WASTE AND VENT APPLICATION

Materials for stormwater installations are discussed in detail in Unit 11.

Ordering materials

Unit 4 provides an example of calculating material quantities.

When ordering and purchasing stormwater pipes and fittings, it is good practice to plan ahead for secure storage in a place where they will not be damaged or stolen. After you have confirmed that they are approved and comply with the standards and local regulations, they can be ordered from plumbing suppliers from your materials list.

Collecting and checking materials

When receiving the delivery of stormwater pipes and fittings, you should count the number and type of items and compare them with your materials list order. Ensure that you have identified all of the items as correct before accepting the order. If you have the wrong materials or damaged items, it may take some time to exchange and order the correct acceptable items.

Sustainable work practices

After collecting the materials, they must be stored so that pipes are straight and supported. This will ensure that they do not become deformed while stored. Fittings should be stored in a way that they are not damaged by other heavy materials and objects.

To minimise costs and waste, always think ahead and plan out what lengths of pipe, fittings and other materials are required. For short lengths of drain, usually it is best to use lengths of pipe that have already been cut.

 COMPLETE WORKSHEET 2

Preparing and excavating the site

The procedures for marking out for installation of stormwater drains are outlined in Unit 11.

Preparing the pits

If the site is carefully prepared and the trenches and pits are neatly excavated, the installation of the drains, sumps and wells will be simplified, resulting in savings on time and materials. To properly prepare and excavate, you should:

- set out the pits and trenches
- determine the depth of the pit
- determine the size of the pit.

The pits should be set out in close proximity to the locations stated on the plan. You must also consider, however, the position of existing underground services, such as electricity, phone and water.

The main types of pits used with a stormwater installation are:

- inlet pits
- stormwater pits.

Inlet pits are installed to permit collection and ingress of ground and surface water to a stormwater drain. They can be installed at low points in roads or paved areas to collect the surface water.

Stormwater pits allow access for junctions, changes in direction and grade of the pipework, as well as for maintenance of the drain in the event of any blockage problems.

The set-out of pits on-site will be determined by several factors. These are:

- depth of the pit
- size of the pit
- type of pit
- changes in direction of pipework
- intersection of main drains
- location of the pit and type of cover required.

Depth of the pit

The depth of a pit can be determined by the depth at which the pipework will be installed. This can be worked out after the levels of the site have been established.

Size of the pit

The minimum size of a stormwater or inlet pit can be determined by the depth at which it will be installed,

and by referring to AS/NZS 3500.3, where a table gives the minimum internal size of the pit based on the depth. For example, based on AS/NZS 3500.3, if a rectangular pit is 700 mm deep, the minimum internal dimensions must be 600 × 600 mm (see Figure 1.1).

AS/NZS 3500.3 PLUMBING AND DRAINAGE. STORMWATER DRAINAGE

Type of pit

The type of pit used will be determined by where water enters the pit, the type of water, and where it is installed (e.g. roadside curb, paved area or in a road surface).

Changes in direction

Pits are used to change the direction of larger pipework and are an access point for inspecting and maintaining the drainage system.

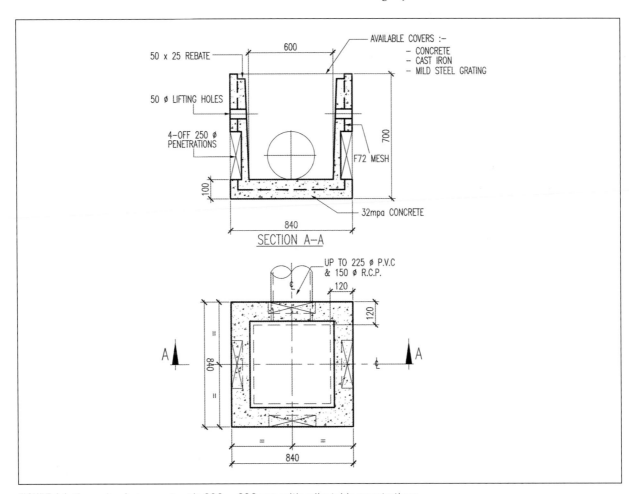

FIGURE 1.1 Example of stormwater pit 600 × 600 mm with adjustable penetrations

Source: Reproduced with permission of Frankston Concrete Ltd.

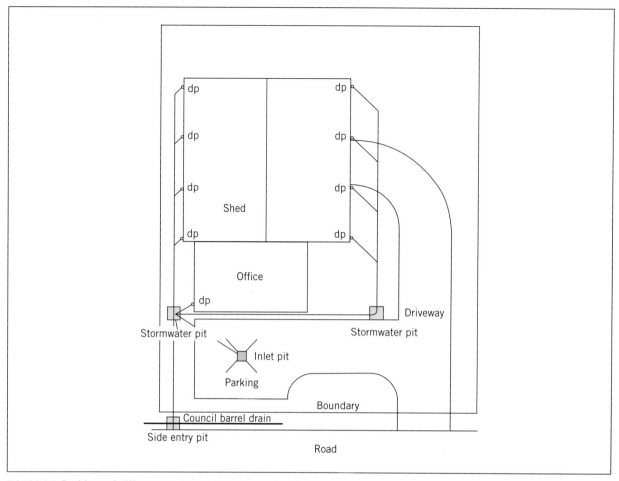

FIGURE 1.2 Positions of different types of stormwater and inlet pits

Source: Department of Education and Training (http://www.training.gov.au) © 2013 Commonwealth of Australia.

Intersection point

It is common practice to have an inspection pit at the intersection of the main stormwater drain to determine who has the responsibility for clearance of blockages.

Type of cover

There are generally three different classes of pit cover. The appropriate cover will depend on the location of the pit within the property. The classes are:

- light-weight: paved and grassed areas
- medium-weight: light vehicular traffic
- heavy-weight: heavy vehicular traffic.

While avoiding clashes with the level and position of existing services, select and mark out on the ground, with lime or marking paint, the most direct route for trenches for the backhoe or excavator to follow. Avoid changes of direction. **Figure 1.2** indicates positions of different types of pits.

Determining the minimum pipe cover

Stormwater drains are generally buried under the ground, so for the protection of the drain it is important to make sure they have the correct depth of cover.

The depth of a drain can be affected by:

- the type of vehicular traffic

- the type of cover over the drain (paved or road surface)
- whether it is in a private or public area.
This information can be obtained from AS/NZS 3500.3.

AS/NZS 3500.3 PLUMBING AND DRAINAGE. STORMWATER DRAINAGE

Determining the minimum gradient

The gradient of a stormwater drain must be sufficient for the drain to be self-cleaning, enabling the flushing out of leaves and debris. The minimum gradient should conform to AS/NZS 3500.3.

AS/NZS 3500.3 PLUMBING AND DRAINAGE. STORMWATER DRAINAGE

As a guide, the minimum gradient for DN 90 to DN 150 stormwater drains in Australia is 1:100, or 1%.

To calculate the fall for any gradient, use the formula:

$$\text{Fall} = \frac{\text{Gradient}}{100} \times \frac{\text{Length}}{1}$$

For the example of a DN 100 drain length of 30 m:

$$\text{Fall} = \frac{1}{100} \times \frac{30}{1}$$

$$\text{Fall} = \frac{30}{100}$$

$$\text{Fall} = 0.3 \text{ m}$$

Taking site levels

Before the set-out of trench depths can be calculated, it is necessary to take site levels to determine whether the block you are working on is sloping in one direction or another. Site levels will also indicate if there is enough cover and fall to run a stormwater drain to an outlet point, or whether you will have to install a sump or well with a pumped system.

This is also a good time to check if the level and position of other underground services is likely to clash with the proposed level and position of the stormwater drain.

Levelling is covered in detail in Unit 3.

Excavation

The excavation of trenches and pits requires levels to be taken. The gradient (fall) of the trenches must be calculated to provide at least the minimum gradients as specified in AS/NZS 3500.3.

AS/NZS 3500.3 PLUMBING AND DRAINAGE. STORMWATER DRAINAGE

The excavation of trenches and pits needs to be planned prior to machinery being brought on-site. Working out how deep to dig the trenches and pits will save time and money.

The following factors need to be considered:

- site levels
- minimum pipe cover
- inlet points
- outlet point
- location of existing services and obstructions
- minimum gradient of the drain
- size of the pits.

Depending on the depth of the trenches and the type of ground, shoring equipment may be required. Trench shoring may be required if the trenches are more than 1.5 m deep or the condition of the ground is unstable and there is a risk the sides may collapse (see Unit 2). Attention must also be given to the types of machinery that may be used near a trench that could vibrate the ground and cause problems of earth movement and potential trench collapse. Trench shoring is used to provide a safe working environment for workers to lay drains and complete associated work.

While the excavation is being carried out, the correct depth of the trench floor in relation to the required gradient in the drain must be checked.

Pipe sizing

Pipe sizing should be calculated by a qualified person to meet load requirements. The minimum diameter sizes are set out in AS/NZS 3500.3. As a guide, where allotments are less than 1000 m², the minimum drain diameter is DN 90.

AS/NZS 3500.3 PLUMBING AND DRAINAGE. STORMWATER DRAINAGE

Installation of stormwater and sub-soil drains

Procedures for installing stormwater drains are outlined in Unit 11.

Sedimentation controls

Erosion and sediment controls must be considered when any excavation or earthmoving takes place. These controls are put in place to address site stability and environmental concerns. Some considerations are listed in AS/NZS 3500.3; however, you will need to comply with local authority requirements and take into consideration rainfall and other site-specific factors, such as slope and surface conditions.

AS/NZS 3500.3 PLUMBING AND DRAINAGE. STORMWATER DRAINAGE

GREEN TIP

The main issues that you must consider are minimising soil erosion and preventing the escape of soil sediment from the worksite.

Temporary drainage and pumping stormwater

Preparation and maintenance of the worksite may require either a permanent or temporary drainage solution. The purpose is to remove water from the site during construction. The preferred solution is to install a permanent drainage system that takes advantage of the natural drainage of the site by gravity; however, you will need to protect against silt-laden water entering the stormwater drain.

When designing a stormwater system, you will need to take levels to verify that drainage of the construction site by gravity is achievable. You will also need to determine whether there are any obstructions that may affect drainage by gravity.

If the worksite cannot be drained by gravity, one solution is to install a pumped system. This may be either a pre-packaged or cast in-situ pump system.

GET IT RIGHT

DRAIN THE WORKSITE

FIGURE 1.3 View of residential backyard in winter. Water is ponding heavily on the site near the new building

In Figure 1.3, the roof eaves gutters are not connected to stormwater drains.

1   Should the stormwater drains have been laid before the roof gutters were installed?

_____

_____

_____

2   Does the placement of soil piles in this situation assist in draining the worksite?

_____

_____

_____

3   You might be surprised that, in this situation, the building slab and walls developed large cracks within three years. This occurred because the worksite was not drained. Do you think it matters whether or not preparation is made for draining the worksite?

_____

_____

_____

Sub-soil drainage

If drainage by gravity below ground is possible, then an alternative to pumping is sub-soil drainage. This is outlined in detail in Unit 11. Sub-soil drains collect the water flowing through the permeable upper soil levels. They are commonly referred to as agricultural, or 'aggie', drains. They usually consist of a slotted or perforated pipe installed in a trench, which is then filled with aggregate and topped off with a thin layer of soil.

Geotextile material is used to limit the amount of fine silt/sand entering the drain. Further information on sub-soil drainage systems can be found in AS/NZS 3500.3.

AS/NZS 3500.3 PLUMBING AND DRAINAGE. STORMWATER DRAINAGE

Wet wells and sumps

Sumps are usually regarded as small, portable-type pump pits, and may be constructed of plastic materials (see Figure 1.4).

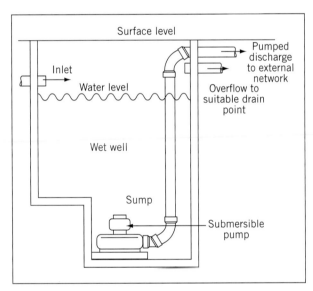

FIGURE 1.4 Example of small stormwater sump with submersible pump

Source: Department of Education and Training (http://www.training.gov.au) © 2013 Commonwealth of Australia.

Wet wells are usually regarded as large, permanent pump pits (see Figure 1.5).

It is preferable that wet wells for pumping are constructed from concrete, because of its strength and ability to be moulded to suit any installation size and shape. The thickness of the concrete can be varied to suit the depth of the installation.

Other benefits of concrete are that it is sound and will generally resist corrosion from contaminated groundwater and soils. For deep installations, concrete

structures have a greater resistance to heave and deformation than some plastic units.

Cast in-situ concrete

Cast in-situ concrete wet wells are constructed on-site and made to suit the capacity of the installation. Arrangements for pumps and pipelines can be included in the construction.

The main disadvantages of in-situ installations are:
- the well is generally at the lowest point of the block, and any rain will flow to this point during the construction
- time is needed to construct the sump or well.

Once the excavation has been completed, the formwork is assembled and fixed before the concrete walls and bottom are poured and formed. The dimensions of the well will vary depending on the capacity requirements (discussed later in the unit).

The base must be constructed on stable ground and consist of compatible materials to the walls. The base should maintain a self-cleaning gradient towards the pump inlet, as required by AS/NZS 3500.3. Any sump or well deeper than 1.2 m must be fitted with an access ladder installed in accordance with AS/NZS 3500.3.

AS/NZS 3500.3 PLUMBING AND DRAINAGE. STORMWATER DRAINAGE

Pre-cast concrete

Pre-cast sumps or wells are manufactured in specified sizes. This means you may be installing in a sump that is over-sized for the installation.

The main problems associated with pre-cast installations are:
- the sump needs to be carted to the site and lowered into position
- a crane is needed to install the sump in the desired location
- the pre-cast sump riser components may not suit the installation depth requirements.

Sizing

The capacity of a permanent wet well is determined by adding the capacity of the pump and storage required. The storage is the wet well volume between the high and low working levels of the well.

The storage must not be less than the volume of the run-off from a storm of average recurrence interval (ARI) of 10 years and duration of 120 minutes, or as otherwise directed by the relevant local authority.

The minimum storage of a permanent stormwater wet well is measured in cubic metres and should be no less than 3 m³.

The storage capacity must be determined in accordance with AS/NZS 3500.3. An example

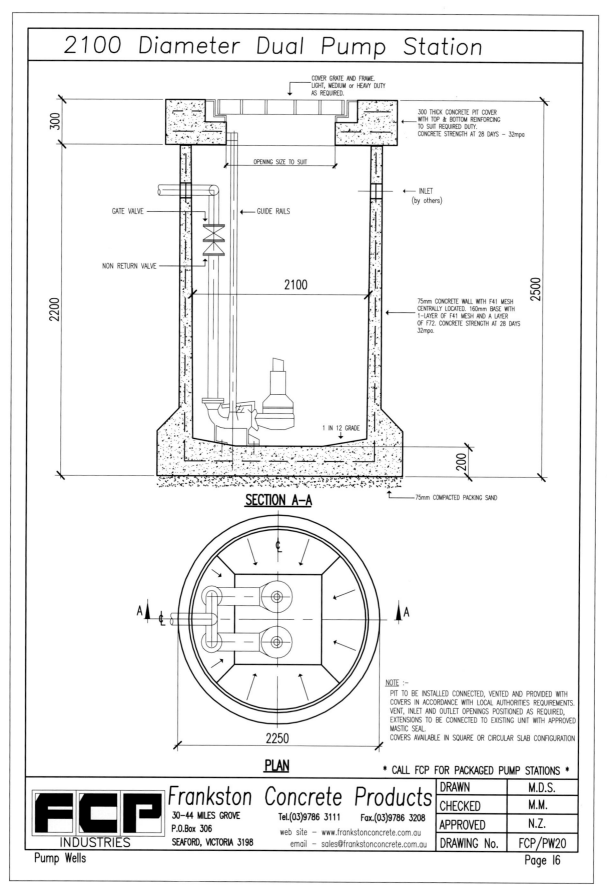

FIGURE 1.5 Example of large concrete stormwater wet well with dual submersible pumps

Source: Reproduced with permission of Frankston Concrete Ltd.

calculation to size a pump system can also be found in AS/NZS 3500.3.

AS/NZS 3500.3 PLUMBING AND DRAINAGE. STORMWATER DRAINAGE

Covers

A cover or grating will be fitted over the wet well at the proposed final surface level. Covers or openings should be constructed of similar material to the sump or well and be large enough to allow access for maintenance purposes.

If the cover is airtight, a breather pipe with a non-corrosive screen should be installed. Most covers for sumps, wells and pits are a cast shell that will be filled with concrete when the installation is completed to finished ground level. The type of cover will depend on the location of the pump well. If the cover is in an area where there is:

- light traffic – that is, walking traffic is allowed – a light-weight cover will be required
- medium traffic – that is, cars can pass over the access cover – a medium-weight cover will be required
- heavy traffic – that is, trucks can pass over the access – a heavy-weight cover will be required.

All of the above situations may require approval from the local authority.

Gratings fitted to admit the flow of water from a paved or graded area should be sized as specified in AS/NZS 3500.3. Where they are installed in a proposed paved area, they should be installed 5 mm below the proposed paving surface level to allow for drainage.

AS/NZS 3500.3 PLUMBING AND DRAINAGE. STORMWATER DRAINAGE

Access covers and grates should comply with AS 3996 Access covers and grates.

AS 3996 ACCESS COVERS AND GRATES

Pumps

Pumps may need to be installed to drain the pump well. Pump installations must comply with AS/NZS 3500.3.

The two types of pump systems commonly installed to drain the pump well are:

1 surface pumps
2 submersible pumps.

More information on pumping can be found in Unit 12.

Surface pumps are installed on the surface and have suction lines running near the base of the well.

The base of the suction line usually has a foot valve encased with a strainer to stop foreign matter from entering the foot valve and pump. The benefit of this type of installation is that the pump is above the ground and can be maintained in position. Also, there is less construction work to be done inside the well.

The disadvantages of this system are:

- the pumps and housing are visible above ground
- the foot valve on a pump's suction assembly can become blocked by foreign matter. This will not allow the foot valve to seal after a pumping cycle, and the pump may lose prime
- the installation process for the surface pump system will rely on two pumps being installed near the edge of the well. Each pump will need a suction line installed to the base of the well, with a foot valve and strainer installed to stop the entry of foreign matter.

Submersible pumps are installed inside the well and are the preferred option for pumped installations. The pump will usually be of the centrifugal type, where the already flooded pump forces liquid through the rising main to discharge at the desired location.

Some pumps may have a grinder mechanism at the pick-up point to assist in breaking up soft solid matter before water travels through the pump. A benefit of this type of pump is that the whole installation is below ground; the only thing visible from ground level should be the alarm system. Another major benefit is that any small pieces of debris will be broken down and pumped away.

The disadvantages of this system are:

- if maintenance to the pumps is required, the pump or pumps will need to be lifted to carry out the work
- there is more work required in the installation of pumps, pipes, electrical system, discharge pipes and location slides for the pump to operate on. Even so, this is the preferred option because this system works better and is visually more acceptable compared with surface pumps.

LEARNING TASK 1.1

Research submersible pumps and find the brand and model number of a single-phase submersible pump that can deliver dirty water and has a float switch.

Connection of pipework

All pipework for the stormwater drains and pump connections must be installed in accordance with the plans and manufacturer's requirements. The relevant authority may require an inspection of the drain and/or rising main before backfilling can commence.

Most stormwater systems fall within the jurisdiction of a controlling authority – a council, water authority or inspection service.

Inspection

The drains may need to be inspected to ensure they are in the location and arrangement as specified in the plan, are laid to the correct grades, and that the correct bedding and backfill materials have been used. Inspection by the authority is not compulsory in all areas, so you will need to research your local authority's requirements.

Testing

It will be necessary to perform a test on the pipe system to check for leaks, which can damage buildings or foundations.

The testing of stormwater drains must take place on all new installations, on repaired or altered sections of drain when they are located in and under buildings, and for rising mains on pumped systems.

There are two main tests that can be used:

1 water test
2 air test.

Testing procedures for stormwater drains and rising mains can be found in AS/NZS 3500.3.

AS/NZS 3500.3 PLUMBING AND DRAINAGE. STORMWATER DRAINAGE

Temporary drainage

While carrying out the installation of a drainage system, wet conditions, waterlogged ground or high water tables may be encountered on a site. A temporary drainage system of pumps and hoses will then need to be set up to drain away the water, which would otherwise impede operations. If the site cannot be drained using gravity, it will be necessary to set up temporary drainage at a low area on the site where any run-off water will gather, and then collect and pump it to a safe location in order to keep the work area dry.

For small-scale temporary site drainage, a submersible pump placed in a trench, sump or well is often adequate. For major water table lowering, specialist equipment is available, comprising a system of wells drilled or jetted into the ground beside where the excavation will take place. A system of header pipes carries the water to the pump and away. Filter material is added to the wells to filter groundwater through suction lines. For more detail, refer to Unit 11.

Preparing the suction inlet

The suction inlet point for the pump should be strategically placed to drain the maximum amount of water from the area of the works. The suction inlet point for the temporary system will need some form of protection, as there will be a lot of debris, such as mud, building debris and rubbish, around the worksite. A form of strainer will be required to protect the inlet point during installation of the temporary pump in the

sump well until the building site and earthworks are finished, and all debris is cleared.

Location of pumps

The location of the pumps may depend on various safety considerations. For example, if a petrol pump is used, there may be exhaust fumes; in that case the area must be well ventilated. If the pump is installed in an open area, some form of protection will be needed to protect it from building traffic, weather and vandals. If the pump relies on electricity, there must be a safe, protected route for the power lead.

Connecting pipes and hoses

A suitable outlet position needs to be selected for the drainage, before the suction and delivery hoses are connected to the pump. These hoses are generally flexible and will need to be positioned so that they are not damaged by traffic during construction on the site. They may be positioned along a fence line, or in an area where construction is not taking place. If the hoses are easily damaged, it is recommended that a temporary form of cover be used to protect the pipes. As this system is temporary, it should be checked regularly to ensure the pipe and outlet points have not been moved for some reason.

COMPLETE WORKSHEET 3

Drainage system operation

This section refers to permanent pumped systems.

Commissioning pumps

After the pumps have been installed, they need to be activated to lower the water level. The pump control system is adjusted to attain the desired water levels in the sump or well. This is a very important section of the installation, because the high and low working levels of the sump determine the capacity of the system.

The floats or switches that turn the pump on and off must be set to the calculated levels. The alarms for high and low levels must be set and checked against the overflow level and the minimum pump operating level.

As part of the commissioning process, it is recommended that the owner of the system be informed about its operation and the potential problems that can occur. Topics discussed with the owner might be the maintenance regime for all components within the system, from cleaning the spouting or guttering, and cleaning the collection pits, to removing any hard, foreign objects from the sump.

The main message is that the system must be regularly inspected and maintained to avoid entry of solids debris that could get into the system and cause it to malfunction.

Pump maintenance

To achieve the reliability required of pumps, they need to be maintained and serviced in accordance with the manufacturer's instructions. Apart from the regular checking of the pumps, the system will need regular monitoring as to the types of solid debris that goes through the system. All sorts of rubbish and debris, collected either from the roof or paved areas, can cause pump problems. Debris such as plastic bags, aluminium cans, dead birds and broken glass can make their way into the pipe system and cause problems with the operation of the pumps. If this is likely to happen, it may be necessary to fit some type of screen or arrestor in the pipework to protect the pumps. These factors will be determined by the location of the installation and can change over time (e.g. as trees grow and drop more leaves, and provide a habitat for birds).

Clean-up

Work area

Worksite preparation and maintenance is required for the health and safety of workers and the public, and to provide an environment where work can be carried out in an efficient manner. Building sites constantly change, and access to various areas is always required by workers and for deliveries. For these reasons, building sites are usually short of space. Site drainage and stormwater drainage work can take up a lot of the building site, as they require excavation and modification to the ground surface around buildings. Adverse weather and changing site conditions may necessitate changing the location of temporary drainage points, but drainage work is necessary to minimise wet ground conditions to make the site accessible and safe.

Most large sites will have site safety procedures and a schedule for cleaning up. Small sites, such as single dwellings, may not have procedures, but the same principles apply as for large sites.

The procedures could include the following:

- The disposal areas for site waste should be identified at site induction. If not, ask your trade supervisor.
- Clean up as you go, if possible, but at least on a daily and weekly basis and on completion of an installation.
- Food and drink waste, wrappers and containers should be disposed of in the correct bin, immediately after use, as build-up attracts vermin and creates associated health risks.

- Packaging should not be allowed to become a windblown, trip or other hazard. Dispose of it in the correct bin as soon as practicable, or take it with you from the site.
- Off-cuts and surplus material that may be used later in the job should be held in a safe storage area until required.
- Off-cuts and surplus material that are not suitable for re-use should be disposed of in the correct bin at the end of each day's work, or sooner if practicable.
- Do not use cupboards, empty rooms or corners to store rubbish and debris.
- After installing temporary drainage, check the site. Remove and store unused materials, and store tools and equipment in the correct location.
- Periodically check and maintain erosion control measures during construction to prevent silting of the stormwater drains, such as temporary or permanent silt traps, hay bales or surface stabilising materials.

Tools and equipment

The procedure for cleaning tools and equipment could include the following:

- Tools and equipment must be cleaned, maintained and stored.
- Tools and equipment may be the responsibility of the apprentice or worker, the contractor, or others such as hire companies.
- The apprentice or worker will be responsible for their own toolkit and PPE.
- The apprentice or worker must maintain a running check on their toolkit and PPE. Check, clean and maintain it at the end of each day.
- Check and maintain company equipment on a regular basis according to company policy, and keep all equipment in good working order.
- Clean, maintain and return hire equipment according to the hire agreement.
- Do not overload or abuse any tools, plant or equipment.
- Dispose of unserviceable tools and equipment that could be unsafe.

Documentation

This section is relevant to documentation required for worksite preparation and maintenance.

Generally, local council and regulatory authority permits are required for projects involving new buildings and building alterations and additions. You will need to research and familiarise yourself with local requirements, which vary around Australia.

In this book, various types of documentation are mentioned, such as:

- plumbing permits
- applications for new sewer and stormwater connections and alterations

- Road Opening Permits
- compliance certificates for completed plumbing work
- as-constructed plans of completed sewerage and/or stormwater works.

Councils or regulatory authorities may require notification when the project is complete in order to carry out an inspection or audit.

COMPLETE WORKSHEETS 4 AND 5

AS/NZS 3500.3 PLUMBING AND DRAINAGE. STORMWATER DRAINAGE

SUMMARY

- Worksite preparation and maintenance are important to protect buildings and other properties and to make working conditions on-site safe and efficient. Among other things, this includes providing for drainage of the worksite.
- When preparing for the work, you will need to have the plans and specifications at hand to plan out where to drain any excess water and where to lay drains and safely store materials and equipment.
- Preparing and excavating the site involves knowing the design building and site surface levels to lay drains with the correct size, gradient and depth of cover. Pits must be of adequate dimensions. Erosion and sediment controls will assist in controlling and diverting surface run-off with clean water to the legal points of discharge.

- When providing temporary drainage and pumping of stormwater, you will need to check site levels and position pumps at a location and depth to be able to effectively drain the worksite. The pump must be suitable and capable of discharging dirty water at adequate flow rates, and it will need to protect against silt-laden water entering any stormwater drains.
- Any drainage system for permanent pumping must be installed to comply with AS/NZS 3500.3 and be correctly commissioned and installed to provide for future maintenance.
- Always ensure that you clean up as you go and dispose of rubbish. Ensure tools are always serviceable, maintained and securely stored.

 **WORKSHEET 1**

Student name: _____

Enrolment year: _____

Class code: _____

Competency name/Number: _____

Task: Review the section 'Preparing for work', then complete the following.

1   Stormwater drainage plans are usually prepared for new building projects. What are five pieces of information that can be found on the plan?

    a _____

    b _____

    c _____

    d _____

    e _____

2   Complete the following: All workers installing plumbing and drainage must be aware of the contract requirements – in particular, the _____ _____ and _____ _____ .

3   Worksites are potentially hazardous for inexperienced persons. What should be done to prevent access to worksites by the general public?

    _____

    _____

4   Name two types of levelling instruments that can be used during worksite preparation and maintenance.

    a _____

    b _____

5   What condition must tools and equipment be in for worksite preparation and maintenance?

    _____

    _____

6   Where should pipe bedding material be positioned when it is delivered to the worksite?

    _____

    _____

WORKSHEET 2

Student name: _____

Enrolment year: _____

Class code: _____

Competency name/Number: _____

Task: Review the section 'Preparing for work', then complete the following.

1   What are six types of sumps and wells?

a   _____

b   _____

c   _____

d   _____

e   _____

f   _____

2   What is the common term for the authority-approved location of discharge for stormwater?

_____

_____

_____

3   Research 'Materials and products' in AS/NZS 3500.3.

a   What Australian Standards are applicable to polyvinyl chloride (PVC) pipes and fittings?

_____

_____

b   What Australian Standards are applicable to fibre-reinforced concrete (FRC) pipes and fittings?

_____

_____

c   What are the limitations on the use of FRC pipes and fittings?

i   _____

ii   _____

d   What is the required composition of cement mortar?

_____

_____

 **WORKSHEET 3**

Student name: _____

Enrolment year: _____

Class code: _____

Competency name/Number: _____

Task: Review the section 'Preparing and excavating the site', then complete the following.

1 Read AS/NZS 3500.0, and define the term 'stormwater pit'.

_____

_____

2 Name the two types of pits used in stormwater installations.

a _____

b _____

3 For a stormwater pit 0.9 m deep, what are the minimum required internal dimensions if it is to be rectangular?

_____

4 What is the minimum cover required for a PVC stormwater drain laid under a driveway without pavement?

_____

5 Calculate the minimum fall required for a DN 90 stormwater drain over a length of 25 m.

_____

6 What is the minimum diameter for a stormwater drain serving a building on an allotment less than 1000 m²?

_____

7 Research 'Erosion and sediment controls' under the Appendix containing General Information in AS/NZS 3500.3. Name four precautions that can be taken to minimise soil erosion and flow of sediment from the site that might occur due to uncontrolled site drainage.

a _____

b _____

c _____

d _____

8   Research 'Subsoil drains' in AS/NZS 3500.3. Name the five requirements for clean-out points for pipes or geocomposite drains

a _____

b _____

c _____

d _____

e _____

9   Read AS/NZS 3500.0 and define the term 'wet well'.

_____

_____

10 What are the requirements for construction of the base in a wet well?

_____

_____

11 Research AS/NZS 3500.3 'Pumped systems' and 'Wet wells'. Name two different types of pump that must be installed in accessible locations.

a _____

b _____

WORKSHEET 4

Student name: _____

Enrolment year: _____

Class code: _____

Competency name/Number: _____

Task: Review the sections 'Drainage system operation' and 'Clean-up', then complete the following.

1 What two levels determine the capacity of the pump system?

a _____

b _____

2 What information should be provided to the pump system owner to ensure the pump system is maintained in a satisfactory manner?

_____

_____

_____

3 What can be done to prevent debris, such as plastic bags and broken glass, causing problems with pumps?

_____

_____

_____

4 When considering maintenance of proper drainage of the worksite, what conditions might necessitate changing the position of temporary drainage points?

_____

_____

_____

5 Why is it necessary to periodically check and maintain erosion control measures?

_____

_____

_____

 **WORKSHEET 5**

Student name: _____

Enrolment year: _____

Class code: _____

Competency name/Number: _____

Task: Practical exercise

All questions in the previous four worksheets must be completed and checked by your trainer/teacher before you attempt the following practical exercise.

Your trainer/teacher should provide preliminary guidance, conduct an induction and ask underpinning knowledge questions.

Install a submersible pump or non-submersible pump to drain water from a pit, trench or excavation containing water to a depth of at least half a metre. The site must remain effectively drained for the duration of the job.

Note: There is flexibility in this exercise to choose the method of pumping, but you must be aware of the implications of running any pump dry (without water). For some pumps, this may cause permanent damage.

2

EXCAVATION AND TRENCH SUPPORT

Learning objectives

This unit provides information needed for safety in trenching operations and trench support.
Areas addressed in this unit include:

- preparation through locating underground services and lodging notices
- signage and traffic management
- inspecting ground conditions
- factors to consider when selecting a support system
- benched and battered excavations
- procedures for installing soldier sets, closed sheeting and shields
- causes of trench collapse.

Introduction

Excavation is necessary for the installation of underground services such as:

- sanitary drains
- stormwater and sub-soil drains
- wet wells
- pre-treatment facilities
- domestic treatment plants
- on-site disposal systems.

While there are many other services installed by plumbers and drainers, the above-listed services may need to be placed in deep excavations. Plumbers and drainers must be competent in shoring a trench to prevent the collapse of trench walls and to provide safe conditions for personnel working in the trench. In addition, a holistic approach must be taken and all risks considered, such as falls prevention and protection against falling objects.

This unit provides broad guidance for the installation of trench support and safe work within trenches. Individual state and territory regulations vary. *At the end of this unit, you will be required to research your local requirements, and apply them in a practical task.*

Before studying this subject, you should have completed training in work health and safety (WHS) as outlined in Chapter 3 of *Basic Plumbing Services Skills*.

Preparing for work

Plans and specifications

When planning and preparing for excavation work, it is necessary to obtain information as follows:

- location of all underground services for example, by contacting Dial Before You Dig (http://www.1100. com.au, or telephone 1100) (note that not all authorities participate with this organisation)
- plans and specifications that indicate the position of the excavation
- safe work method statement (SWMS) for any high risk construction work (HRCW)
- WHS regulations
- authority notification forms (if applicable)
- trench support, and other equipment manufacturer instructions
- organisation quality procedures (if applicable)
- relevant Australian Standards.

Safety and environmental requirements

The safety requirements for excavation and trench support centre on adopting a holistic approach. This means that you must be aware of different aspects of the work environment, and the movement of personnel in and around that environment, and then consider the associated risks before implementing control measures to prevent injuries.

Each state and territory has WHS regulations that require employers and self-employed people to identify hazards, and assess and control risks, at the workplace in consultation with their workers. All hazards must be identified before excavation and installation of trench support commences.

This unit will provide you with an awareness of both the hazards and safe work practices. For example, never climb on any trench support system. Only industrial grade portable ladders should be used to gain access to the trench floor.

The environmental requirements for excavation and trench support centre on protecting the environment from any contamination that might be caused by excavation operations. Occasionally, it is necessary to dewater trenches, as described in Unit 11. Only clean water may be discharged to the stormwater drainage system.

GREEN TIP

Some sites have contaminated or filled ground, and the nature of the contamination must be considered, not only for obvious reasons of health and safety, but also because contaminated water, soil and excavation refuse material can only be disposed of in a manner, and to a location, approved by the relevant state/territory authorities.

Signage and traffic management

When excavation involves disruption to vehicular or pedestrian traffic, whether on public or private property, it is necessary to plan and provide for safety and control of traffic.

Personnel involved in traffic control should have completed a recognised training course, and this may be compulsory, depending on the relevant local regulations.

When carrying out excavations in public areas, it will always be necessary to consult with the relevant local authority to determine the extent of disruption to vehicular traffic and pedestrians, and, where applicable, to have an approved plan for traffic control. As a guide, any work that requires opening of a road or pathway and disruption of pedestrian or vehicular traffic will require notification to the local authority.

Often a Road Opening Permit is required by the local (council) authority, and this must be obtained through lodgement of a fee prior to the commencement of work. The authority may also require evidence of adequate insurance before issuing any permit to work in a public area.

At the time of obtaining a Road Opening Permit, it is important to ascertain any special requirements and whether a traffic control plan is needed. Traffic control plans are required particularly where there will be disruptions to heavy vehicle or night traffic.

Special requirements may include notification to affected businesses, residents, and rail and road traffic authorities, and occasionally must include public newspaper advertisements.

In all cases where roadways and pathways are to be excavated, a job safety analysis (JSA) worksheet should be completed in consultation with all workers.

It is recommended that the following issues are addressed in preparing the JSA, as appropriate:

- The relevant regulations or Australian Standards providing guidance for traffic control are considered and an arrangement diagram is prepared.
- All workers to be involved with the job should acknowledge and sign the JSA.
- Hazards that may cause injury to those involved in the tasks or to other persons in the vicinity are identified.
- For each hazard, assess the level of risk, and then list the control measures required to eliminate or reduce those risks.
- Identify the personnel responsible for implementing the control measures.

The following are some examples of risks and control measures that should be considered prior to commencing work:

- Are employees capable and qualified to operate mobile plant? (National Certificates of Competency may be required for the operation of plant.) If not, training and experience is required before the job starts.
- Has the plant and equipment planned for use been serviced/calibrated, and is there evidence of servicing and calibration? If not, servicing and calibration are required before the job starts.
- Do pedestrians need to be isolated from the job and mobile plant? If so, barricading, fencing, witches hats, reflective tape, restricted zones or lockouts may prevent access.
- Are there distracting unrelated activities? If so, use job programming to keep unrelated activities apart.
- Are there existing high risks on-site, such as overhead power lines or excessive slopes? If so, eliminate the risks, or make plant operators aware of the risks.
- Are there traffic management issues? Control the risks associated with movement of plant and pedestrians (see Figure 2.1).

FIGURE 2.1 Pedestrian traffic control

Where the local authority requires a traffic management plan (see Figure 2.2), the following information may be required:

- the location and duration of the job
- the relevant regulations or Australian Standard providing guidance for traffic control, including a specific arrangement diagram
- the work area layout
- traffic volumes
- traffic destinations
- traffic prioritisation
- identification of potential 'black spots'
- traffic flow (i.e. shared, one way, divided way, priority traffic)
- separation of roadways, walkways, pathways and aisles
- signage and signalling
- speed limitations
- flaggers and spotters for controlling traffic in localised areas where a high risk exists, and equipping of flaggers with whistles, reflective vests, flags, torches, lights and other communication equipment
- reflective vests for all workers when working near powered mobile plant.

FIGURE 2.2 Road traffic control

Jobs in progress using the JSA and/or traffic control plan

Regardless of whether the municipal authority has requested a traffic control plan, it should be noted that the actual provision of resources and performance of the work must be in accordance with relevant Australian Standards and local authority regulations.

At any given time, the JSA and/or traffic control plan must be up-to-date, with any new employees signed as inducted, date changes listed and authority notifications recorded.

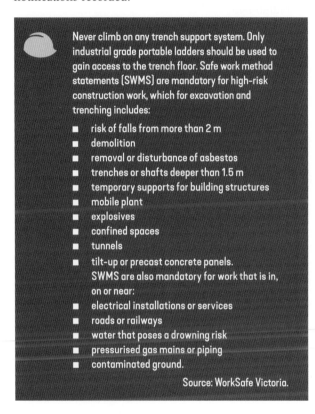

Never climb on any trench support system. Only industrial grade portable ladders should be used to gain access to the trench floor. Safe work method statements (SWMS) are mandatory for high-risk construction work, which for excavation and trenching includes:

- risk of falls from more than 2 m
- demolition
- removal or disturbance of asbestos
- trenches or shafts deeper than 1.5 m
- temporary supports for building structures
- mobile plant
- explosives
- confined spaces
- tunnels
- tilt-up or precast concrete panels.
  SWMS are also mandatory for work that is in, on or near:
- electrical installations or services
- roads or railways
- water that poses a drowning risk
- pressurised gas mains or piping
- contaminated ground.

Source: WorkSafe Victoria.

Quality assurance

The critical aspects of quality assurance in safe excavation and trenching operations are:

- safe operation of excavating machinery
- correct selection and assembly of shields and trench support systems.

To obtain a licence or certificate to do excavation work, you must complete one of the following qualifications:

- RII30815/RII30813/RII30809 Certificate III in Civil Construction Plant Operations
- BCC30607 Certificate III in Civil Construction (Plant Operation)
- BCC30603 Certificate III in Civil Construction (Plant Operation)
- BCC30198 Certificate III in Civil Construction (Plant)

- Qualification 2555 Certificate III in Civil Construction (Plant Operator).

To find organisations that deliver nationally recognised training, go to training.gov.au and search via the course name.

If you are an apprentice or trainee, you may not be able to apply for a licence or certificate, so check with your state or territory authority before operating an excavator.

There are very real risks of serious injury or death due to the lack of knowledge and skills in safe excavation and trenching operations. The following case study is just one documented case of injury.

Worker falls in trench (21 October 2019)

While putting pickets and bunting around an excavation on a construction site in Marsden Park, a 30-year-old labourer fell three metres into the trench, suffering fractures to his leg.

The investigation:

- SafeWork NSW inspectors responded to the incident
- SafeWork NSW commenced an investigation to determine the cause and circumstances of the incident.

Safety information:

Consider 'reasonably practicable' control measures to manage the risks of falling into an excavation at a workplace, including:

- developing a plan – in consultation with your workers and before work begins – on managing the risks of falls in and around excavations
- considering barrier fencing to further minimise accidental encroachment – perimeter barriers should be effective
- considering a suitable location for inspection, and access and egress for emergency situations
- consulting with an engineer to determine distances where shoring, benching and battering should be used – and monitor distances, particularly after rain
- using a landing platform – or scaffold towers for big excavations – to 'in-fill'
- preparing a safe work method statement when there is a risk of falling more than two metres – and ensure workers are inducted and supervised in its implementation
- backfilling or covering the excavation.

Source: Incident Information Release – Worker Falls in Trench, SafeWork NSW, 21 October 2019, https://www.safework.nsw.gov.au/

To avoid incidents such as that described in this case study, and to achieve an assurance of quality in safe excavation and trenching operations, each state and territory has strict regulations, and provides guidance for complying with the regulations.

Lodging documentation before commencing deep excavations

Some state/territory authorities require that a notice be lodged before commencing a deep excavation. For the purposes of this unit, a deep excavation may be regarded as a trench if the excavated depth is more than 1.5 m, or a shaft if the excavated depth is more than 2 m. It should be noted, though, that there are conditions and variations that determine whether notices are necessary.

At the end of this unit, you will be required to research your state/territory requirements.

Notification may not be required if the excavation will be part of building work for which a building permit has been issued and is in force.

Plant, tools and equipment

Safety equipment

An important safety consideration when carrying out excavation and trench support activities is the correct selection and use of safety equipment.

Types of personal protective equipment (PPE) that are used include:
- hard hat
- overalls
- boots
- safety glasses or goggles
- ear plugs or ear muffs
- gloves
- dust masks or respirators.

Individual state and territory regulations vary. Under some conditions, such as when connecting to a live sewer, a trench may be classified as a confined space, and in these circumstances you may be required to have confined space training and special safety equipment.

Tools

Tools and equipment typically used for excavation and trench support activities include:
- excavator/backhoe
- levelling equipment (see Unit 3)
- pick, shovel, crowbar
- measuring tape
- shoring systems (often available on hire)
- ladders and scaffolding
- wood panel saw and mitre box
- cordless drill with a range of bit tools for self-drilling screws
- communications equipment (e.g. two-way radio)
- screwdrivers
- shifting spanners
- string line
- plumb-bob
- spirit level
- claw hammer and duplex nails
- jemmy bar.

When moving trench support members, you might be required to use lifting or load-shifting equipment such as:
- cranes
- rollers
- slings and chains
- trolleys
- hydraulic jacks.

Special equipment – selection and operation

Special equipment may be required to assemble and install trench support members. Suppliers of proprietary trench support systems provide special equipment that must be used in accordance with their installation instructions. Examples include:
- special tools (e.g. lifting attachments)
- proprietary hydraulic jacking equipment.

 COMPLETE WORKSHEET 1

Excavation

Dangers of excavation

Excavation by machine must only be done by persons who have completed the appropriate training as required by individual state and territory regulations. It is mandatory that the operator must be free of alcohol, drugs or any condition that might put other workers at risk.

Excavation methods

Excavation methods include:
- trenching
- tunnelling
- shafts
- bulk excavations.

The nature of the excavation work being undertaken will affect the selection of excavation method and the safe system of work.

GET IT RIGHT

RECOGNISE DANGERS

FIGURE 2.3 Unsafe trench

The photograph in Figure 2.3 shows a man who could be digging his own grave. He is working at the bottom of a 4 m-deep trench. His hard hat provides very little protection. Excavated soil and rock has been piled near the edge of the trench, and rocks could easily fall, causing him permanent injury or death. Increasing the danger are ground vibrations from the excavator operating nearby. There is no ladder provided as a means of escape. Also, there is nothing to protect the worker from a cave-in; no sloping back of the trench walls, no shoring of the walls and no shielding of the worker. Unfortunately, this is a common situation in excavation work, where such dangers can be avoided by using safe excavation methods.

The trench is 4 metres deep. Why would you need to prepare an SWMS?

_____

_____

_____

_____

_____

Within metropolitan areas, ground conditions can vary considerably from suburb to suburb. Therefore, the potential hazards also vary. While new areas may have hazards such as rock or unstable ground, some older suburbs have hazards such as supporting foundations, tanks, basements and corroded pressurised pipes.

Trench

The following definitions for trench and tunnel are derived from SafeWork NSW January 2020 *Code of Practice Excavation Work*, p. 67.

A horizontal or inclined way or opening:

- the length of which is greater than its width and greater than or equal to its depth
- that commences starts at and extends below the surface of the ground, and
- that is open to the surface along its length.

Tunnel

An underground passage or opening that is approximately horizontal and starts at the surface of the ground or an excavation.

Design of tunnels

Safe tunnel construction depends on an adequate pre-construction engineering investigation of the ground and site and accurate interpretation of the information obtained.

Designers should:

- obtain or be provided with all available relevant information
- be advised of gaps in the information for planning and construction
- undertake or be involved in data acquisition for the site investigation program, and
- have on site involvement during the engineering investigation.

The information obtained from the engineering investigation and the anticipated excavation methods should be considered in preparing a tunnel design. The design should include:

- details on the tunnel dimensions and allowable excavation tolerances
- temporary and final support and lining requirements for each location within the tunnel
- details of expected tunnel drive lengths and shaft location, and
- other requirements for the finished tunnel.

Designers must also give the business that commissioned the design a written report that specifies the hazards relating to the design of the tunnel that, so far as the designer is reasonably aware:

- the hazards and design do not create a risk to the health or safety of persons who are to carry out any construction work on the tunnel, and

- the design is unique and not derived from other designs of the same type of structure.

The design should also include information on the excavation methods and ground conditions considered in the design. This will allow the design to be reviewed if another excavation method is chosen or the ground conditions differ from that expected as the excavation proceeds.

The design also needs to take into account the construction methods used to construct the tunnel so that a safe design for construction purposes is achieved.

Tunnelling hazards and risks

Common hazards and risks in tunnel construction generally relate to the confines of working underground including:

- tunnel stability – rock or earth falls and rock bursts
- changing ground conditions–strata and stress fluctuations
- limited space and access, with possible confined spaces involved
- air contamination or oxygen depletion
- fire or explosion
- the use of fixed and powered mobile plant
- the interaction of people and powered mobile plant
- temporary electrical supplies and circuits including loss of power for lighting and ventilation
- compressed air use and high pressure hydraulics
- large scale materials and equipment handling
- overhead seepage, ground and process water
- uneven and wet or other slippery surfaces
- falls of people or objects
- contaminated groundwater
- ground gas and water in-rush
- noise
- vibration
- heat and humidity
- ground loss or settlement at surface level, and
- hazardous substances.

Control measures include:

- ground support, for example tunnelling shields, mesh, rockbolts and shotcrete
- fall protection, for example temporary work platforms
- plant and vehicle traffic management systems
- regular plant maintenance
- pumps or dewatering systems to remove groundwater
- mechanical ventilation to control airborne contaminants and air temperature/humidity
- dust extraction
- plant fitted with water scrubbers
- plant fitted with catalytic converters, and
- providing breathing equipment when a hazardous atmosphere is present and cannot be effectively ventilated by external means.

Using ground support designed for the unique circumstances of the work is essential to control the risk of a collapse or tunnel support failure. All excavation for tunnelling should be supported.

Shafts

Shafts are often constructed to provide access or ventilation to a tunnel.

Comparatively shallow shafts can be sunk for investigating or constructing foundations, dewatering or providing openings to underground facilities.

Shafts vary greatly in design and construction technique, depending on their purpose and the local conditions. They may be vertical or inclined, lined or unlined, various shapes, and excavated using various techniques.

Shaft sinking involves excavating a shaft from the top, with access and spoil removal from the top. Other construction methods include raise-boring, which is a method of constructing a shaft or raise where underground access has already been established. Raised bored shafts can be from the surface or from one horizon to another underground. The method can be remotely executed, not requiring people to enter the shaft.

Access to shaft openings should be controlled by using a secure cover that is lockable and accessible only by a designated person. An alternative means is to use a suitable guardrail and toe board with a gate for access and supporting the sides by steel frames or sets of timber. In special cases support can also be provided by installing precast concrete or steel liners.

Shafts can have special features so design and construction advice should be obtained from a competent person, for example an engineer before excavation and installation. In some cases, special ventilation facilities may be required.

Common hazards and risks involved in shaft construction include:

- shaft dimensions limiting work space, possibly including confined space work
- the potential for ground instability for lifting and removing spoil
- falls and falling objects including fine material and water from the shaft wall
- hoisting equipment such as a winch, ropes and hooks
- hoisting and winching people, materials, spoil and plant
- water in-flow/in-rush and dewatering
- airborne contaminants and ventilation
- confined space
- manual tasks
- hazardous materials
- fire or explosion
- inadequate communication systems

- mobile plant
- noise, and
- emergency exits.
  Control measures include:
- stabilising the ground at the head of the shaft and removal of spoil
- continuously lining or supporting the shaft
- providing fall protection, for example temporary work platforms
- providing and maintaining hoisting equipment
- installing dewatering systems
- installing mechanical ventilation to control airborne contaminants and air temperature/humidity
- isolating access to moving parts of plant and equipment
- guiding the working platforms and material
- avoiding overfilling material kibbles and cleaning kibbles before lifting
- closing shaft doors before tipping, and
- cleaning the spillage off doors, stage and steelwork.
  Further guidance on confined spaces is available in the *Code of Practice: Confined spaces*.

Bulk excavations

Bulk excavations are often carried out when the construction design specifies large underground spaces (e.g. parking lots and basements – see **Figure 2.4**).

FIGURE 2.4 Bulk excavations

Common hazards and risks involved in bulk excavation work include:

- ground collapse
- undermining of the ground supporting neighbouring structures such as buildings, footpaths and roadways
- damage to buried pipes, conduits and drains (e.g. gas and sewerage pipes)
- falling into excavation trenches and holes (both people and equipment)
- drowning if the excavation floods due to open stormwater drains.

Control measures include engaging a suitably qualified person (e.g. a geotechnical engineer or civil engineer) to determine:

- ground structure
- type of ground support or retention systems
- methods of work for the installation of the ground support or retention system.

Control measures also include ensuring that:

- the work is managed by a competent person
- the engineer's design and procedure for using the ground support is followed
- all workers involved in the work are trained and have signed the SWMS
- work is not carried out in an area that ground support is not installed
- workers are aware of the emergency procedures
- the public is prevented from accessing the edge of the excavation or the work site.

Whilst the bulk excavation is open, the excavation area, ground support system and site security is regularly inspected. Rainfall and ground movement can affect the safety of an excavation.

Reducing the risk of ground collapse

The following section is based on SafeWork NSW 2020 *Code of Practice Excavation Work*, p. 49.

Ground collapse is one of the primary risks to be controlled in excavation work. Ground collapse can occur quickly and without warning, giving a worker virtually no time to escape, especially if the collapse is extensive.

Trench collapses of this nature can cause fatal injuries. A buried worker is likely to die from suffocation before help arrives, either because their head is buried, or their chest is so restricted by the weight of ground that the worker can no longer breathe.

Figure 2.5 shows a typical example of ground failure where material collapses onto a worker pinning them against the wall of a trench.

When planning the work and selecting appropriate excavation methods and control measures, it is important to consider:

- if the need for people to enter the excavation can be eliminated
- the type and strength of the material to be excavated, for example whether the ground is natural and self-supporting or has been previously backfilled
- the moisture content of the soil
- if the ground is level or sloping
- if groundwater is present
- if there are discontinuities or faults in the strata
- if there are other nearby watercourses, drains or runoff that might affect the stability of the excavation
- the work area and access or operational limitations

- the planned height of the excavated face
- if vehicle traffic and powered mobile plant will operate near the excavation
- if there will be other construction activity nearby that may cause vibration
- other loads adjacent to the planned excavation, for example buildings, tanks, retaining walls, trees, and
- underground essential services.

The ground conditions will have a significant impact on selecting an excavation method and the control measures implemented.

Ground conditions

In their natural condition, soils have varying degrees of cohesive strength and frictional resistance. Examples of materials with virtually no cohesive strength are dry sand, saturated sand and gravels with minimum clay content. Ground encountered in excavations can generally be categorised as one of three types:

- hard, compact soil
- soil liable to crack or crumble, and
- loose or running material.

Of these materials, hard compact soil is the type that can cause the most trouble because the face 'looks good' and this often leads to risks being taken. Loose or running material is often the safest, because the need for safety precautions is obvious from the start.

Soil liable to crack or crumble is doubtful and should be given careful consideration before the treatment to be given is determined. Useful information can often be obtained from local authorities.

Non-cohesive faces can be very hazardous. With the right amount of moisture they can look safe and solid. A little loss of water by evaporation from the face or an increase in water content from rain or other causes can make the soil crumble. The stability of any excavated face depends on the strength of the soil in the face being greater at all times than the stresses it is subjected to.

The following situations all increase soil stresses in an excavated face and may lead to failure under adverse weather conditions, additional load or vibration:

- deep cuts and steep slopes, by removal of the natural side support of the excavated material
- loads on the ground surface near the top of the face, for example excavated material, digging equipment or other construction plant and material
- shock and vibration, which could be caused by pile-driving, blasting, passing loads or vibration-producing plant
- water pressure from groundwater flow, which fills cracks in the soil, increases horizontal stresses and the possibility of undermining, and
- soil saturation, which increases the weight and in some cases the volume of the soil.

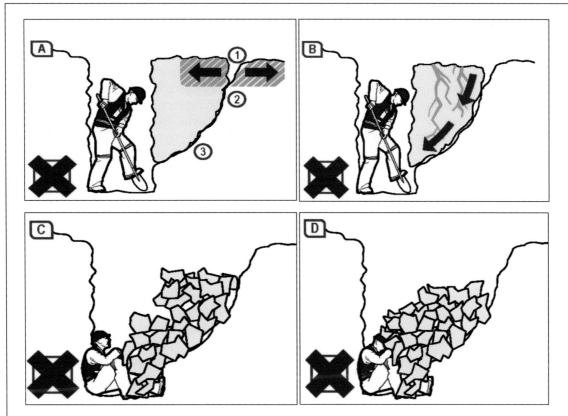

Figure A

This is a very dangerous situation, requiring ground support. No worker should be in a trench > 1.5 m deep unless support has been installed.

1. Area of tension, as all starts to collapse
2. Slipping plane
3. Seepage along the slipping plane further reduces the stability of the wall. Water seeping into the excavation, tension cracks on the surface and building side walls are all signs of imminent collapse.

Seepage in trench bottom may not be obvious until the actual collapse.

Figure B

Shear plane failure along the seepage (slippage) plane.

Figure C

Worker trapped and crushed against the trench wall by the quick collapse.

Figure D

Worker badly injured and probably smothered after being crushed against the opposite wall by the collapsing ground. The weight of a wedge of sand over a one metre length of trench two metres deep is about three tonnes.

FIGURE 2.5 Trench collapse and associated ground forces

Source: SafeWork NSW. Code of Practice Excavation Work, January 2020, p. 49.

The following may reduce soil strength and stability:

■ excess water pressure in sandy soil which may cause boils and saturate the soil and increase its plasticity
■ dryness of the soil may reduce cohesion in sandy soil and soils high in organic content which then crumble readily

■ prolonged stress, may cause plastic deformity (squeezing or flowing), and
■ prolonged inactivity at an excavation site.

A soil evaluation should be carried out before work restarts.

Controlling the risk of ground collapse

The following sections are based on SafeWork NSW *Code of Practice Excavation Work* January 2020 pp. 51 – 62.

There are three main types of ground collapse control measures that can be used where ground collapse may occur:

1 shielding – shields do not ensure ground stability but they protect workers inside the shield from ground collapse by preventing the collapsing material from falling onto them
2 benching and battering, and
3 positive ground support, for example shoring.

Trench shields

A shield is a structure, usually manufactured from steel, which is able to withstand the forces imposed by a ground collapse and to protect workers within it. Shields can be permanently installed or portable and designed to move along as work progresses. Many different shield system configurations are available for hire or purchase. **Figure 2.6** shows a typical trench shield.

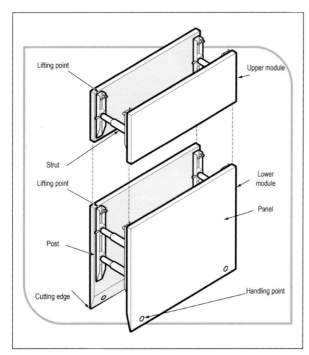

FIGURE 2.6 Typical trench shield

Source: SafeWork NSW, *Code of Practice Excavation Work*, January 2020, p. 62.

Shields and boxes used in trenches are often referred to as trench shields or trench boxes, and are designed and constructed to withstand the earth pressures of particular trench depths and ground types. They incorporate specific lifting points for installation and removal.

Trench shields and boxes differ from shoring as shoring is designed to prevent collapse, while shielding and boxes are only designed to protect workers if a collapse occurs.

Trench shields and boxes are useful where it is not reasonably practicable to install other forms of support.

They are mainly used in open areas where access is available for an excavator or backhoe to lower and raise the boxes or shields into and out of a trench. They are generally not suitable where access is difficult and ground conditions prevent the use of lifting equipment.

Steel boxes for trench work can be light or heavy duty construction depending on the depth of the trench and ground conditions. Trench shields and boxes should be designed by a competent person, for example an engineer and be pre-manufactured to job specific dimensions.

Used correctly, shields and boxes can provide a safe workspace for workers needing to enter an excavation. Trench shields and boxes should be maintained or they may fail unexpectedly, particularly if they have been abused or misused. The manufacturer's instructions for the installation, use, removal and maintenance of shields and boxes should always be followed.

Trench boxes should not be subjected to loads exceeding those which the system was designed to withstand. Earth pressures are reduced when correct benching and battering practices are used.

Shields and boxes should be stored and transported in accordance with the manufacturer's instructions. Heavy duty equipment may require disassembly for transport.

Boxes should be regularly inspected for damage. They should never be altered or modified without the approval of a qualified engineer.

Benching and battering

One fairly simple way of controlling the risk of ground collapse is to bench or batter the excavation walls. An excavated slope is safe when the ground is stable. That is, the slope does not flatten when left for a considerable period, there is no movement of material down the slope, and the toe of the slope remains in the same place.

If excavation work is planned to be carried out without positive ground support (that is, shoring), the continuing safety of the excavation will depend on the conditions arising during construction. If the conditions during construction are not as expected, or if conditions change during the course of the work, for example different soils, heavy rain/flooding, action should be taken immediately to protect workers, other persons and property. Implement control measures, for example temporarily suspending work until the ground is stable or, if necessary, providing positive ground support.

Benching is the creation of a series of steps in the vertical wall of an excavation to reduce the wall height and ensure stability (see **Figure 2.7**). Benching is a method of preventing collapse by excavating the sides of an excavation to form one or more horizontal levels or steps with vertical surfaces between levels.

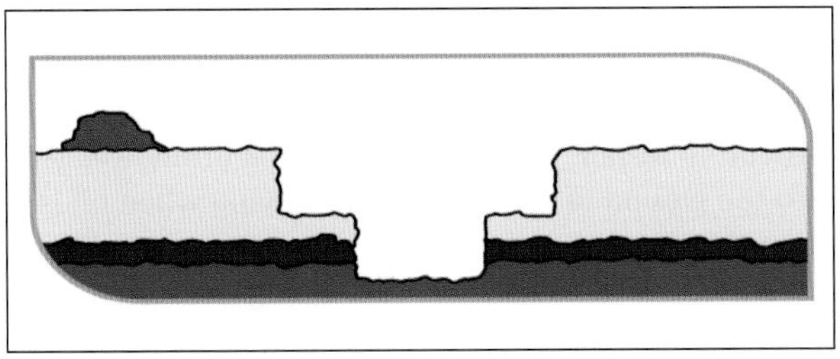

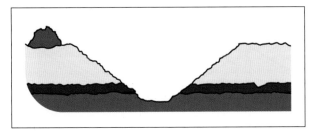

FIGURE 2.7 Benching

Source: SafeWork NSW, *Code of Practice Excavation Work*, January 2020, p. 51.

Battering is where the wall of an excavation is sloped back to a predetermined angle to improve stability (see **Figure 2.8**). Battering prevents ground collapse by cutting the excavated face back to a safe slope. Battering should start from the bottom of the excavation, and in some circumstances, it may be appropriate to use a combination of the two methods on an excavation (see **Figure 2.9**).

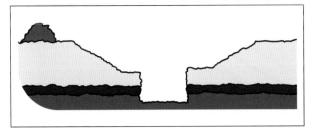

FIGURE 2.8 Battering

Source: SafeWork NSW, *Code of Practice Excavation Work*, January 2020, p. 52.

FIGURE 2.9 Combination of benching and battering controls

Source: SafeWork NSW, *Code of Practice Excavation Work*, January 2020, p. 52.

Benching and battering of excavation walls can minimise the risk of soil or rock slipping into the excavation. Control measures should be designed by a competent person, for example a geotechnical engineer, and be relative to the soil type, the moisture content of the soil, the planned height of the excavated face and surcharge loads acting on the excavated face.

It is not necessary to bench or batter the face of excavations which a competent person determines are in stable rock or has assessed there is no risk of collapse. When benching or battering the walls of an excavation, an angle of repose of 45 degrees should not be exceeded unless designed by a competent person and certified in writing.

Benches should be wide enough to stabilise the slopes and to prevent material from the top falling down to the working area. They should also be sloped to reduce the possibility of water scouring. The size and type of earthmoving machinery to be used, and related haul routes, should be considered when designing the face slopes and widths of benches.

Shoring

Shoring is a positive ground support system that can be used when the location or depth of an excavation makes battering and benching impracticable. It should always be designed for the specific workplace conditions by a competent person, for example an engineer.

Shoring is the provision of support for an excavated face or faces to prevent the movement of soil and therefore ground collapse. It is a common method of ground support in trench excavation where unstable ground conditions are often encountered, for example soft ground or ground liable to be wet during excavation, for example sand, silt or soft moist clay.

Where ground is not self-supporting and benching or battering are not practical or effective control measures, shoring should be used. Shoring should also be used when there is a risk of a person being buried, struck or trapped by dislodged or falling material which forms the side of, or is adjacent to, the excavation work.

Where such a risk also exists for those installing shoring, other control measures must be in place to ensure the health and safety of people entering the excavation. Shoring the face of an excavation should progress as the excavation work progresses. Where earthmoving machinery is used risk assessment should be used to determine whether a part of the trench may be left unsupported.

The system of work included in the SWMS should ensure workers do not enter a part of the excavation that is not protected. They should not work ahead of the shoring protection if it is being progressively installed.

The basic types of shoring are hydraulically operated metal shoring and timber shoring. The most common shoring used consists of hydraulic jacks and steel struts, walls and sheeting. Sometimes aluminium or timber components are used.

The use of metal shoring has largely replaced timber shoring because of its ability to ensure even distribution of pressure along a trench line, and it is easily adapted to various depths and trench widths.

Some of the common types of shoring are:

- closed sheeting and side lacing
- steel sheet piling
- steel trench sheeting
- timber systems, for example soldier sets
- hydraulic systems
- ground anchors.

Closed sheeting

Closed sheeting (see Figure 2.10) is where vertical timber or metal members are used to fully cover and support a trench wall and which are in turn supported by other members of a ground support system. The closed sheeting may also be supported by the other members of the ground support system, such as toms and walers.

This method of ground support is used where there is a danger of the ground running or collapsing.

Side lacing

Side lacing is a form of closed sheeting used primarily to ensure worker safety by preventing soil from slipping by the placement of fill behind timber boards or steel plates (see Figure 2.11). Side lacing is used in all types of ground, and is particularly useful where long or large diameter pipes are to be installed and in variable ground conditions where steel or timber supports are difficult to install. Side lacing should be firmly wedged into the ground to prevent it from moving when fill is placed against it.

When closed sheeting or side lacing is used to prevent ground collapse, workers should not:

- enter the excavation prior to the installation of the sheeting or lacing
- work inside a trench, outside the protection of sheeting or lacing
- enter the excavation after sheeting has been removed, and
- enter an area where there is sheeting or lacing, other than by a ladder.

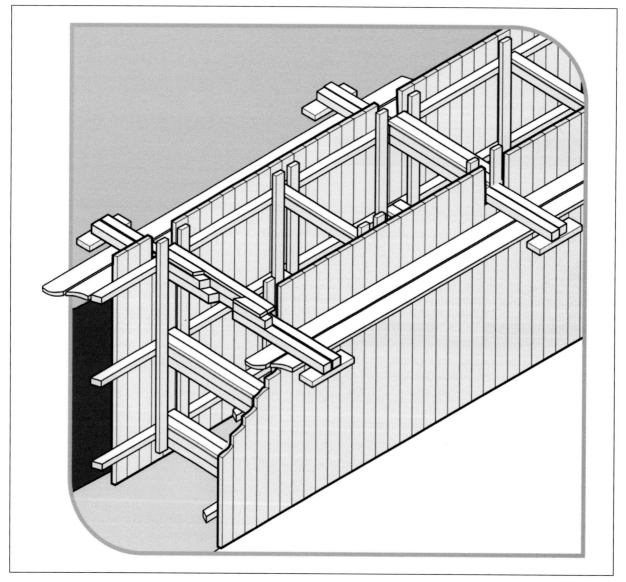

FIGURE 2.10 Example of closed sheeting

Source: SafeWork NSW, *Code of Practice Excavation Work*, January 2020, p. 59.

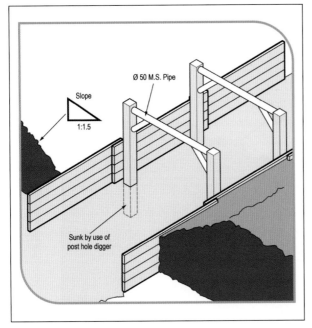

FIGURE 2.11 Side lacing in sand trench

Source: SafeWork NSW, *Code of Practice Excavation Work*, January 2020, p. 60.

Steel sheet piling

Steel sheet piling is generally used on major excavations, for example large building foundations or where large embankments are to be held back and can be installed prior to excavation work commencing. It is also used where an excavation is in close proximity to adjoining buildings (see **Figure 2.12**).

Sheet piling may be used when the ground is so unstable that side wall collapse is likely to occur during excavation, for example in loose and running sand. In such cases, sheet piling should be installed before excavation starts.

Steel trench sheeting

Other methods of excavation may require the use of steel trench sheeting or shoring. It is positioned and pneumatically driven in to final depth. Toms and walings are placed into position as the soil is excavated. Although timber can be used it is more efficient to use adjustable jacks or struts (see **Figure 2.13**).

Steel trench sheeting is lighter weight than normal sheet piling and in some circumstances may be driven by hand-held pneumatic hammers or electrical operated vibrating hammers. The potential for manual handling injuries to occur in this operation is very high, as is the risk of lacerations due to sharp metal protrusions. These risks should be addressed before the driving of the steel sheet starts. Projections on the underside of the anvil of jackhammers should be removed to prevent damage to the driving cap and potential injury to the operator.

During driving operations, if it is likely workers may be exposed to noise levels in excess of the exposure standard, a method of controlling the noise exposure is required.

Further information on exposure to noise levels is available in Safe Work Australia *Code of Practice: Managing noise and preventing hearing loss at work*. Steel shoring and trench lining equipment should be designed by a competent person.

Soldier set systems

The soldier set system is a simple form of excavation support set, which can be formed with steel or timber. This system is mostly used in rock, stiff clays and in other soil types with similar self-supporting properties.

Unlike closed sheeting sets, soldier sets retain the earth where there may be a fault in the embankment. Soldier sets only provide ground support at regular intervals and do not provide positive ground support to the whole excavated face. Open soldier sets are only suitable for use in stable soil types (see **Figure 2.14**).

Hydraulic systems

Hydraulic support systems are commonly used to provide temporary or mobile ground support while other ground supports are being installed (see **Figure 2.15**).

Ground pressures should be considered prior to installing hydraulic supports. The hydraulic support system should be designed by a competent person in consultation with the geotechnical engineer.

The hydraulic capacity of the temporary ground support system must be designed to resist the expected ground pressures and potential for collapse. Hydraulic support systems may become unreliable if not properly maintained and properly used. Frequent inspections of pressure hoses and rams are necessary to detect abrasion, fatigue or damage, for example bent or notched rams.

When a trench has been fully supported the hydraulic support systems should be dismantled to prevent costly damage. The hydraulic supports should be inspected, repaired if necessary and carefully stored prior to re-use.

Ground anchors

A ground anchor is a tie-back to the soil behind the face requiring support and is typically used with steel sheet piling (see **Figure 2.16**).

Ground anchors may be installed in either granular or clay soils. The design of ground anchors should be carried out by a competent person, for example a geotechnical engineer.

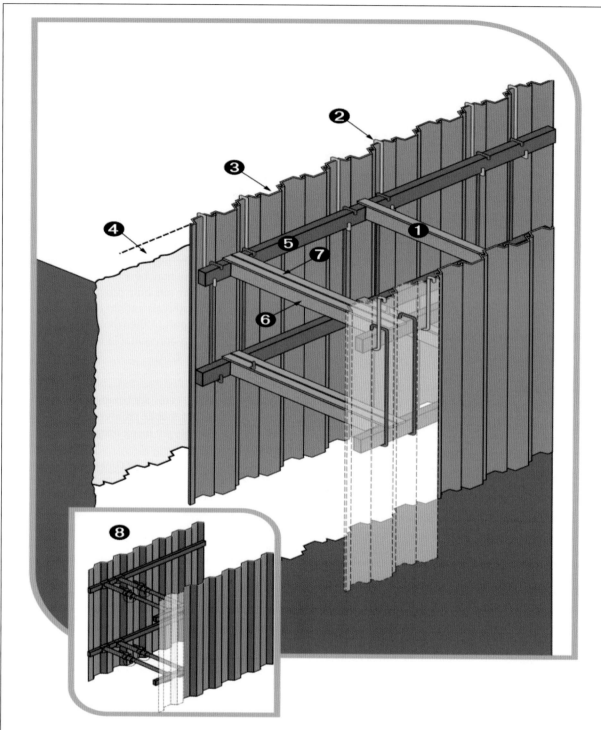

1. Centre capped single tom
2. Hanging bar
3. Sheet piling
4. Minimum height of sheet piling above surface: 300mm
5. Waling
6. Twin toms
7. Twin capping
8. Twin steel jacks should be used where extra strength is required due to heavy loading

FIGURE 2.12 Steel sheet piling

Source: SafeWork NSW, *Code of Practice Excavation Work*, January 2020, p. 56.

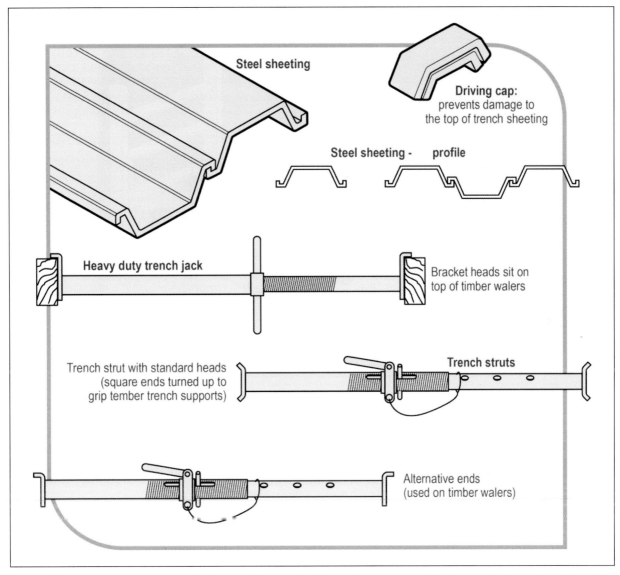

FIGURE 2.13 Steel trench sheeting and jacks

Source: SafeWork NSW, *Code of Practice Excavation Work*, January 2020, p. 57.

In granular soil, the anchorage zone is usually a plug of grout located behind the active soil limit line. This plug resists the tension force induced in the stressing cables, due to the shear and cohesion forces developed along its length.

These forces can be due, in part, to the overburden. Removal of soil above installed ground anchors should only be carried out after approval has been received from a competent person.

Removal of the soil between the retaining wall and the active soil limit line may cause sheet piling to bend. This bending will release the load in the stressing cable, and render the ground anchor useless and dangerous to workers in the excavation area.

The ground anchor may not develop its original load carrying capacity on replacement of the soil. The anchorage of the stressing cable at the face of the sheet piling may be also dislodged or loosened. This depends on the type of stressing cable and the respective anchoring systems. While the ground anchoring system is operative, periodic checks with hydraulic jacks and pressure gauges are used to assess anchor behaviour over long periods.

Installing and removing ground support systems

An excavated area is potentially unstable when the ground supports are being installed or removed. Any support systems must be installed and removed in a way that protects workers from collapsing ground or structures. Before installing or removing ground support systems, temporary structural supports must be installed as necessary to ensure that the work can be carried out safely.

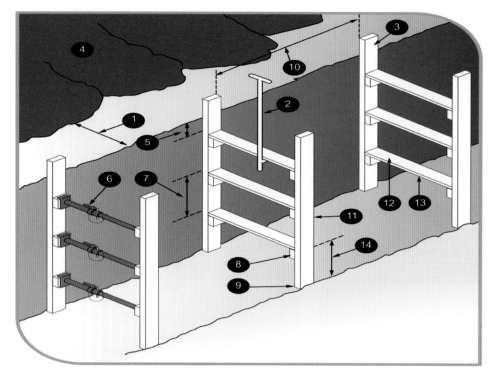

1. Spoil heap at least 1000 mm clear of excavation allows access along the side of the trench top and prevents material from the heap rolling into the trench.
2. Toms placed from surface with special timbering tongs.
3. Soldiers protrude 500 mm above the top of the trench.
4. Spoil heap or pile
5. Top tom no lower than 300 mm from the trench top
6. For added side support, steel jacks may replace timber toms.
7. Maximum spacing of toms no more than 750 mm
8. Cleats securely nailed to soldiers before placing soldiers in trench.
9. Soldier resting securely on trench bottom.
10. Maximum spacing between soldier sets: 1.5 metres
11. Soldier: minimum size 150 mm × 38 mm
12. Tom: minimum size 150 mm × 38 mm
13. Tom should be long enough to force soldiers firmly against trench sides. To prevent excessive bowing of soldiers against irregular trench sides, wood packing, between the trench wall and the soldier, may be used.
14. Space between the bottom tom and trench floor should be sufficient to allow installation of a pipe—normally, no more than 1000 mm.

FIGURE 2.14 Timber soldier sets

Source: SafeWork NSW, *Code of Practice Excavation Work*, January 2020, p. 58.

Installing ground support systems

During the erection of support systems, as more material is excavated, the walls of the support system will be under increasing load.

Workers must not enter an excavated area until permanent supports have been correctly installed. Always ensure that temporary protection in the form of timber supports or trench shields is used as required to protect workers from ground collapse, structural collapse or falling rocks.

Removing ground support systems

Shoring and all support systems should be removed in a way that protects workers from ground collapse, structural collapse or being struck by structural members. Before removal begins, temporary structural members may need to be installed to ensure worker safety.

When removing shoring, the support system should be extracted or dismantled in reverse order to its installation. Persons performing the work in the excavation should not work outside the protection of the ground support system. No part of a ground support system should be removed until the trench is ready for final backfill and compaction.

Regular inspection

The following section is based on SafeWork NSW, *Code of Practice Excavation Work* January 2020, p. 63.

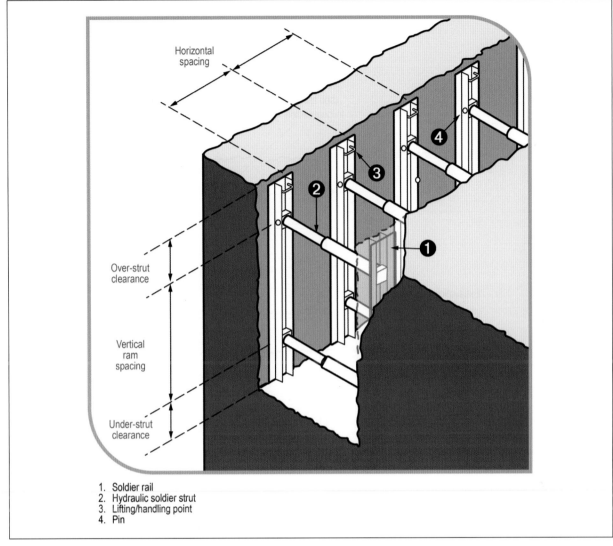

1. Soldier rail
2. Hydraulic soldier strut
3. Lifting/handling point
4. Pin

FIGURE 2.15 Hydraulic shoring (soldier set style)

Source: SafeWork NSW, *Code of Practice Excavation Work*, January 2020, p. 55.

The condition of soil surrounding excavations can change quickly due to the soil drying out, changes in the water table or water saturation of the soil. The soil condition and the state of shoring, battering and trench walls should be frequently checked by a competent person for signs of earth fretting, slipping, slumping or ground swelling. Where necessary, repair the excavation or strengthen the shoring system from above before allowing work below ground to continue.

COMPLETE WORKSHEET 2

Clean-up

Work area

After the trench support system has been removed, the trench must be backfilled. The requirements for backfill

vary, but as a minimum they must comply with AS/NZS 3500 Plumbing and drainage. If the excavation is in a public area or roadway, the local council will normally have a specification for backfilling.

AS/NZS 3500 PLUMBING AND DRAINAGE

The clean-up may necessitate reinstatement of surfaces and clean-up of the work area. Clean-up includes:
- ensuring backfill is level and trip hazards are removed
- where applicable, soil is seeded/planted as required
- where applicable, clean-up of any wastewater spills
- where applicable, pick-up and disposal of waste (e.g. Duplex nails)

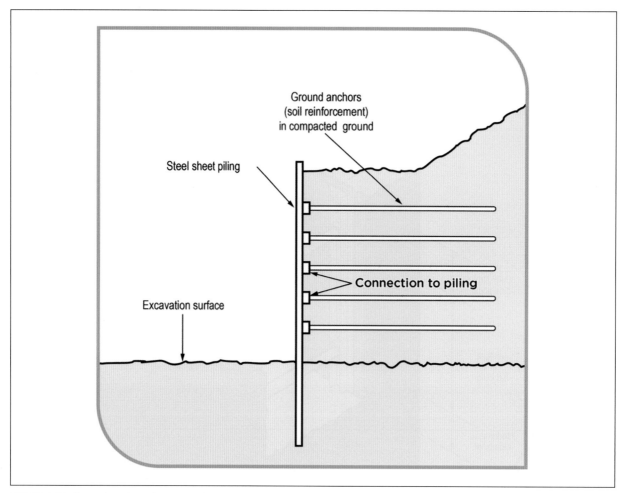

FIGURE 2.16 Ground anchors for supporting steel sheet piling

Source: SafeWork NSW. *Code of Practice Excavation Work*, January 2020, p. 61.

- where applicable, removal and disposal of contaminated fill in an approved manner after the contaminated fill has been assessed in accordance with the relevant Environmental Protection Agency requirements
- restoring any soiled building surfaces to a clean condition
- restoring the worksite where possible to its original condition.

Tools and equipment

The procedure for cleaning tools and equipment could include the following:

- Excavating equipment should be removed and serviced to maintain it in good condition.
- Tools and equipment must be inspected, cleaned, maintained and stored correctly to prevent corrosion.

- Unserviceable and unsafe tools and equipment should either be repaired or disposed of.
- Clean-up and return of hire equipment in accordance with the hire agreement.

Documentation

The documentation for excavation and trench support is discussed earlier in this chapter. It is important not only to lodge applications, but also to finalise documentation, such as notification of completion of work, as required by the relevant local authorities (e.g. Road Opening Permits).

COMPLETE WORKSHEETS 3 AND 4

SUMMARY

- When preparing to excavate, research local regulations so that you comply with legislation regarding excavator plant operation, traffic management and lodging applications for Road Opening Permits and notices for excavations.
- Plant, tools and equipment considerations include the correct selection and use of safety equipment, PPE, and tools needed for excavation and trench support activities, including for assembling, installing and removing trench support members.
- Be aware of the dangers of excavations and the excavation methods for trenching, tunnelling, shafts and bulk excavations. To reduce the risk of ground collapse, you will need to understand the ground conditions and factors that increase soil stresses and likelihood of soil movement. To control the risk of ground collapse, you will need to understand the difference between ground support systems and shields. It is important to be able to safely transport, install and remove shields.
- If you are required to install a ground support system, then the shoring system should be designed by a qualified engineer. Benching and battering, when carried out correctly, removes the requirement for trench shields and ground support systems. When excavating, do not undermine buildings and other structures.
- Always clean up properly and keep tools and equipment in a safe and serviceable condition.

 **WORKSHEET 1**

To be completed by teachers	
Student competent	☐
Student not yet competent	☐

Student name: _____

Enrolment year: _____

Class code: _____

Competency name/Number: _____

Task: Review the section 'Preparing for work', then complete the following.

1   When planning and preparing for excavation work, it is necessary to obtain information. Name three pieces of information that are required.

a _____

b _____

c _____

2   Research your local requirements for traffic control. When involved in traffic control in a public area such as a road, is it necessary to have completed a recognised training course?

_____

_____

_____

3   Research your local authority requirements for excavating an open trench across a roadway. You may be able to find information on the internet. Obtain a Road Opening Permit application from your local authority, and keep it, as it will be used in a later task. If the form is completed online, simply complete the research activity.

a What is the name of your local authority?

_____

b What is the cost of lodging a Road Opening Permit application?

_____

c What is the type and value of insurance required to obtain a Road Opening Permit?

_____

d Under what conditions is a traffic control plan required?

_____

_____

_____

4   What type of ladder should be used to gain access to a trench?

_____

_____

5 Research your local requirements for lodging notices before commencing trench excavations deeper than 1.5 m and longer than 2.5 m.

a Are notices required to be lodged?

_____

b How long before excavation work must notices be lodged?

_____

c Are notices required for excavation of trenches on new building construction sites?

_____

d Are notices required for excavation of trenches for drain repair replacement work?

_____

6 Research your local requirements for entering confined spaces. When does a trench become classified as a confined space?

_____

_____

_____

WORKSHEET 2

Student name: _____

Enrolment year: _____

Class code: _____

Competency name/Number: _____

Task: Review the sections 'Excavation', 'Reducing the risk of ground collapse' and 'Controlling the risk of ground collapse', and complete the following.

1 What are the four main methods of excavation?

   a _____

   b _____

   c _____

   d _____

2 What are three ground collapse prevention methods?

   a _____

   b _____

   c _____

3 What are five considerations in the design of a tunnel?

   a _____

   b _____

   c _____

   d _____

   e _____

4 What is an example of a risk control to provide fall protection at the top of a shaft?

_____

5 What is an example of a project requiring bulk excavation?

_____

6 While regular inspections of excavations are important, what are two examples of events that could affect the safety of an excavation?

   a _____

   b _____

7   What is one example of an injury that could be caused to a worker trapped due to ground collapse?

_____

8   Name three categories of ground generally encountered in excavations.

a   _____

b   _____

c   _____

9   What are four factors that may reduce soil stability?

a   _____

b   _____

c   _____

d   _____

10 What are three main types of ground collapse risk controls?

a   _____

b   _____

c   _____

11 In what conditions are trench shields not suitable?

_____

_____

12 What distance should a shield be extended above the ground surface where it is needed to provide fall protection?

_____

_____

13 What type of connections should be avoided when lifting trench shields?

_____

_____

_____

_____

14 Define 'benching'.

_____

_____

15 What is the minimum distance that a spoil pile should be placed from the edge of a trench?

_____

16 Define 'angle of repose'.

_____

_____

17 When is trench shoring used?

_____

_____

18 The following diagram shows trench shoring installed in a trade school sandpit. Name the type of support system being used.

_____

WORKSHEET 3

Student name: _____

Enrolment year: _____

Class code: _____

Competency name/Number: _____

Task: Review the section 'Clean-up', and the 'Glossary' at the end of the book, and then complete the following.

1   Provide definitions for:

   a   Tom

   _____

   b   Waler

   _____

   c   Soldier

   _____

2   Where is it permitted to dispose of contaminated soil and fill?

   _____

   _____

3   Use the Road Opening Permit obtained in completing the previous task in Worksheet 1. Complete the form for an imaginary address at 69 Spring Street, Smithfield. There will be a trench more than 1.5 m deep and 6 m long across the unsealed public street. (This is a theoretical task – do not lodge the form.) Note: If the form is completed online, simply complete the research activity and state the backfill requirements for the unsealed public street.

WORKSHEET 4

Student name: _____

Enrolment year: _____

Class code: _____

Competency name/Number: _____

Task: Practical exercises

All questions and exercises in the previous three worksheets must be completed and checked by your trainer/teacher before you attempt the following practical exercises.

The following provides broad guidance on the minimum requirements. Individual trainers and teachers have flexibility to create and combine exercises from this book.

Your trainer/teacher should provide preliminary guidance, conduct an induction and ask underpinning knowledge questions.

1 Complete a safe work method statement (SWMS) for the installation of trench support.

2 Research your state/territory regulations for excavation and installation of trench support.

3 Excavate and install trench support for two projects in trenches deeper than 1.5 m and longer than 2.5 m. The trench support must be installed, approved by your trainer/teacher and then safely removed.

3 LEVELLING

Learning objectives

This unit provides information needed for levelling. Areas addressed in this unit include:

- levelling definitions, abbreviations and expressions
- methods for calculating fall and gradient
- procedures for using spirit level, hydrostatic level, boning rods and laser level
- handling and caring for levelling equipment
- differential levelling from benchmarks.

Introduction

Levelling is one of the fundamental skills required in plumbing, as it can affect the operation and appearance of plumbing systems, fixtures and appliances. Plumbers must be able to select and correctly use levelling equipment to establish, record and apply levels to plumbing work.

In this unit, we will study the use and operation of levelling equipment to read, record, establish and check:

- horizontal and vertical levels for the placement of pipe
- levels at specific points along a set-out
- levels in drainage excavations and plumbing operations.

The primary focus of this book is sanitary plumbing and drainage, and from this perspective, levelling equipment is most commonly used during drain-laying operations and installation of sanitary plumbing. However, levelling equipment is broadly used elsewhere in plumbing, such as in the installation of plumbing fixtures, roof gutters and downpipes.

Before studying this subject, you should have completed training in work health and safety (WHS), plan reading and plumbing quantities, and levelling equipment as outlined in Chapters 3, 4 and 6 of *Basic Plumbing Services Skills*.

Preparing for work

Determining job requirements

The job requirements are broadly determined by:

- building plans
- specifications
- plumbing regulations and standards
- WHS regulations
- site inspection.

Table 3.1 lists sources and types of information.

TABLE 3.1 Sources and types of information

Source of information	Type of information
Building plans	Layout and shape of buildings Location of important walls Property boundaries Floor levels
Specifications	Installation requirements
Plumbing regulations and standards	Pipe gradient Pipe location in relation to buildings, and general installation requirements
WHS regulations	Safe work in trenching operations Safe work in confined spaces
Site inspection	Site access Site layout and elevations Obstructions

Safety and environmental requirements

General

Each state and territory has WHS regulations that require employers and self-employed people to identify hazards, and to assess and control risks at the workplace in consultation with their workers before undertaking any work in the construction industry.

 When working in trenches with levelling equipment, you must comply with your state or territory WHS regulations, and be aware of the potential for trench collapse (see Unit 2).

Laser levels

Lasers are divided into different classes. The type generally used in plumbing is a Class 1 laser. A Class 1 laser is regarded as not dangerous to the operator. Some classes of laser levelling equipment may be harmful to people, particularly by looking directly into the light.

Class 2 and 3 lasers may only be operated by qualified people. You should read AS 2397 Safe use of lasers in the building and construction industry, and manufacturers' specifications, before using laser levels.

 AS 2397 SAFE USE OF LASERS IN THE BUILDING AND CONSTRUCTION INDUSTRY

When using any levelling instrument, the work area must be safe for the equipment and users. In some areas – for example, where there is public access – it may be necessary to fence or barricade the work area.

Quality assurance

The quality assurance requirements are determined by the quality policies and procedures of the plumbing business.

Quality assurance in levelling is mainly determined by the:

- qualifications and training of plumbers using levelling equipment
- care and protection of levelling equipment
- maintenance and calibration of levelling equipment in accordance with manufacturer's specifications.

It is important that any measurements taken in the process of levelling are as accurate as possible. For pipe gradient, an acceptable fall tolerance is ± 5 mm over a 6 m length. When working to plans that specify the design pipe levels, you must be competent in levelling.

Tools and equipment

Tools and equipment that may be required when levelling include:

- hammer or mallet
- wooden pegs

- string line
- shovel
- tape measure
- telescopic staff
- pen or pencil and paper to work out and record measurements.

There may be other tools required to service and adjust levelling equipment.

Levelling equipment and procedures

Levelling benchmarks

The datum is the point from which the heights of other points are reckoned or measured. Survey markers are used as an official point from which other levels can be taken.

Figure 3.1 illustrates a survey marker, which states: 'This is an observation station within the primary mapping control network of Australia and was established by the Lands and Survey Branch of the N.T. Administration in 1958. Elevation 28.54 metres (A.H.D.)'.

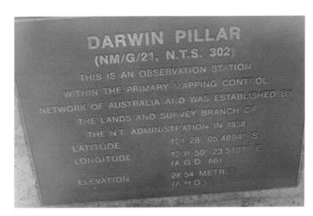

FIGURE 3.1 Survey marker

The abbreviation 'AHD' stands for 'Australian Height Datum'. This is the theoretical datum for altitude measurement in Australia, and is the average sea level around the coast of the Australian continent. AHD is the control datum to which mapping and surveying functions must be referred.

Further information on levelling benchmarks and definitions follows later in this unit.

Plumb-bob

A plumb-bob is a weight that is suspended from a vertical string line (see **Figure 3.2**). The weight is pulled towards the earth by gravity. The plum-bob is commonly used to transfer the centre from a high point to a low point, such as determining the position for a drain connection from a stack branch or discharge pipe.

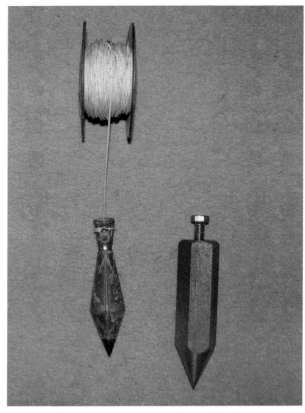

FIGURE 3.2 Two different plumb-bobs and string line

Spirit level

The spirit level is a straight edge that is fitted with a glass or plastic cylinder almost filled with liquid and sealed at both ends. Within the cylinder is an air bubble that rises to the highest part in the liquid. When the bubble in the cylinder is at centre, it is said to be 'level' when the cylinder is truly horizontal, or perpendicular to gravity.

Generally, spirit levels are fitted with two cylinders – one to show vertical level and the other to show horizontal level. They are made in different lengths, with 600 mm being the most common size. Electronic spirit levels are also available, which can be set to measure inclination in degrees, percentage and millimetres per metre.

For plumbers, the effective length of a spirit level may be increased by sitting the level onto a straight edge.

Using a straight edge and spirit level

Two basic levelling tools are the straight edge and the spirit level. A straight edge may be purchased and can be made of different materials, including aluminium. Straight edges of this type have a rectangular or box-shaped profile. The fairly common straight edge is a length of timber, which can be of differing lengths, depending on the job. When choosing a piece of timber, you should sight along the lengths of timber and choose one that is straight.

When using a spirit level in conjunction with a straight edge, it is important that the level is sitting flat against the straight edge, with a clean surface between them to create a close fit. To make it easier to handle the two tools, one method is to tape the two together.

From one given point we can transfer a level to a new point a short distance away. This is done by making sure the bubble in the level is exactly centre when the procedure is carried out (see Figure 3.3). Figure 3.4 shows using a spirit level and straight edge.

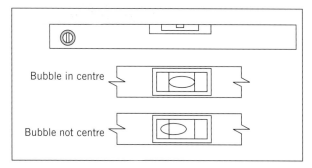

FIGURE 3.3 Reading a spirit level

Source: Department of Education and Training (http://www.training.gov.au)
© 2013 Commonwealth of Australia.

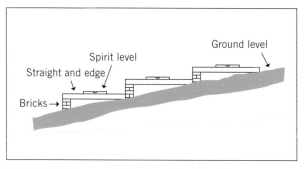

FIGURE 3.4 Using a spirit level and straight edge

Source: Department of Education and Training (http://www.training.gov.au)
© 2013 Commonwealth of Australia.

Using a spirit level in drain laying

When using a spirit level in conjunction with a straight edge during drain laying, the straight edge can be tapered, so that the spirit level shows level for a specified gradient. For example, to measure a common gradient 1.65% (1:60) with a 600 mm level, attach a piece of tapered timber 600 mm long that is 10 mm deeper at one end than the other. The level and timber can be screwed together (see Figure 3.5).

Testing the accuracy of a spirit level

To test the accuracy of a spirit level, find a straight level surface, lay the spirit level flat and note the position of the bubble in the cylinder. Now turn the level 180°. The bubble should be in the same position. If it is not, the level is not accurate, and it may be possible to fix this by loosening and adjusting the position of the cylinder cradle. This procedure applies for both vertical and horizontal surfaces.

Some electronic spirit levels can be calibrated in the same way, while using the electronic settings to confirm the accuracy of the instrument. Figure 3.6 shows an electronic spirit level.

Hydrostatic (water) level

The hydrostatic, or water, level is a simple device that consists of a clear plastic tube filled with water (see Figure 3.7). The diameter of the tube is not important, but 10 mm to 15 mm is commonly used. The length of tube may vary also, but 10 m would be a useful length. This level provides fast and accurate levelling from one point to another, even around corners.

This method can be used for obtaining levels for many different pipework installations. It can also be used in roof plumbing to set fall in eaves gutters. You usually need two people to operate a water level, with one person at each end. If necessary, you may use

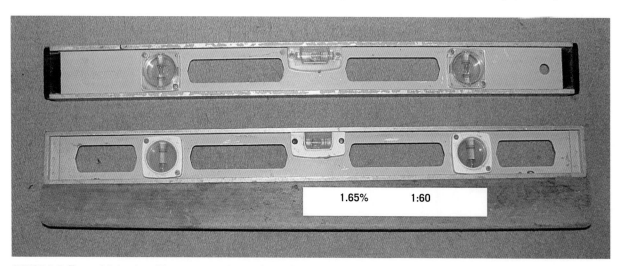

FIGURE 3.5 Two spirit levels, with one screwed to a piece of tapered timber to measure a gradient of 1.65%

FIGURE 3.6 Measuring drain gradient using an electronic spirit level

the water level by yourself. This is done by attaching a clamp to hold one end in position. The ends of the water level must be open to the atmosphere, and you must ensure there is no air trapped in the hose. It is also a good idea to add food colouring to the water, to make it easier to see.

Boning rods

Boning rods are used to achieve an even gradient along a trench floor. Each boning rod consists of two pieces of timber or metal fixed together to form a 'T' shape. The top of the 'T' is about 400 mm wide and must be fixed at right angles to the vertical section, which is usually about 1.2 m in length. Boning rods are used in sets of three, with all three being exactly the same length. The rods are set at the top and bottom ends of the trench at the desired gradient. While one person sights the top of a pair of fixed boning rods, another person moves the third rod along the trench. The objective is to sight along the tops of all three boning rods for alignment. Painting the top of each 'T' a different colour makes this easier.

Using boning rods

To use boning rods, first dig two holes, one at each end of a proposed trench. The two holes should be to the required depth of the trench, allowing for the pipe gradient, pipe and bedding.

It is best practice to hammer a peg into the ground at the bottom of each hole for the rods to sit on. This is done so that, if the soil moves, there will still be a fixed point to work from. Stand the boning rods on the pegs, and securely fasten them so they are vertical and will not fall.

To find how much soil is to be removed and to obtain an even gradient between the rods, sight over the rods while an assistant moves the remaining (third) rod along between the two ends, adjusting the base of the trench as they go. This process may be undertaken as the excavator digs the trench. Alternatively, a profile may be set up at each end of the excavation at the required height, with a boning rod used as previously indicated to obtain the correct depth and grade.

Repeat these steps until an even trench floor is achieved and the tops of all three boning rods are in common alignment. This method of using boning rods can be a very accurate way to achieve an even trench floor. See Figures 3.8 and 3.9.

Laser levels

A laser level (Class 1; see Figure 3.10) is simple to use and accurate. The instrument is attached to and supported by a tripod. This type of level is either rechargeable or powered by batteries. When switched on, a continuous horizontal sweeping beam through 360° is emitted from the instrument. The beam being transmitted is picked up by an electronic receiver, which is attached to a staff. The receiver will sound intermittent beeps when it is not level with the laser beam. When the receiver is moved closer to level, the beeps will sound more frequently. When the receiver is exactly level with the laser beam, the receiver will emit

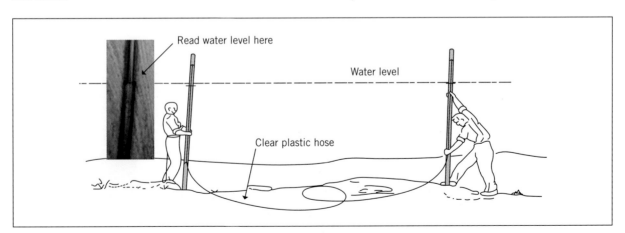

FIGURE 3.7 Using and reading a hydrostatic (water) level

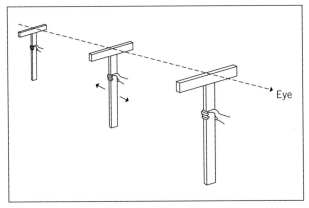

FIGURE 3.8 Using boning rods to achieve even gradient along a trench floor

Source: Department of Education and Training (http://www.training.gov.au) © 2013 Commonwealth of Australia.

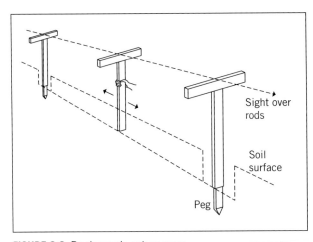

FIGURE 3.9 Boning rods set on pegs

Source: Department of Education and Training (http://www.training.gov.au) © 2013 Commonwealth of Australia.

FIGURE 3.10 Laser level

a constant tone. The operator is then able to record levels from the beam over a large radius.

One big advantage that the laser rotating level has over other levels is that it can be operated by one person. This instrument is ideal for many areas of a plumber's work, including the setting of gradients in drainage work.

Pipe laser levels can be used in trenches and preset to measure gradients in drains.

The staff

To operate a laser level successfully, you require a telescopic staff. Telescopic staffs are available in different lengths. Each staff has a series of distinct major and minor markings, and new operators should take the time to understand how to read these markings (see Table 3.2).

TABLE 3.2 Staff markings

Major markings	Numbers that indicate the vertical distance in metres from the bottom of the staff. They are marked every 0.1 m (100 mm).
Minor markings	Left and right facing 'Es', which are 50 mm high with 10 mm-thick legs: • the bottom of the E is the 0.1 m or 100 mm mark • the right-facing E indicates the decimetre number will be even • the left-facing E indicates the decimetre number will be odd.

On a staff, the white spaces between the black spaces are 10 mm high; and the black marks, whether they are the leg of an E or a black square, are 10 mm high. The smallest graduation on the staff is therefore 10 mm. For accurate work, the operator must estimate the single millimetre graduations between the marks. See Figure 3.11.

Using a laser level

The laser level consists of a transmitting unit containing a laser diode and an intermittent electronic beam, which continually rotates through a 360° horizontal plane. Because the beam is electronic, it is sometimes called an electronic level.

To set up the laser level, choose a position from where the beam may transmit to the points to be checked. Make sure the area chosen is safe and clear of machinery. It may be necessary to fence or barricade the area to ensure the safety of the instrument.

Spread the legs of the tripod and press them firmly into the ground, ensuring the top platform on the tripod is approximately level (see Figure 3.12). Fit the level to the tripod by tightening the tripod screw into the level. Adjust the levelling screws until the bubble is centred in the circular spirit level. Turn on the level, so that it will transmit the electronic beam. Connect the receiver to the staff, which must be held vertical, and switch on the receiver (see Figure 3.13).

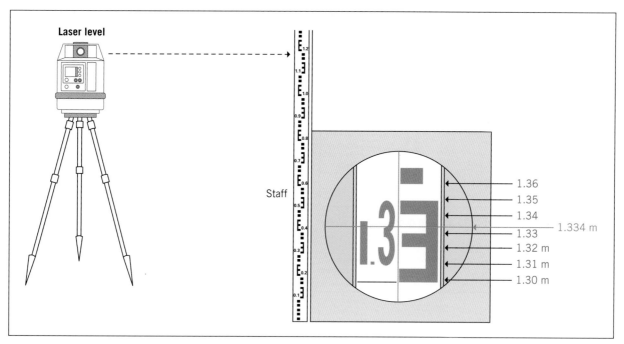

FIGURE 3.11 Reading a staff using a laser level

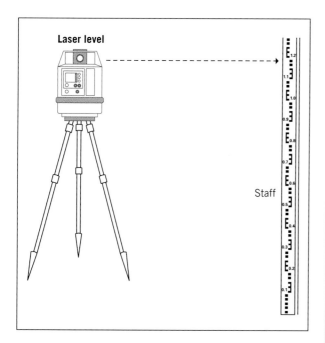

FIGURE 3.12 Taking level measurements using a laser level and staff

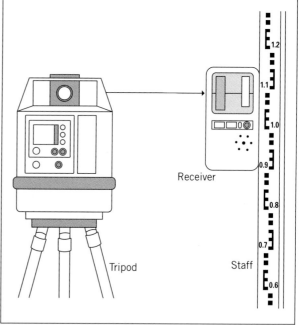

FIGURE 3.13 Laser level with staff and receiver

Source: Department of Education and Training (http://www.training.gov.au)
© 2013 Commonwealth of Australia.

Note: Most modern laser levels are self-levelling and it is not necessary to adjust the levelling screws. Self-levelling pipe lasers are also available, which can be set to measure a predetermined gradient.

You are now ready to take measurements at different points as necessary.

Handling and caring for levelling equipment

Levelling devices, such as laser levels, are precision equipment and must be handled correctly and with care. Table 3.3 provides some tips for handling and caring for this type of equipment.

AS 2397 SAFE USE OF LASERS IN THE BUILDING AND CONSTRUCTION INDUSTRY

GET IT RIGHT

CARE FOR EQUIPMENT

FIGURE 3.14 Pipe laser lying in trench

The pipe laser in **Figure** 3.14 has been left in a trench while the trench is being backfilled.

1   Is this an appropriate place to leave a levelling instrument?

_____

_____

_____

2   What aspects of this location might be detrimental to the pipe laser?

_____

_____

_____

3   What problems might arise from a damaged pipe laser?

_____

_____

_____

TABLE 3.3 Handling and caring for levelling equipment

Setting up	Take care when removing the level from its box.
	Make sure the level is securely fixed to the tripod before using it.
	Do not use excessive force when adjusting the position for level.
Relocating to another area or position	Carry the instrument against the shoulder in a vertical position.
	Fold the legs of the tripod inward when carrying it over obstacles.
	Hand the instrument to another person when crossing rough ground or over fencing.
Windy conditions	Make sure tripod legs are spread wide apart and pushed firmly into the ground.
	On surfaces such as concrete, tie the tripod down with a weight.
	In extremely windy conditions, do not leave the instrument unattended.
Dust and dirt	Avoid leaving lasers in dust and dirt, which can accumulate and damage control buttons and lenses.
Wet weather	Cover the instrument when not in use, to protect the lenses.
	If necessary, remove any moisture with a cleaning cloth.
	If moisture penetrates the level, it must be cleaned out because fungus may grow inside it.
The tripod	Do not use excessive force on screws and clamps.
	If cleaning is necessary, use a soft brush and cloth.
	Apply a small amount of oil to the clamps and screws if necessary.
Storing equipment	Store equipment in the box or case provided.
	When transporting in a vehicle, store in a place that is safe for equipment (i.e. separate from tools).

Levelling definitions and abbreviations

Table 3.4 lists definitions and abbreviations commonly used on plans and for recording levels.

TABLE 3.4 Levelling definitions and abbreviations

FFL	Finished or final floor level. After floor treatment has been applied – i.e. carpets, tiles, slate, etc.
FL	Floor level. Floor level of poured concrete.
RL/EL	Reduced level/Elevation level. Height above mean sea level.
HI	Height of instrument. The elevation of the line of sight established by the instrument. It is calculated by adding a known elevation (RL or TBM) and the staff reading.
BS	Back sight. Also known as 'plus sight'; the reading on the staff from a known or assumed elevation. Used to establish HI.

FS	Forward sight. The last reading taken in a block or group of readings where the elevation is to be determined.
IS	Intermediate sight. All other readings taken in the block or group of readings other than BS and FS.
TBM	Temporary benchmark. A movable object that has a known RL/EL established from a PBM by a surveyor.
PBM	Permanent benchmark. A fixed level relative to mean sea level determined by surveyors.
Ch.	Chainage. The cumulative distance from the centre of the instrument to the staff.
NSEL	Natural surface elevation level.
MSL	Mean sea level. The midway level between high tide and low tide. Used as a basis for barometric pressure readings.

Examples of symbols and expressions of levels

When levelling, contour lines are used to connect points of equal elevation or height and are usually drawn on a site plan to express:

- direction of fall or slope in land
- inclination (steepness) or grade of land.

Plumbers can use this information to plan a route for drains.

Other expressions of levels are shown in Table 3.5.

TABLE 3.5 Examples of symbols and expressions of levels

Element	Symbol
Levels and gradients	
FINISHED FLOOR LEVEL	F.F.L. 69.300
REDUCED LEVEL OF CONTROLLING PLANE	R.L. 69.300 ◯
LEVELS Existing Level New Level Note – Levels to be expressed to record 5 mm	(a) 69.045 (b) 69.345
LEVELS Level should be prefixed by RL or FFL where confusion with other dimensions may occur	RL 69.345
JOB DATUM LEVEL	100.00
SPOT LEVELS Alternate methods Existing Requires	69.045 69.345 69.045 69.345

Element	Symbol	
Levels and gradients		
CONTOUR LINES Alternate methods (a) Existing (b) Requires	(a) 69.000 68.000 (b) 69.000 68.000	
INDICATION OF LEVEL ON SECTION OR ELEVATION PLAN	FFL LO25	
THE GRADIENT OF A SLOPE	Y X GRADIENT = $\frac{Y}{X}$	
BENCHMARK		

Source: Department of Education and Training (http://www.training.gov.au)
© 2013 Commonwealth of Australia.

FROM EXPERIENCE

Did you know that 'plumb' means a ball of lead or other heavy object attached to the end of a line and used for determining the vertical on an upright surface? To be a plumber, it is critical that you have the skills to accurately establish levels and install pipes to very specific gradients.

 **COMPLETE WORKSHEET 1**

Differential levelling

Differential levelling is commonly used in drain laying. The procedure also can be used to calculate the simple difference between two levels. For example, it is necessary for an overflow relief gully to be at least 150 mm in level lower than the lowest fixture connected to a sanitary drain. If the lowest fixture is a shower, you can use the procedure in differential levelling to confirm that there is at least 150 mm between the top of the overflow gully riser and top surface level of the shower grate.

Procedure example

Before following this example, study the definitions and abbreviations shown earlier in Table 3.4.

Imagine that you are taking staff readings and need to calculate the RLs. First, you must determine HI. This is the height of the instrument from a known benchmark, plus the RL at the benchmark (HI = TBM + BS).

Assume there is a known elevation at TBM of 45.050 (see Tables 3.6 and 3.7).

After levelling the instrument, obtain and record a back sight (also known as plus sight) reading. In this example, we have calculated that the HI is 46.850.

In the next step, we take IS (intermediate sight) and FS (forward sight) readings and transfer them to the chart.

The individual RLs are calculated by using:

$$RL \text{ (elevation)} = HI - (IS \text{ or } FS)$$

If we are given individual peg RLs, we can also calculate the staff readings by using:

$$(IS \text{ or } FS) = HI - RL$$

TABLE 3.6 Differential levelling – calculating HI (height of instrument)

BS	IS	FS	HI	RL	Ch.	Remarks
1.800			46.850	45.050		TBM
						Peg 1
						Peg 2
						Peg 3
						Peg 4
						Peg 5
						Peg 6

TABLE 3.7 Differential levelling – calculating RLs (reduced levels)

BS	IS	FS	HI	RL	Ch.	Remarks
1.800			46.850	45.050		TBM
	1.600			45.250		Peg 1
	1.587			45.263		Peg 2
	1.523			45.327		Peg 3
	1.515			45.335		Peg 4
	1.498			45.352		Peg 5
		1.450		45.400		Peg 6

While calculating levels in this procedure can be done manually, if you are proficient in using computer spreadsheets such as Microsoft Excel®, the process will be more efficient, and the spreadsheet function can be expanded to calculate fall and gradient in drainage.

Calculating the fall and gradient

The primary reason for gradient and fall in drainage is to achieve a self-cleaning velocity. This is the ability of liquid containing solids to completely drain out from the system.

Gradients for pipes may be expressed as a percentage or a ratio. For example, AS/NZS 3500.2 Sanitary plumbing and drainage states the minimum

gradient for DN 100 drainage pipe is 1.65%, which may also be expressed as the ratio 1:60.
- To convert 1.65% to a ratio, divide 100 by 1.65 = 60.6 (1:60 approximately).
- To convert 1:60 to a percentage, divide 100 by 60 = 1.66 (1.65% approximately).

AS/NZS 3500.2 SANITARY PLUMBING AND DRAINAGE.

Refer to the conversion table in AS/NZS 3500.2 for these and other conversions.

Calculating fall

To calculate the fall of a drain, you will need to know the minimum depth requirement from the plan and the depth of the connection to the stormwater or sewer system. These levels are usually taken from the sewer/stormwater point of connection and may be found on the authority plan. It should be noted that depths marked on authority plans were taken at the time of the sewer installation, and these measurements can and do change due to many factors, such as the owner/developer altering the land surface levels.

To calculate the fall in a proposed drain, you need to use simple calculations, as in Example 3.1.

EXAMPLE 3.1

Calculate the fall required in a drain 54 m long.

To find the fall for a drain, you first need to know the length of the proposed drain and the gradient at which it is laid. In Figure 3.15, the length is 54 m with a gradient of 1.65%.

To calculate the fall, use the formula
Fall = Gradient × Length.

Fall = Gradient × Length

Fall = 1.65% × 54 m

$$Fall = \frac{1.65}{100} \times \frac{54}{1}$$

Fall = 0.891 m

Fall = 891 mm

This means that the bottom of the trench at the lower end will be 891 mm below the bottom of the trench at the top end of the drain. (It has a fall of 891 mm; see Figure 3.16.)

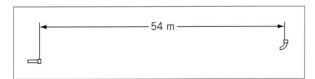

FIGURE 3.15 Example, calculating fall for drainage

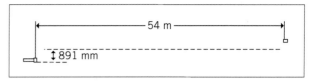

FIGURE 3.16 Example, fall calculated for drainage

Calculating gradient

Sometimes it is necessary to calculate the available gradient of a trench or ground surface to establish if there is enough gradient for the pipe to be laid. To calculate gradient, you will need to know the length and the fall.

Gradient percentage may be calculated by using the formula Gradient = Fall ÷ Length × 100.

Example: If a DN 100 drain is to be laid with a fall of 0.675 m and a length of 37 m, would the drain have the adequate (minimum) gradient, being 1.65% for a DN 100 drain? By working through the following calculation, you can see the answer is 'yes' – that is, 1.82% is greater than 1.65%.

$$Gradient = \frac{Fall}{Length} \times 100$$

$$Gradient = \frac{0.675}{37} \times 100$$

$$Gradient = 0.0182 \times 1000$$

$$Gradient = 1.82\%$$

Practical application

When you understand levelling terminology, the procedures for differential levelling and how to calculate fall and gradient, you will be able to apply what you have learnt to set out and install drainage to the design. At first glance, some plans and specifications can appear complicated, but if you study them carefully, your understanding will follow. For example, Figure 3.17 shows a longitudinal section drawing of a stormwater retention system, and pit schedule. You will see that the scales are 1 in 50 vertically and 1 in 200 horizontally. This drawing is purposely out of proportion to present levels and gradients in a format that can easily be seen. All of the information is there, with levels and gradients for pipes and pits. The drawing would simply be read in conjunction with the drainage plan.

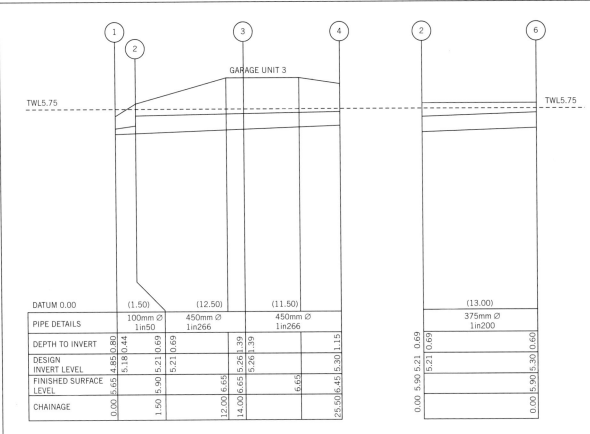

LONGITUDINAL SECTION – STORMWATER RETENTION SYSTEM
SCALE: HORIZONTAL 1:200
VERTICAL 1:50

PIT SCHEDULE									
PIT NO	TYPE	INTERNAL DIM. L × W	INLET		OUTLET		FINISHED TOP	DEPTH TO INVERT	REMARKS
			DIA.	RL	DIA.	RL			
1	JUNCTION PIT	900 × 600	100	5.18	Ex 150	4.85	5.63	0.78	CONSTRUCT JUNCTION PIT AS PER STANDARD DRG No SD215 FITTED WITH "TERRA FIRMA" LID AS PER SD207
2	RETENTION PIT	900 × 600	375 450	5.21 5.21	100	5.21	5.90	0.69	SEE DETAIL
3	JUNCTION PIT	900 × 600	450	5.26	450	5.26	6.65	1.39	FIT "TERRA FIRMA" LID AS PER STANDARD SD 207 AND STEP IRONS
4	GRATE PIT	900 × 600	150	6.00	450	5.30	6.45	1.15	FIT HEAVY DUTY TRAFFICABLE GRATE LID AND STEP IRONS
5	GRATE PIT	600 × 600	DP's		150	7.21	7.66	0.45	FIT HEAVY DUTY TRAFFICABLE GRATE LID
6	GRATE PIT	450 × 450	DP's		375	5.30	5.90	0.60	FIT GRATE LID

FIGURE 3.17 Example longitudinal section and pit schedule for stormwater retention system

LEARNING TASK 3.1

Look at Figure 3.17 and calculate:
1. the difference in reduced levels between the inlet to Junction Pit 1 and outlet of Retention Pit 2
2. the chainage between the inlet to Junction Pit 1 and outlet of Retention Pit 2.
   Do these match the detail for the design gradient?

Clean-up

Work area

Cleaning up includes clearing the worksite of trip hazards, debris and unused materials before and after levelling work.

Tools and equipment

You should also clean and maintain your tools and equipment, especially levelling equipment, before returning them to be stored. This will also prolong the life of the equipment.

The type of work undertaken while using levelling equipment can be messy, especially drainage work. The work area may be exposed to dust, wind and rain. In these environments, levelling instruments can become damaged, which can cause costly delays while the level is being repaired.

If a level is found to be faulty, it should be fixed as soon as possible by a person with the necessary technical knowledge (usually the manufacturer), so that it is working for the next job. Whenever the lens of a level needs to be cleaned, be sure to use only a cloth specifically designed for this job.

GREEN TIP

Laser levels must be maintained in clean, reliable working condition, ensuring that both base unit and receiver have adequate charge/power for the job. Laser levels must also be periodically checked (or calibrated) to ensure continued accuracy. Future work may depend on the instrument being in first-class working condition.

Documentation

The information and records from levelling will depend upon the job specification, and the quality assurance procedures of the business. For large projects, it may be necessary to provide as-constructed details of drainage installations, including level information.

 COMPLETE WORKSHEETS 2 AND 3

SUMMARY

- Prepare for your work by always having plans and specifications before the work starts, and inspect the job site to find out whether there are services and obstructions that might affect the design location and levels of pipes. Ensure that you are aware of the accuracy requirements so that all pipes, pits, etc. will be installed at the correct design levels and gradients.
- Always care for equipment and ensure that levelling instruments are serviced and calibrated for accuracy. Study levelling definitions, abbreviations, symbols and expressions of levels so that you can understand plans and specifications.
- Always carry out differential levelling to a procedure where levels are progressively recorded and checked for accuracy.
- When you have finished using levelling equipment, ensure that it is cleaned and safely stored to prevent damage. Ensure that powered levelling instruments and electronic receivers are fully charged and, if necessary, have spare batteries available for the next job.

 **WORKSHEET 1**

Student name: _____

Enrolment year: _____

Class code: _____

Competency name/Number: _____

Task: Review the section on 'Preparing for work', then complete the following.

1 Identify four pieces of information that can be obtained from building plans.

a _____

b _____

c _____

d _____

2 What is the standard for safe use of lasers in the building and construction industry?

_____

_____

_____

3 What is an acceptable fall tolerance for pipe gradient?

_____

_____

_____

4 Under what circumstances might it be necessary to fence or barricade the work area?

_____

_____

5 What class of lasers is acceptable for use in the building and construction industry?

_____

_____

_____

6 What type of staff is used in levelling?

_____

_____

_____

WORKSHEET 2

Student name: _____

Enrolment year: _____

Class code: _____

Competency name/Number: _____

Task: Review the section on 'Levelling equipment and procedures', and then complete the following.

1   Explain the term 'AHD (Australian Height Datum)'.

_____

2   Explain how the accuracy of a spirit level can be tested.

_____

3   When using boning rods, why is it best practice to place the end boning rods onto fixed pegs?

_____

4   What is the smallest graduation on a levelling staff?

_____

5   On the staff shown in the following diagram, what is the reading taken at the arrow?

_____

Source: Department of Education and Training (http://www.training.gov.au)
© 2013 Commonwealth of Australia.

6   When setting up a laser level that is not self levelling, why is it necessary to ensure that the bubble in the small circular spirit level is centred?

_____

7   Why must the staff be held vertical?

_____

8 When using a laser level and receiver, what does it mean when there is an intermittent beep sound?

_____

9 What is the main advantage of a laser level?

_____

10 Explain how you would protect a laser level instrument during windy conditions.

_____

11 Explain the following abbreviations:

HI _____

BS _____

TBM _____

FS _____

12 What are contour lines?

_____

13 Calculate and record the staff readings in the tables. The reading at TBM (BS) is 1.335.

RL	Remarks	RL	Remarks
42.655	Invert 1	45.437	Invert 4
42.600	Invert 2	45.413	Invert 5
42.579	Invert 3	45.387	Invert 6

FS	IS	BS	HI	RL	Ch.	Remarks
				45.050		TBM

14 Calculate and record the RLs given in the staff readings below. The reading at TBM (BS) is 2.335.

Staff reading	Remarks	Staff reading	Remarks
1.600	Peg 1	1.515	Peg 4
1.587	Peg 2	1.498	Peg 5
1.523	Peg 3	1.450	Peg 6

				10.250		TBM
						Peg 1
						Peg 2
						Peg 3
						Peg 4
						Peg 5
						Peg 6

15 Calculate the fall of a 25 m drain with a gradient of 1.65%.

_____

16 Calculate the fall of a 65 m drain with a gradient of 2.00%.

_____

17 Calculate the fall of a 40 m drain with a gradient of 2.50%.

_____

18

a   Calculate the gradient for a section of drain 45 m long and with a fall of 600 mm. Show all
calculations.

_____

_____

_____

b   After finding your answer to (a) above, is your answer adequate for a DN 100 sanitary drain?
(Note: The minimum required gradient for a DN 100 sanitary drain is 1.65%.)

_____

_____

_____

WORKSHEET 3

Student name: _____

Enrolment year: _____

Class code: _____

Competency name/Number: _____

Task: Practical exercise

All questions and exercises in the previous two worksheets must be completed and checked by your trainer/teacher before you attempt the following practical exercise.

The following exercise must be completed to demonstrate competency in levelling.
Equipment options:
Preparation equipment: a) Boning rods; b) String line
Levelling instruments: a) Pipe laser; b) Rotary laser

Using one of the preparation equipment options and one of the levelling instrument options, lay a DN 100 drain below ground level at a gradient of 1.65%. The drain must be at least 10 m long. The accuracy required is ±5 mm. This exercise must be documented. Your trainer/teacher will provide you with a TBM, then complete the following chart.

BS	IS	FS	HI	RL	Ch.	Remarks
						TBM
						Peg 1
						Peg 2
						Peg 3
						Peg 4
						Peg 5
						Peg 6

Your trainer/teacher may choose to combine the above exercise with the practical exercise at the end of Unit 10, 'Installing below-ground sanitary drainage systems'. The section of drain must be 10 m long.

4
PLANNING THE LAYOUT FOR A RESIDENTIAL EXTERNAL SANITARY DRAINAGE SYSTEM

Learning objectives

This unit provides the basic principles, knowledge and procedures for planning the layout of a residential sanitary drainage system. Areas addressed in this unit include:

- preparation through obtaining information from plans, specifications, Australian Standards and regulations
- understanding various requirements, such as connection methods, depth of cover, gradient, sizing, ventilation and surcharge protection
- developing a plan for a drainage system layout through following an example project
- determining suitable materials
- developing a list of pipes, fittings and other materials needed to complete a project.

Introduction

In this unit you will study the requirements for planning the layout of an external sanitary drainage system for a residential building. Where it is practical to do so, sanitary drains should be located externally to buildings.

A sanitary drainage system is an assembly of pipes and fittings that conveys discharges from a sanitary plumbing system and fixtures to an approved point of discharge or on-site disposal system. In this unit, we focus on connection to an approved point of discharge. On-site disposal systems are covered in Unit 13.

A sanitary drainage system can be constructed above or below ground. In this unit, we study the requirements for external (below-ground) sanitary drainage (see Figure 4.1). Above-ground drainage is covered in Unit 5.

It is important to develop skills in planning the layout of a sanitary drainage system, so that you will be able to install a system that complies with job specifications, regulations and standards.

While studying this subject, you will be required to respond to questions and complete exercises referring to AS/NZS 3500.2 Sanitary plumbing and drainage.

AS/NZS 3500.2 SANITARY PLUMBING AND DRAINAGE

Before studying this subject, you should have completed training in work health and safety (WHS) as outlined in Chapter 3 of *Basic Plumbing Services Skills*.

Preparing for work

When planning the layout of a sanitary drainage system, you will first need to prepare for the work. The issues you will need to consider are:

- excavating equipment, which should be removed and serviced to maintain it in good condition
- plans and specifications
- WHS and environmental requirements
- quality assurance
- planning and sequencing of tasks
- tools and equipment
- preparation of the work area.

Plans and specifications

Plans (or design drawings) and specifications will help you to make decisions about the layout of the drain. Where any sanitary drain is to be installed, altered or extended, an application will need to be made to the local sewerage authority before work commences. After reviewing the application, the authority might provide additional conditions and specifications that will further determine the layout. Other factors can also determine the layout, such as the location of existing underground services and obstructions. So, you can see that proper preparation – by obtaining information *for every job* – is critical, because you will only want to do the job once.

Most large projects have specifications and drawings that will determine both the plumbing fixtures and the types of drain materials to be used. In such cases, the licensed plumber/drainer will have

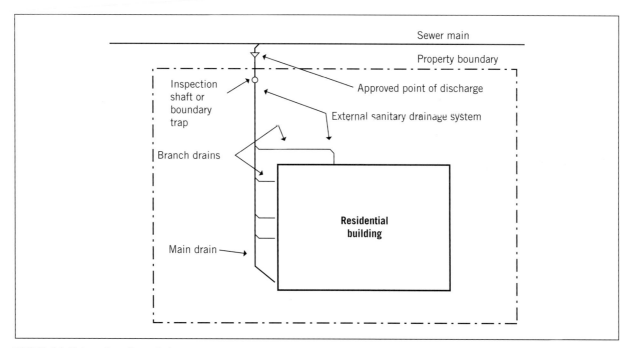

FIGURE 4.1 External sanitary drainage system

Source: Department of Education and Training (http://www.training.gov.au) © 2013 Commonwealth of Australia.

little control over what materials are to be used, and will need to purchase the drain materials and construct the drain in accordance with the specifications and drawings. Regardless of any specification or drawing, the licensed plumber/drainer can only construct a sanitary drainage system in accordance with plumbing regulations, codes and standards. Licensed plumbers and drainers must have a good working knowledge of the plumbing regulations, and must keep abreast of changes.

For small residential building projects, the specifications are unlikely to determine the materials to be used, other than 'approved materials' that are specified in the Australian Standards. In these projects, the plumber/drainer will determine the pipe materials to be used and the layout of the residential sanitary drainage system.

Responsibilities

The plumbing auditor/inspector's responsibility is to:
- ensure compliance with standards
- ensure compliance with regulations
- ensure compliance with conditions of permit/consent
- protect public health
- protect the sewerage authority's assets.
  The plumber's responsibility is to:
- comply with standards and regulations
- comply with the terms of the sewerage authority permit/consent
- protect public health
- protect the sewerage authority's assets
- comply with health and safety regulations
- protect the environment from pollution
- comply with quality assurance requirements.
  The plumber should also:
- check the site soil classification to determine whether the drain will need to be installed with special provisions for movement, such as swivel and telescopic joint fittings
- verify with the property owner or agent that the layout of the proposed drainage installation is acceptable before commencing work
- verify the position of any easements, existing services and obstructions before commencing work
- take precautions to prevent damage to any easements and services belonging to any authority
- rectify any work that does not comply with the regulations as directed by the relevant authority
- submit an 'as completed' plan if required by the authority
- notify the authority on the completion of work
- fill out and issue a Certificate of Compliance where required by the authority. The Certificate of Compliance is a statement confirming that all work is completed and meets regulatory requirements and standards, and provides a warranty. Note

that many authorities require that a copy of the certificate be issued to both the authority and the property owner/consumer.

Authority permits and fees

Prior to the commencement of any building work, the local government authority (usually the council) must inspect and approve all building plans. Before the plans are approved, the applicant must pay the required fees.

When the plans have been approved (usually stamped), the applicant then submits them to the local sewerage authority for their approval and pays the applicable fees or charges.

Each sewerage authority has different fees and charges, and these can amount to hundreds or thousands of dollars. It is important, therefore, to clarify the responsibility for payment of these fees and charges with the owner/builder. It is also good practice to confirm the responsibility for fees and charges with the owner/agent in writing.

Each regulatory authority has a system for receiving and processing applications for sanitary drainage works. The permit (or consent) provides details of the property, approved point of discharge and conditions of authorisation.

Before commencing work, the plumber/drainer should ensure that:
- the plans have been approved and stamped by the local council
- all appropriate fees and charges have been paid
- the plumbing permit (or consent) is obtained from the sewerage authority. Unless carried out as an emergency (e.g. a blocked drain), the licensed plumber must obtain a plumbing permit/consent before work commences
- information relating to the location of all underground services is obtained (e.g. by contacting Dial Before You Dig: http://www.1100.com.au, or telephone 1100)
- the worksite will be safe for the duration of works (e.g. fencing is installed to prevent public access to open trenches).

Emergency work

Occasionally, emergency work is necessary to:
- prevent the overflow of sewer/waste water
- prevent damage to property
- clear obstructions and blocked pipes
- protect public health and safety.

In these cases, the licensed plumber may carry out work without first obtaining a permit or consent. However, when the emergency work is complete, the plumber must inform the sewerage authority of completed work for issue of a permit.

It also may be necessary for the licensed plumber to issue a Certificate of Compliance, as described earlier under 'Responsibilities'. *At the end of this unit, you will be required to research your local authority requirements in relation to applications and permits.*

Depth of cover

Plans and specifications may indicate that the drain requires protection against mechanical damage and deformation due to loadings from vehicles. There is different minimum cover for buried piping, and this depends upon the type of cover and vehicular loading.

For example, if you see that the building plan for the site indicates the location of a residential driveway where a PVC (polyvinyl chloride) sanitary drain is to be laid, and you are unsure when there might be brick or concrete paving, you must lay the drain with at least 500 mm of cover between the top of the drain and the natural surface level.

Approved point of discharge

Generally, the licensed plumber is responsible for locating the approved point of discharge. Sewerage authorities will provide a plan that indicates the location and depth of the point.

When planning to lay the sanitary drain, it is good practice to first locate the approved point of discharge, as there may be slight variations in the actual depth and location that will affect the layout. You should be prepared to excavate by hand to locate the approved point of discharge. Machinery must not be used to excavate within 600 mm of an authority sewer main. Table 4.1 outlines the minimum cover for external sanitary drains.

TABLE 4.1 Minimum cover for external sanitary drains

Location	Minimum depth of cover in mm		
	Cast iron and ductile iron	PVC and other materials	All materials where there is at least 50 mm cover
Light vehicle traffic	300	500	75 mm-thick brick or concrete paving*
Heavy vehicle traffic	300	500	100 mm-thick reinforced concrete*
Elsewhere (such as garden area)	Nil	300	50 mm-thick brick or concrete paving*
*Paving to cover at least the full width of the trench.			

If the point of connection cannot be located, you should excavate 1 m either side of the point indicated on the plan and also 1 m below if it is a deep connection. If the connection point still cannot be located, contact the sewerage authority for further direction. Occasionally, it is found that the point of connection has not been installed or is in the wrong location. In that case, the authority will arrange for it to be inserted at the point indicated on the plan.

The approved point of discharge provided by the sewerage authority is usually DN 100, or 150 mm, which can be in the form of a junction fitting or inlet into an inspection chamber. The connection point may be located either inside or outside the property, in the road or in the footpath. If you will need to work outside the property boundary to gain access to the approved point of discharge, then you will need to seek the permission of the relevant landowner or council.

The sewerage authority plan will also indicate whether there are any special lot conditions, such as a surcharge area, and whether you will need to install a boundary trap or inspection shaft.

Soffit level

When you have the sewerage authority plan and approved building plans, you should be able to ascertain the difference in levels between the soffit (or obvert) of the approved point of discharge and the lowest point of discharge, which is usually the overflow relief gully. The flood rim of the lowest point of discharge must be of sufficient height above the soffit of the sewer to reduce the possibility of sewer reflux or surcharge on the property.

Australian Standards do not specify the minimum difference in levels, and sewerage authorities may have differing specifications.

While soffit level requirements vary between sewerage authorities, as a general rule there should be at least 1200 mm vertical separation between the soffit of the sewer main and the lowest point of discharge (see Figure 4.2). If there is any less vertical separation than this, then advice should be sought from the sewerage authority. Refer to the relevant authority for more information.

Generally, if the above requirements cannot be met, a reflux valve must be installed.

Where minimum soffit requirements cannot be achieved, fixtures connect by means of:

- an ejector
- pump, or
- reflux valve.

The above must be installed in accordance with AS/NZS 3500.2.

AS/NZS 3500.2 SANITARY PLUMBING AND DRAINAGE

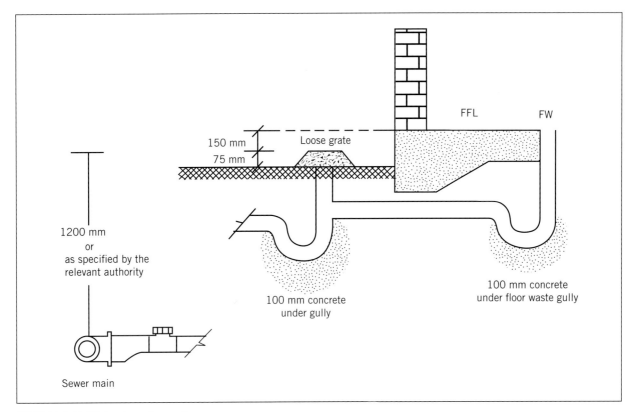

150 mm
75 mm
Loose grate
FFL
FW

1200 mm
or
as specified by the
relevant authority

100 mm concrete
under gully

100 mm concrete
under floor waste gully

Sewer main

FIGURE 4.2 Soffit level requirements

Source: Gary Cook.

Prior to installing a reflux valve, approval must be obtained from the sewerage authority for its use and location.

Safety and environmental requirements

Preparing for WHS will ensure that it is not a *second thought* after the job has commenced or, worse, after a workplace accident.

Each state and territory has WHS regulations that require employers and self-employed people to identify hazards, and assess and control risks, at the workplace in consultation with their workers.

While planning the layout of a sanitary drainage system, you will observe issues that attract your attention, and you should note any special requirements so that you have all of the necessary WHS issues covered before the job commences. For example, you might see that a long run of drain is to be laid, and this might necessitate a deep excavation. From Unit 2, you will remember that trench support is required for a deep excavation, and so you will need to prepare and plan for sufficient trench support.

Planning the layout of a sanitary drainage system is not only about the drain, but also about planning how long the job will take to complete, and allowing for trench support if it is required to carry out the work safely.

You will also need to consider any environmental requirements. As previously mentioned, when the approved point of discharge has been located, you will need to be prepared to provide protection to the point and prevent the entry of any stormwater, sand, silt or rubbish. If you are working in wet conditions, any water from the trench must not be allowed to enter the authority sewer. Trenches must only be dewatered by pumping the trench water out in an approved manner to *an approved point of discharge for stormwater*.

Quality assurance

In planning the layout of a sanitary drainage system, there may be requirements for quality assurance. Normally, a regulatory authority will inspect all, or a percentage, of your drainage work as a form of quality assurance of your work for the sewerage authority and property owner. On larger building sites, it is common to see additional requirements for quality assurance, such as sign-off on portions of your work for progress payment, and additional tests and inspections.

All pipes and fittings used in a sanitary drainage system must comply with the *National Construction Code (NCC) Volume 3, Plumbing Code of Australia* and bear WaterMark and Australian Standards certification.

Preparation for quality assurance requirements may involve 'hold points', where your work must stop in order to allow time for the necessary inspections and tests (see **Table 4.2**).

TABLE 4.2 Examples of tests and inspections

Examples of tests	Examples of inspections
Hydrostatic or air testing	Pipe bedding material and concrete support
	Depth of cover
	Correct 45° junction fittings

Some contractors view inspections and tests as a waste of time, but when something goes wrong, the full value of inspection and testing is realised.

Sequencing of tasks

In planning the layout for a residential sanitary drainage system, it is important that tasks are sequenced so that the building project will run smoothly. Each trade on a building site has a job to do. Understanding the project schedule, and communicating with others, is fundamental to the success of the plumbing business. For example, it was previously mentioned that when preparing for work, you might see on the building site plan that a driveway is to be constructed over the sanitary drain. If this is to be concreted, you will need to know the concreting date so that you will have time to install, test and backfill the drain before that date.

Another example might be that scaffolding is to be erected where you propose to lay the sanitary drain, and so you will need to know when the scaffolding is to be erected and removed.

Proper planning and sequencing of work will assist you in understanding the dates and times of work on one or more projects, so that you can best utilise your time for the successful operation of the plumbing business.

Tools and equipment

To plan the layout of a sanitary drainage system, tools and equipment are needed for measuring, adding quantities, drawing and site inspection. It is best practice first to inspect the site.

Examples of useful tools and equipment are:
- shovel (e.g. to expose inspection covers)
- 30 m wind-up measuring tape
- compass
- torch
- note pad and pencil
- digital camera
- personal protective equipment (PPE), such as sunhat and sunscreen.

Second, off-site measurement and inspection is carried out to determine the quantities and layout of the drainage system. Useful equipment includes:
- drawing board
- scale ruler, with multiple scales, and must include 1:100, 1:200 and 1:500
- pencils, to draw the layout of the sanitary drain
- note pad
- eraser
- calculator
- computer and printer.

Preparing the work area

The efficient planning of sanitary drainage system layouts is achieved by having a clean and well-organised work area. The options for this work area will depend on your level of proficiency in different methods of planning.

In this unit, we will focus on basic drawing skills and methods for determining material quantities. As you progress further, you may decide to develop your skills in using computer software for planning and estimating quantities.

COMPLETE WORKSHEET 1

Planning the system layout

Site inspection

After all the plans, specifications and permits have been obtained, a site inspection can be carried out. Information from the site inspection will assist you in determining any site-specific requirements for the layout of the sanitary drain.

The site visit should be carried out with the building plans, the sewerage authority plan and any information obtained on the location of existing services.

First, determine the location of the approved point of discharge. If the job is for a new residential building, then it is likely that the connection will be to an existing point near a boundary of the property. If the job is for an extension or addition to an existing property drain, then you will need to determine the best location for the 'cut-in', considering the suitability of depth. This is discussed in detail later in this unit.

Second, determine whether there are any existing obstructions in the path of the drain, such as brick fences, foundations, rocks and trees, and the location of existing services, such as power, telecommunications and stormwater.

Third, look at the site levels, the gradient and the level of proposed fixtures. Occasionally, buildings are cut into a hillside. This might restrict the amount of fall available to achieve the necessary gradient in the drain, or might restrict the cover available over the drain.

The above factors – that is, the level at the point of discharge/cut-in; existing obstructions and services; and available fall and cover – are all factors that must be determined from the site inspection, which will assist in planning a layout for the sanitary drain.

Determining the quantity, location and type of fixtures

Quantity
From the building plan/s and specifications, determine the number and location of fixtures that need to be connected to the drain. One method of doing this is by using a highlighter pen to mark all of the internal fixtures.

Location
The location of the fixtures will partially determine the layout of the sanitary drain, because the main line of the sanitary drain should be as short as possible. If the fixtures are, or are to be, located on a first floor, then you will need to know where the drain is to be located in order to connect stack/s.

Type of fixtures
The type of plumbing fixtures will determine the size of drains; therefore, they must be identified at the planning stage.

The main line of the sanitary drain (main drain) must be a minimum of DN 100 from the approved point of discharge or on-site disposal system to the connection of the upstream drainage vent. Connections to this main drain are known as branch drains, and normally range in size from DN 65 to DN 100. The branch drain must be at least the same size as the fixture outlet; however, multiple fixtures connected to the same branch drain can determine its size. This is discussed in greater detail later in this unit.

Developing the system layout in accordance with plans and standards

In this section, we will develop a system layout in accordance with plans and standards, working from the approved point of discharge to upstream. We will take the example of a two-storey residential building on concrete slab construction. The building plan shown in **Figure 4.3** will be used in this exercise.

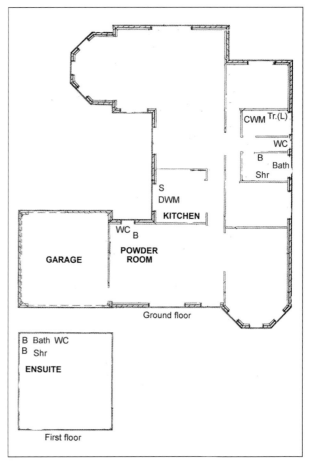

FIGURE 4.3 Example project building plan

Development of example project: As you can see in **Figure 4.3**, we have already found the quantity, location and type of plumbing fixtures and marked them on the building plan. The correct abbreviations have been used, and these are identified in AS/NZS 3500.2. While we are looking at the correct abbreviations for fixtures, it will be useful to examine fixture unit ratings.

AS/NZS 3500.2 SANITARY PLUMBING AND DRAINAGE

Fixtures and fixture unit ratings
When planning the layout of a sanitary drainage system, you need to identify the fixtures, and understand the basics of how the flow from fixtures

will determine drain sizing. Table 4.3 indicates the standard abbreviations for some residential fixtures and their fixture unit rating. These abbreviations must be used when you are preparing a plan drawing for connection of fixtures to the sanitary drainage system.

TABLE 4.3 Fixture abbreviations and fixture unit ratings for residential fixtures

Fixture	Fixture abbreviation	Fixture unit rating
Basin	B	1
Bath	Bath	4
Bidet/Bidette	Bid	1
Clothes-washing machine	CWM	5
Dishwashing machine	DWM	3
Shower	Shr	2
Sink	S	3
Trough	Tr. (L)	5
Water closet pan (with cistern)	WC	4
Bathroom group of fixtures (including bath, basin, shower and water closet in same room)		6
For further information, refer to AS/NZS 3500.2.		

The fixture unit rating for each fixture in Table 4.3 is used to determine the size of the section of the sanitary drain that will connect that fixture. If a group of fixtures is to be connected, the fixture unit ratings are added together to determine the total fixture unit loading on the section of drain that will connect that group of fixtures. This is discussed in greater detail later in this unit. (Read 'Definition', 'Fixture unit' in AS/NZS 3500.0 Plumbing and drainage, 'Glossary of terms'.)

AS/NZS 3500.2 SANITARY PLUMBING AND DRAINAGE

Location and size of approved point of discharge

Once you have the building plans and have identified the quantity, location and type of plumbing fixtures, the approved point of discharge should be located to ensure it is at an adequate depth for drainage by gravity.

Development of example project: On the example sewerage authority plan (see Figure 4.4), we have marked the proposed location of our example building, and will then locate the approved point of discharge. (Note: This arrangement is not approved by all authorities.) The connection at the approved point of discharge for the main drain may be either DN 100 or

DN 150 (depending on the local sewerage authority). The connection can be in the form of:
- a junction
- an inlet into an inspection chamber (sometimes referred to as a 'manhole'), or
- a property connection sewer (sometimes known as a 'side line', where the sewer main is located within another property).

The connection point may be located:
- inside the property
- at the property boundary, or
- outside the property in the footpath or road.

If you work in different parts of Australia, you will find that sewerage authorities use slightly different terminology and different plans; however, all authorities can provide information on the location of their sewer mains and the approved point of discharge for every property. Generally, the approved point of discharge will have been designed and constructed at a level that will allow plenty of scope for connection by gravity, but you must plan to lay drains at the correct minimum gradient.

Development of example project: In the example property sewerage plan, the approved point of discharge is located by referring to 'TIE', 'NSL', 'IL' and 'DIA', where:
- 'TIE' is the distance measured along the property boundary (or fence) line between the boundary corner and centreline of the approved point of discharge
- 'NSL' is the natural surface level
- 'IL' is the invert level, or the level at the lowest point of the internal surface of the point of discharge
- 'DIA' is the diameter of the point of discharge.

To find the depth of the approved point of discharge in the example, simply subtract the IL from NSL – that is, 104.73 minus 102.39 = 2.34 m deep. Remember that the IL is the lowest point of the internal point of discharge, so if the DIA (diameter) is 100 mm, then in our example the depth to the top of the pipe will be approximately 2.24 m.

The top of the approved point of discharge is referred to as the 'soffit'. The soffit is the highest point of the internal surface of pipe. So, we have now found for our example project that the location of the approved point of discharge is:
- 2.24 m depth of cover over the soffit
- 4.2 m measured along the property boundary (or fence) line between the boundary corner and centreline of approved point of discharge
- DN 100 size drain connection at the approved point of discharge.

Types of connection to approved point of discharge

When planning to connect to the approved point of discharge, you should be aware that there are two types

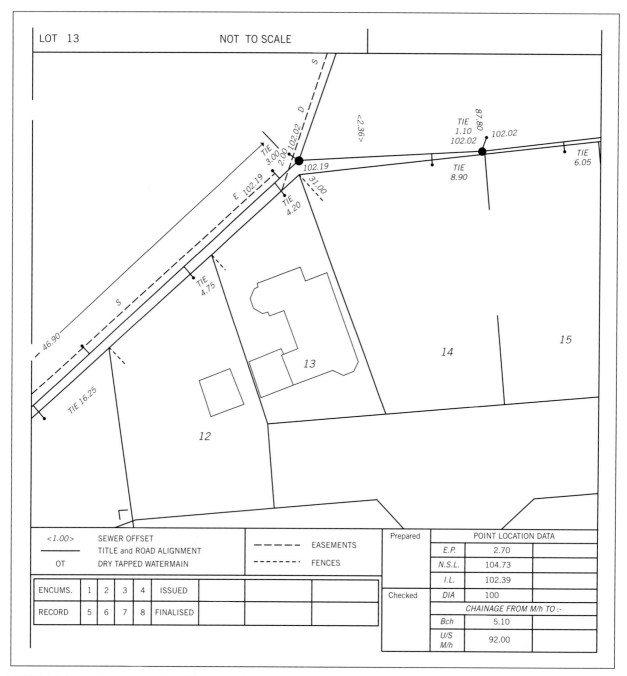

FIGURE 4.4 Property sewerage plan with proposed building

of connections that the property drain can make to the sewer main. They are:

1 boundary trap connection
2 direct connection.

The local sewerage authority determines when a boundary trap is required. Generally, when a boundary trap is required, the sewerage authority plan will have an abbreviated mark such as 'BT' or 'WITH BT'. If there are no such abbreviations on the plan, the connection required will usually be the 'direct' type. If you are unsure, enquire at the sewerage authority when obtaining the sewerage plan.

Boundary trap connection

The purpose of a boundary trap is to disconnect the property main drain from the sewer main. By disconnect, we mean that a water seal is provided to act as a barrier, preventing sewer gases from entering the property drain from the authority's sewer main. Boundary traps are specified for property drains to avoid the possibility of noxious gases or odours from venting sewer mains through property drains.

When a boundary trap is specified, the property drain will require both upstream and downstream ventilation (see **Figure 4.5**).

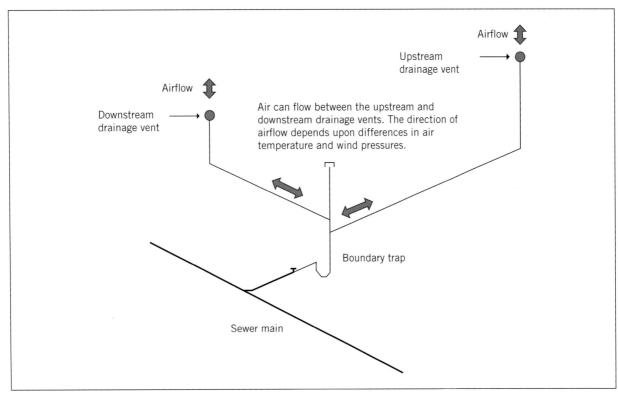

FIGURE 4.5 Boundary trap connection and airflow

Where a boundary trap is located inside a building or in a traffic area, a ground vent cannot be installed at the top of the boundary trap shaft. In these cases, downstream ventilation must still be provided in accordance with AS/NZS 3500.2.

Note that not all authorities permit a vent at the top of a boundary trap shaft, as shown in Figure 4.6.

A boundary trap should always have an inspection shaft extended to the surface level. This provides for inspection and allows access to clear any blockages in the boundary trap or sewerage connection point. Also, a boundary trap must have 100 mm of concrete placed under the trap. This concrete will support the trap material during maintenance and clearance of blockages.

Other conditions for the installation of boundary traps are outlined in AS/NZS 3500.2.

AS/NZS 3500.2 SANITARY PLUMBING AND DRAINAGE

Direct connection

Regardless of whether the connection type is boundary trap or direct, open ventilation is required in the main drain of all property drains. Previously, we found that when a boundary trap is installed, you must provide for both upstream and downstream ventilation in the property sewerage drain. Where boundary traps are not required, downstream ventilation is not needed because ventilation is from the sewer main. Upstream ventilation in property drains assists in the ventilation

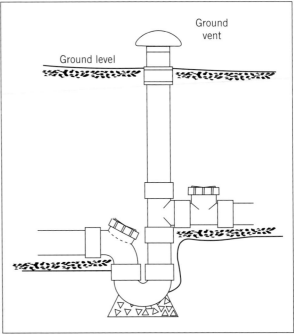

FIGURE 4.6 Boundary trap arrangement with downstream vent fitted to the top of the boundary trap shaft

Source: Department of Education and Training (http://www.training.gov.au) © 2013 Commonwealth of Australia.

of both the main sewer and the property drain. Where boundary traps are not required, the sewer gases are not regarded by the sewerage authority as noxious; therefore, the gases may pass from the sewer main into the property drain and out of the property drain upstream vent (see Figure 4.7).

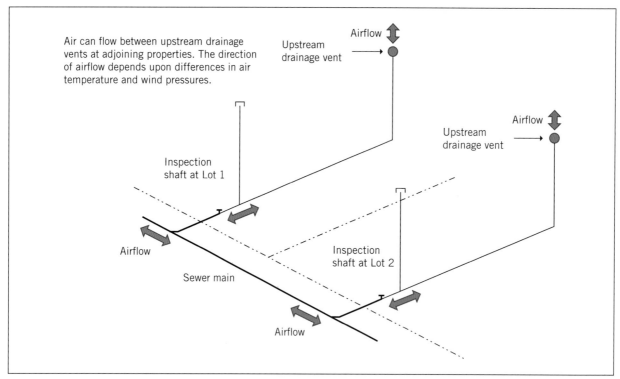

Air can flow between upstream drainage vents at adjoining properties. The direction of airflow depends upon differences in air temperature and wind pressures.

Airflow

Upstream drainage vent

Airflow

Upstream drainage vent

Inspection shaft at Lot 1

Inspection shaft at Lot 2

Airflow

Sewer main

Airflow

FIGURE 4.7 Direct connection and airflow: inspection shaft access

On every property drain where a boundary trap is not required, an inspection shaft must be installed within the property, as close as practicable to the authorised point of discharge, in an accessible position and clear of easements. This provides for inspection and allows access to clear any blockages in the sewerage connection point. Inspection shafts must have 100 mm of concrete placed under the square junction, or around and under the bend forming the jump-up. This concrete will support the fitting material during maintenance and clearance of blockages.

An inspection shaft must be provided where a boundary trap is not required via a:
- square junction (see **Figure 4.8**), or
- jump-up with riser shaft extended to surface level (see **Figure 4.12** later in the unit).

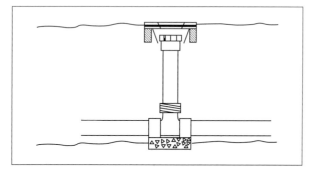

FIGURE 4.8 Inspection shaft riser from square junction

Other conditions for the installation of inspection shafts are outlined in AS/NZS 3500.2.

AS/NZS 3500.2 SANITARY PLUMBING AND DRAINAGE

Summary of requirements for boundary trap shafts and inspection shafts

Boundary trap shafts and inspection shafts must:
- be located wholly within the property, close to the property boundary, close to the authorised point of discharge, clear of easements and accessible
- be visible
- terminate at the surface level with a removable airtight cap (except where a boundary trap shaft is fitted with a ground vent as permitted by the sewerage authority).

Development of example project: In our example project, it can be seen that there are no markings on the property sewerage plan indicating that a boundary trap is required; therefore, our project will require an inspection shaft.

Depth of cover

Earlier in this unit, we discussed the importance of obtaining information in relation to the required depth of cover. If the layout will necessitate the drain being laid under a driveway, and you are unsure if there will be paving, then the depth must be greater than in a garden. AS/NZS 3500.2 provides the minimum depths of cover that must be considered.

AS/NZS 3500.2 SANITARY PLUMBING AND DRAINAGE

Generally, there are three factors that determine depth of cover:

1 the minimum depth of cover specified in AS/NZS 3500.2, considering the location and drain material
2 existing services or obstructions that necessitate a change in drain level
3 the level of existing drain connections or fixture outlets.

Development of example project: In the example project, we will assume that the site is level and there is no vehicular traffic at the rear of the garage. Therefore, the minimum cover over the drain at the furthermost distance from the approved point of discharge must be 300 mm. This is relevant, because on a level site it is important to establish the minimum depth of drain cover at the 'head' of the drain (the top end of the longest length of drain). This will determine the minimum excavation required. Note that the example project is on concrete slab construction, and it is possible that under-slab drains will have been laid to the external foundation. This has been disregarded for the example project. If there are under-slab drains, they must be located first, as they may determine the minimum depth of the external sanitary drain. Drains laid under or through concrete foundations must have a minimum of 25 mm clearance around the pipe.

Line of drain

In planning the layout for sanitary drainage, the drain needs to be located external to the building, where possible. This is to allow future access to the drain for maintenance and alterations. The alignment of the drain should be straight, with minimum continuous even gradient, minimum bends, and clear of buildings and easements. Where the drain is in close proximity to the building, care must be taken to ensure that trenching required for the drain does not compromise the structural strength of the building (see AS/NZS 3500.2).

Depending on soil composition, the distance from the bottom side of the trench to the base of the building footings can range from a ratio of 8:1 (stable rock) to 1:4 (silt). For example, if a drain is to be laid in an area that has stable rock, a ratio of 8:1 is to be used (see Figure 4.9). In this ratio, 1 is 'L' (relative distance from the edge of the trench to the edge of the footing) and 8 is 'H' (relative vertical height from the base of the trench to the base of the footing – i.e. rise:run).

When excavating for the drain, it might not be possible or sensible to dig near the footing to determine the footing depth. In such cases, it is best to refer to the building drawings, which specify the footing design depth.

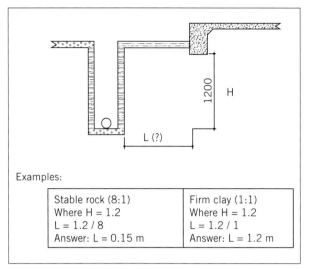

Examples:

Stable rock (8:1)	Firm clay (1:1)
Where H = 1.2	Where H = 1.2
L = 1.2 / 8	L = 1.2 / 1
Answer: L = 0.15 m	Answer: L = 1.2 m

FIGURE 4.9 Calculating the required distance from trench to footing

Source: Department of Education and Training (http://www.training.gov.au) © 2013 Commonwealth of Australia.

In any situation where the required distance from the footing cannot be achieved, you should consult a qualified engineer to determine the necessary requirements.

FROM EXPERIENCE

Plan to keep excavations away from building foundations. The protection of building structures is critical.

Development of example project: In the example project, it can be seen that all drains could be kept external of the building and well clear of footings, except at the side of the building adjacent to the bathroom, where there is only 1 m between the footings and the property boundary. The distance from the centre of the bathroom to the rear corner of the building is 6.5 m (see Figure 4.10).

The minimum depth of the trench at the top end (head) of the drain will be minimum cover + pipe diameter + bedding. The minimum depth in the example is 300 mm + 100 mm + 75 mm = 475 mm. Over a distance of 6.5 m, and with a drain gradient of 1.65% (1:60), our trench will fall approximately 107.25 mm (1.65/100 × 6500 mm). Therefore, the depth of our trench will be 565.75 mm (475 + 107.25), or approximately 585 mm at the end of the building.

If we assume that the soil conditions are firm clay (ratio 1:1), and the footing depth is 500 mm, then the vertical distance from the base of the footing to the closest bottom edge of the trench near the end corner of the building will be only 85 mm. Therefore, the edge of the trench must be at least 85 mm away from the footing. Remember, this is only a small distance, but the measurement is taken from the edge of the trench, not the drain.

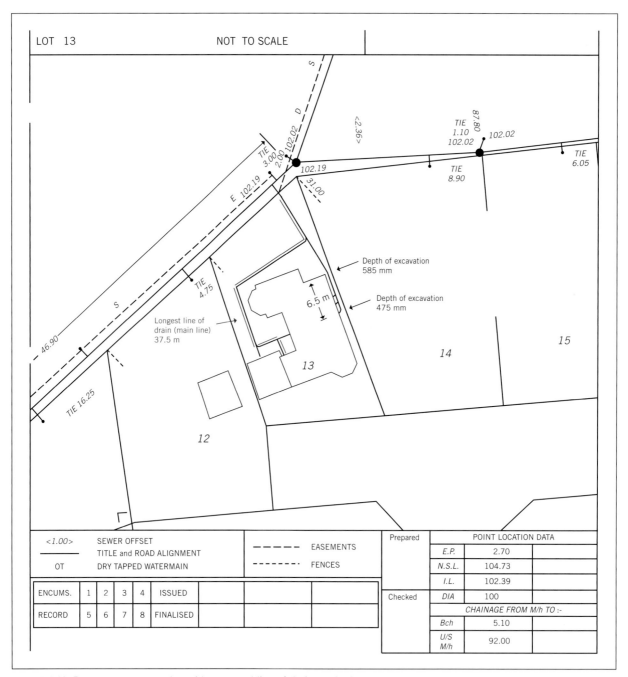

LOT 13   NOT TO SCALE

Depth of excavation 585 mm

Depth of excavation 475 mm

6.5 m

Longest line of drain (main line) 37.5 m

13

14

15

12

	<1.00>	SEWER OFFSET			— — — —	EASEMENTS	Prepared		POINT LOCATION DATA	
		TITLE and ROAD ALIGNMENT						E.P.	2.70	
	OT	DRY TAPPED WATERMAIN			- - - - -	FENCES		N.S.L.	104.73	
								I.L.	102.39	
ENCUMS.	1	2	3	4	ISSUED		Checked	DIA	100	
RECORD	5	6	7	8	FINALISED			CHAINAGE FROM M/h TO :-		
								Bch	5.10	
								U/S M/h	92.00	

FIGURE 4.10 Property sewerage plan with proposed line of drain marked

In our example case, we can see that our drain can easily be laid between the building and property boundary fence without encroaching on the restricted zone below the line of footing.

Drain gradient

Drains must be laid at a sufficient grade to maintain self-cleansing velocity. This means solids will not accumulate in the drainage system, causing blockages. Table 4.4 indicates the maximum number of fixture units permitted to pass through drains laid at various gradients. The 'nominal' minimum grades at which that drain may be laid are in bold.

The figures in brackets indicate the maximum fixture unit loadings for drains at reduced gradients; however, laying drains at reduced gradients is not recommended, unless it is absolutely necessary. There are also minimum fixture unit loadings for drains laid at reduced gradients (refer to AS/NZS 3500.2).

AS/NZS 3500.2 SANITARY PLUMBING AND DRAINAGE

TABLE 4.4 Maximum fixture unit loadings for vented drains

Gradient (%)	Nominal size of drain		
	DN 65	DN 80	DN 100
	Maximum fixture unit loadings		
5.00	60	215	515
3.35	36	140	345
2.50	**25**	100	255
1.65	np*	**61**	**165**
1.45	np	(50)	(140)
1.25	np	(42)	(120)
1.00	np	np	np

*'np' indicates 'not permitted' for this size of drain.

Notes: (a) The minimum size of a branch drain receiving discharges from a water closet pan is DN 80. (b) The minimum size of a main drain is DN 100. (c) The minimum size of a branch drain is DN 65. (Refer to AS/NZS 3500.2.)

Development of example project: At this point, we could measure and calculate whether the available fall will satisfy the gradient requirements. We have assumed a level block, and found that the depth of cover over the soffit at the approved point of discharge is 2.24 m. Also, the minimum cover required over the drain at the rear of the building garage is 300 mm. Therefore, the available fall for the main line of drain on this level block is 1.94 m (2.24 – 0.3 m = 1.94 m). Now, we can use the property sewerage plan or site plan to mark and measure a theoretical main line of drain. Remember to consider the location of any other services and obstructions, and to keep the drain well clear of buildings and easements (see Figure 4.10).

As can be seen on the property sewerage plan, the proposed longest line of drain (main line or main drain) is between the authorised point of discharge and the downstairs powder room. The plan in Figure 4.10 is not to scale. For the exercise, accept that the line of drain along the red line is approximately 37.5 m. The required gradient for a DN 100 drain is 1.65%. Therefore, the required fall is approximately 0.619 m (1.65/100 × 37.5 = 0.619 m).

Since the available fall is 1.94 m, and the required fall is 0.619 m, we have excessive fall, and can lay the drain at 1.65% and install a jump-up at the inspection shaft.

Jump-ups

Vertical jump-ups can be constructed at any point along the line of the drain. They can be used to:

■ reduce the amount of excavation required
■ change the level of the drain to avoid existing services and obstacles.

A jump-up has a bend at the base of the vertical section that is supported by a 100 mm-thick concrete footing (see Figure 4.11). At the top of the jump-up, a bend incorporating a full-size inspection opening (IO) is used. Alternatively, a junction fitting that is fitted with an inspection opening may be used instead of an IO bend.

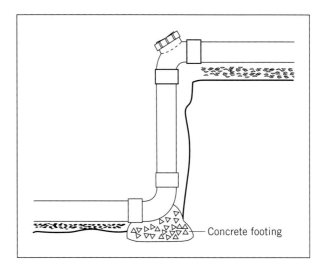

FIGURE 4.11 Jump-up constructed in PVC

Source: Department of Education and Training (http://www.training.gov.au) © 2013 Commonwealth of Australia.

An inspection shaft jump-up is simply an extension of a normal jump-up. The inspection shaft jump-up is often used for 'direct' connections to the authority-approved point of discharge where the connection point has depth in excess of that required to achieve the minimum gradient.

Instead of a bend at the top of the jump-up, a sweep junction is used. An IO may be fitted to the lead-off drain (see Figure 4.12). The top of the inspection shaft is terminated with an airtight cap. An inspection

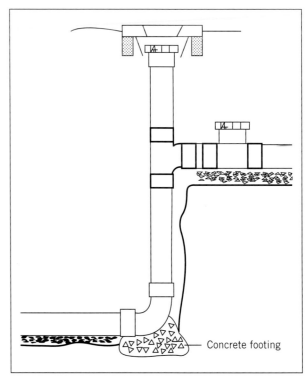

FIGURE 4.12 Inspection shaft jump-up constructed in PVC

Source: Department of Education and Training (http://www.training.gov.au) © 2013 Commonwealth of Australia.

shaft cover is located over the top of the shaft and independently supported so that if there is any loading or traffic on the cover, that loading is not transferred to the shaft.

Development of example project: In the example project, an inspection shaft jump-up (as illustrated in Figure 4.12) is used at the connection to the approved point of discharge (point) because the available fall is 1.94 m and the required fall is 0.619 m. Therefore, the difference in soffit levels between the drain at the base of the jump-up and lead-off drain will be 1.321 m (1.94 – 0.619 = 1.321 m). By installing an inspection shaft jump-up, it is possible to save 1.321 m in excavation depth over the entire length of the main line of drain.

Note: The inspection shaft jump-up should always be constructed last, as services and obstructions (such as rock) are often encountered in the process of laying the drain and this can alter the layout and drain levels.

Inclined drains and inclined jump-ups

Another means of reducing excavation is to install an inclined jump-up (see Figure 4.13). Any drain that is laid on a grade of more than 20% (1:5) is referred to as an 'inclined drain' or 'steep grade drain', and must have anchor blocks installed. The anchorage points ensure the drain is not subject to movement because of the forces of soil movement due to gravity, and inertia applied by hydraulic forces within the drain.

Inclined drain anchor blocks are installed at the bed or junction, at the top and bottom of the inclined drain, and at intervals not exceeding 3 m in long sections of steep-graded drains. An inspection opening is installed immediately above every jump-up. (Refer to AS/NZS 3500.2.)

AS/NZS 3500.2 SANITARY PLUMBING AND DRAINAGE

Proximity to other services

Service pipes and conduits, such as gas, stormwater, communication or electricity services, are likely to be encountered on a building site, and you will need to read information obtained from authorities (e.g. through Dial Before You Dig), read the building plans and communicate with other trades on-site to determine whether allowances need to be made when planning to lay out the sanitary drain. This will ensure that damage to other services or the sanitary drain does not occur during the building project, or in the future when maintenance or alterations are carried out.

In general, you are required to provide a separation of 100 mm between the drain and protected gas pipes and electrical conduits, while unprotected services must be provided with a separation of 600 mm.

When laying the sanitary drain in close proximity to a stormwater drain exceeding 100 mm in size, you will need to provide a clearance of 300 mm.

You should research the requirements for clearance from services by reading AS/NZS 3500.2.

AS/NZS 3500.2 SANITARY PLUMBING AND DRAINAGE

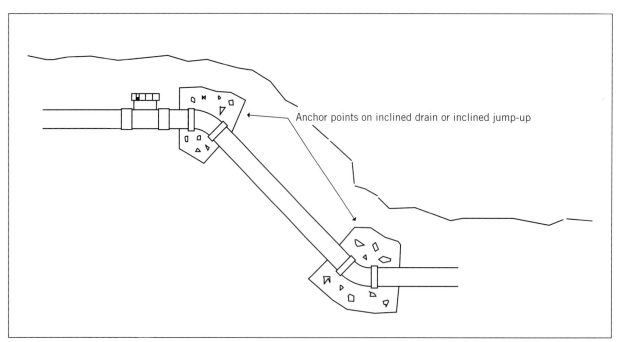

Anchor points on inclined drain or inclined jump-up

FIGURE 4.13 Inclined drain or steep grade drain with anchor blocks

Source: Department of Education and Training (http://www.training.gov.au) © 2013 Commonwealth of Australia.

Size of drains

Main drain

Earlier in this unit we found that the minimum size of the main drain is DN 100. The main drain could be defined as the longest section of DN 100 property drain that provides upstream ventilation.

If the DN 100 main drain is to be laid at the minimum gradient of 1.65%, then, as indicated in Table 4.4 (see earlier in the unit), the maximum number of fixture units that can pass through this drain is 165.

In some developments, the number of fixture units may exceed 165 and a larger main drain might therefore be required. However, where a steeper gradient is available, the fixture loading can be increased without increasing the size of the drain. (Refer to AS/NZS 3500.2.)

Branch drains

Branch drains are connected to the main drain, and may be the same size or smaller than the main drain. Branch drains receive the discharges from individual fixtures or groups of fixtures.

Branch drains may be vented or unvented, and while the gradient requirements are the same for both, vented branch drains can receive a much greater number of fixture units (fixture unit loading). For example, for a 65 mm branch drain the maximum number of fixture units is:

- vented 65 mm drain laid at 2.5%: 25 fixture units
- unvented 65 mm drain laid at 2.5%: 5 fixture units.

When planning the layout of a sanitary drain, you should be aware of whether the drain will be vented or unvented. There are restrictions on unvented drains.

While larger drains are permitted to carry a greater number of fixture units, the drain must not be increased in size to take advantage of lesser gradients. Large drains carrying low fixture unit loadings are more prone to blockages.

Maximum distances in unvented branch drains

Unvented branch drains are commonly used to receive discharges from a fixture or small group of fixtures. The allowable fixture unit loading on the drain is considerably less than that for a vented drain, but the benefit is that a vent is not required. There are also disadvantages to unvented branch drains. For example, the maximum length of piping measured along the line of pipe between any fixture trap and the vented drain must not exceed 10 m.

Why are there restrictions on unvented drains? All fixtures are provided with a trap, which retains a 'water seal'. Loss of the water seal would result in foul and offensive gases entering the building. The air in the main vented drain contributes some ventilation to unvented connections; however, when the unvented branch drain is too long, the protection of trap water seals is limited. Lack of drain ventilation can result in loss of trap water seal/s by syphonage. (See Unit 5 for more information on syphonage.)

If the distance of 10 m must be exceeded, a vent must be installed. As mentioned earlier, branch drains may be vented. The benefit of vented branch drains is that they can receive a much greater number of fixture units, and they can be run for an unlimited distance, providing that the drain has the correct minimum gradient. (Read the section 'Unvented branch drains' in AS/NZS 3500.2.)

AS/NZS 3500.2 SANITARY PLUMBING AND DRAINAGE

Development of example project: In the example project, we can see that there are three branches from the main drain. Moving upstream from the point of discharge, there is one branch to the laundry and bathroom fixtures, another branch to the rear of the garage to connect the first-floor fixtures, and another branch to the kitchen sink. Will these branches be vented or unvented?

The plan in Figure 4.10 is not to scale. For the exercise, accept the measurements as follows:

- The branch to the laundry and bathroom fixtures is approximately 12.5 m; therefore, this branch must be vented.
- The branch to the rear of the garage is less than 2 m; therefore, on first appearance, this branch could be unvented. However, later we will learn that it must be vented, as it will connect a stack to first-floor fixtures.
- The branch to the kitchen sink is only 2 m; therefore, this branch could be unvented.

Ventilation of drains

When connecting drains of different sizes, the top (soffit) of the drain must be in continuous alignment. This will ensure that air is free to flow along the top inside of the drain, maintaining ventilation within the drainage system.

Vents are installed to serve four main functions. These are to:

1 stop the loss of water seals through syphonage
2 maintain equilibrium of pressures in the drainage system
3 reduce objectionable and corrosive gases accumulating in drains
4 allow the drain to dry out to assist self-cleansing. Vents are required at:

- both ends of a main drain that incorporates a boundary trap
- the upstream end of any main drain to a direct connection (no boundary trap)
- the upstream end of any branch drain that has a fixture trap or floor waste gully where the distance

from the weir of the trap to the vented drain exceeds 10 m

- the upstream end of any branch drain connected to a drainage trap where the distance from the weir of the drainage trap to the vented drain exceeds 10 m
- the upstream end of any DN 100 branch drain where three or more water closet pans are connected
- the upstream end of any DN 80 branch drain where two water closet pans are connected. (Read AS/NZS 3500.2.)

At this point, you should now be aware of two types of drainage connections: boundary trap and direct. The drainage ventilation requirements for each of these connections are different.

Upstream ventilation

Upstream ventilation must be provided on every drain, whether the connection type is boundary trap or direct.

An upstream vent must be provided in one of the following ways:

- the total length of unvented pipework connecting the furthermost fixture trap must not exceed 10 m in length, or
- as the vent extension of a stack at or near the upstream end of the main drain. (Read AS/NZS 3500.2.)

The upstream vent may be connected as a separate branch to the main drain (see Figure 4.14) or to a fixture discharge pipe (see Figure 4.15).

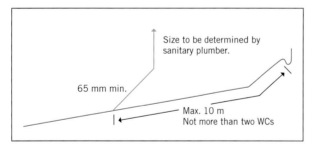

FIGURE 4.14 Upstream drainage vent extended as a separate branch off the drain

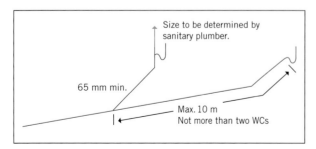

FIGURE 4.15 Upstream drainage vent extended from a fixture discharge pipe connected directly to the drain

Any section of drain acting as a vent must be no less than 65 mm. The size of the vent is determined by the total number of fixture units discharging through the drain. Regardless of the number of fixture units discharging through the drain, the minimum size of any upstream main drain vent is DN 50. (Read AS/NZS 3500.2.)

Downstream ventilation

Where a boundary trap is fitted, the property sewerage drain must be fitted with a downstream vent. This is because the boundary trap water seal prevents the passage of air and gases from the sewer main to the property drain; and the property drain requires air circulation to function properly.

Downstream ventilation is achieved by installing a vent at the lower end of the drain close to the boundary trap, allowing air to circulate through the main drainage system between the upstream and downstream vents.

Where a boundary trap has been installed, downstream ventilation is provided in property drains in one of the following ways:

- boundary trap ground vent (see Figure 4.16) (some authorities do not permit this type of vent)
- downstream vent. (Read AS/NZS 3500.2.)

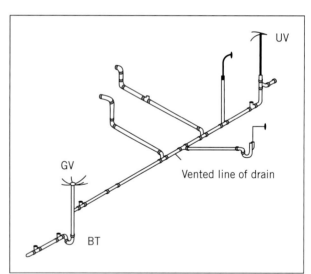

FIGURE 4.16 Boundary trap with downstream ground vent and upstream vent

Where a ground vent is fitted to the top of a boundary trap shaft, it must be clear of driveways and pathways, and must be located more than 3 m from any opening into a building and 5 m from an air duct intake. This rule is necessary to prevent foul air from the drain entering the building. There are also other requirements for the location of vents. (Read AS/NZS 3500.2.)

Some sewerage authorities do not permit a downstream ground vent to be fitted to the top of a boundary trap shaft. The alternative is to fit a junction to the top part of the boundary shaft and extend a drain from this junction to terminate elsewhere with a vent.

AS/NZS 3500.2 SANITARY PLUMBING AND DRAINAGE

Ground vents must be installed so that the inlet of the vent is not less than 150 mm above ground level, to prevent the ingress of stormwater.

If it is not permitted or possible to install a ground vent at the top of the boundary shaft, then the downstream ventilation must be provided within 10 m of the boundary trap shaft, providing there is no other connection between the boundary trap riser and the vent connection (see Figure 4.17). (Read AS/NZS 3500.2, and if you are unclear about this, discuss it with your teacher/trainer.)

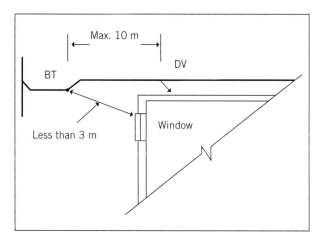

FIGURE 4.17 Plan view: boundary trap shaft less than 3 m from an opening to a building, and downstream vent branch connected to the main drain within 10 m

Source: Department of Education and Training (http://www.training.gov.au) © 2013 Commonwealth of Australia.

The minimum size of any downstream vent is 50 mm for up to 30 fixture units. The size of downstream ventilation is determined based on the fixture unit loading on the boundary trap. (Read AS/NZS 3500.2.)

Air admittance valves

Air admittance valves may be used as trap vents, group vents and stack vents. They can also be used to ventilate branch drains. Air admittance valves have the same general purpose as vents in protecting trap water seals.

Air admittance valves allow air to enter the drain; however, gases within the drain cannot escape the drainage system through an air admittance valve. For this reason, they must not be used for ventilation of the main property drain. (Read AS/NZS 3500.2.)

Development of example project: Now that we have learnt about ventilation, we can apply it to our example project.

- The branch to the laundry and bathroom fixtures is approximately 12.5 m; therefore, this branch will be over 10 m to the furthermost fixture trap on that section of drain and so must be vented. This is a branch from the main drain; therefore, it does not require a main drain vent. An air admittance valve could be installed (see Figure 4.18).
- The branch to the rear of the garage is less than 2 m. On first appearance, this branch could be unvented. However, we have learnt that it must be vented, because it will connect a stack to first-floor fixtures. Remember that upstream ventilation can be provided through the vent extension of a stack.
- Now that we can achieve upstream ventilation through the vent extension of a stack, we can redefine our main drain. It will now be the main line of drain between the authorised point of discharge (point) and the stack.
- The branch drain from the main drain connecting the sink is approximately 6 m, and the branch drain from the main drain to the powder room is approximately 6.5 m. As both measurements are less than 10 m, these drains shall be classified as unventilated branch drains. Remember that the 10 m measurement must include allowance for fixture discharge pipes. If you are unsure, read AS/NZS 3500.2 and discuss with your teacher/trainer.

AS/NZS 3500.2 SANITARY PLUMBING AND DRAINAGE

Surcharge protection

All residential buildings must be protected from surcharge from the authority sewer main. Surcharge is an overflow caused by overloading on the sewer main, usually caused by wet weather infiltration or inflow.

Surcharge from sanitary fixtures can cause damage to the building, and can create unsanitary conditions. Therefore, it is important that the lowest opening on each property is not internal within a building. In most cases, an overflow relief gully (ORG) is satisfactory as the lowest opening and is the best protection against sewer surcharge.

Overflow relief gully

The purpose of the ORG is to prevent surcharge from entering residential buildings. Each property sewerage

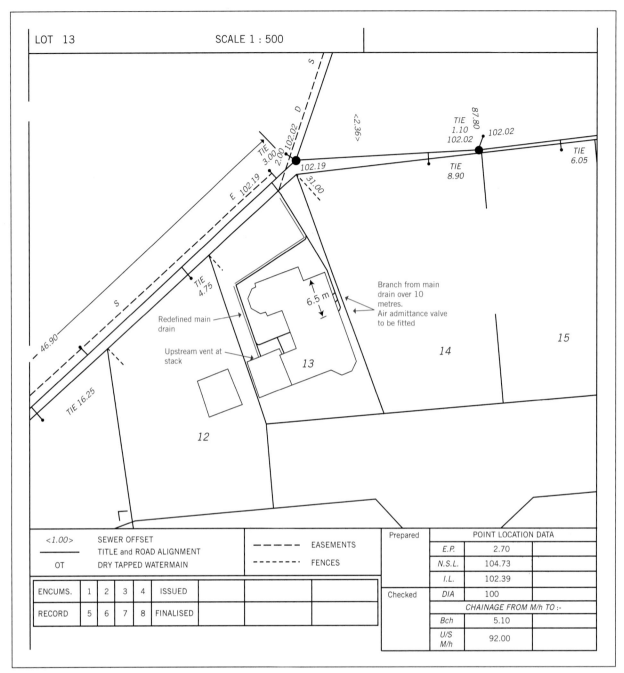

LOT 13   SCALE 1 : 500

Branch from main
drain over 10
metres.
Air admittance valve
to be fitted

6.5 m

Redefined main
drain

Upstream vent at
stack

13

14

15

12

<1.00>	SEWER OFFSET					- - - - -	EASEMENTS		Prepared	POINT LOCATION DATA		
	TITLE and ROAD ALIGNMENT					- - - - -	FENCES			E.P.	2.70	
OT	DRY TAPPED WATERMAIN									N.S.L.	104.73	
										I.L.	102.39	
ENCUMS.	1	2	3	4	ISSUED				Checked	DIA	100	
										CHAINAGE FROM M/h TO :-		
RECORD	5	6	7	8	FINALISED					Bch	5.10	
										U/S M/h	92.00	

FIGURE 4.18 Development of example project: drain ventilation

drain must be fitted with at least one ORG, except where omitted as permitted in AS/NZS 3500.2.

A disconnector gully can be constructed as an ORG, providing that it meets the necessary vertical clearance requirements (see **Figure 4.19**). *The spill level of the ORG must be at least 150 mm below the lowest fixture.* This is measured differently for different fixtures, but usually the measurement will be at least 150 mm between the spill level of the ORG and the top surface level of the shower grate.

The spill level of an ORG must also be *75 mm above natural ground level.* Where the gully surface fittings are located in a pathway or a paved area, the spill level of the fittings must be above paving level to prevent the ingress of stormwater/surface water. (Read AS/NZS 3500.2.)

AS/NZS 3500.2 SANITARY PLUMBING AND DRAINAGE

ORGs must be charged – that is, the water seal in the gully must be maintained by a source of water/waste. This is usually achieved by connecting a waste fixture to the ORG, and in this case the ORG

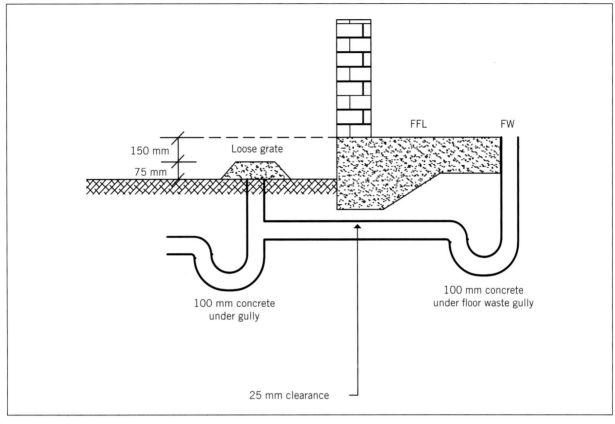

FIGURE 4.19 Vertical clearances: ORG and fixtures

Source: Gary Cook.

must be located as close as possible to the wall of the building near the plumbing fixture/s. (Read AS/NZS 3500.2.)

Occasionally, there may be some problems achieving the necessary vertical separation for the ORG surface fittings. In these cases, there may be some benefit in installing the ORG out from the building, and charging the ORG with a tap, and not by a fixture (see Figure 4.20).

Earlier in this unit we discussed the requirements for vertical separation between the soffit of the sewer main and the lowest fixture outlet. Remember that this soffit level requirement varies between sewerage authorities; however, as a general rule, there should be at least 1200 mm vertical separation between the soffit of the sewer main and the lowest fixture outlet. When an ORG is fitted, this requirement applies to the vertical distance between the soffit of the sewer main and the spill level of the ORG (see Figure 4.21). If there is any less than 1200 mm, then advice should be sought from the sewerage authority.

The surface collar fitting of the ORG must incorporate a loose-fitting domed grate. The purpose of the domed grate is to discourage the owner/occupier from placing pots and other objects on the grate,

because this may prevent the grate from lifting in the event of surcharge. For the same reason, the grate must not be a screwed type or be otherwise fastened to the collar fitting. The ORG should function in a way that allows the sewer surcharge to escape, with the upward pressure of liquid containing solids lifting the grate.

In most domestic installations, the size of an ORG is DN 100; however, if the drain is greater than DN 150, the size of the ORG is increased to DN 150.

The location of the ORG must be:

- within the property boundary, and
- generally external to the building, and
- the grate must be accessible and surcharge noticeable, and
- there must be 2 m clear access above the grate for maintenance.

An ORG can be installed internal to the building, providing it meets special conditions. (Read AS/NZS 3500.2.)

AS/NZS 3500.2 SANITARY PLUMBING AND DRAINAGE

Reflux valve

Generally, reflux valves are installed where it is not possible or practical to install an overflow relief gully.

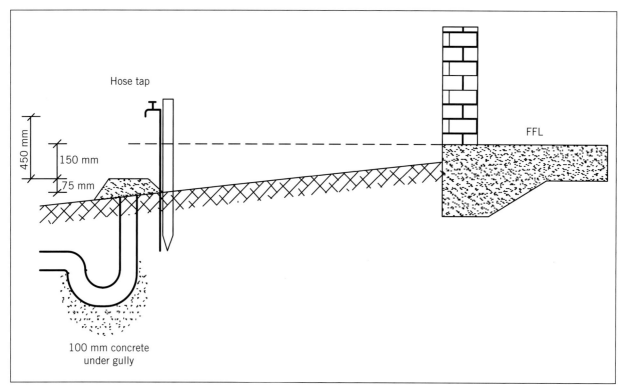

FIGURE 4.20 Achieving vertical separations for ORG fittings

Source: Gary Cook.

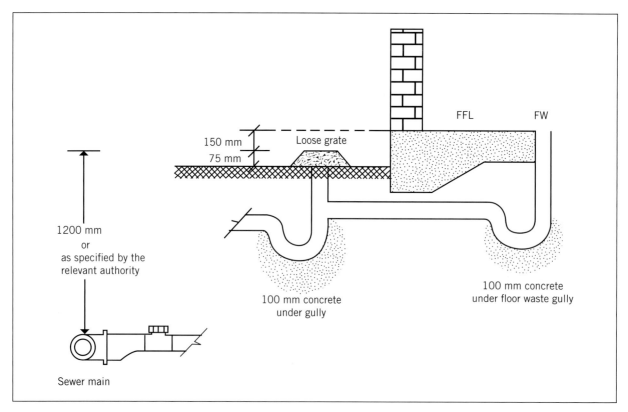

FIGURE 4.21 Vertical clearances: ORG and soffit of sewer main

Source: Gary Cook.

Reflux valves are used to prevent property damage from sewer surcharge where:

- the minimum required vertical separations for ORG surface fittings, as mentioned earlier, cannot be achieved
- the property is located in an area that is highly susceptible to surcharge, or
- there are fixtures that are at a lower level than the ORG.

Reflux valves are designed to allow flow in one direction only. The flow under normal conditions lifts the flap off its seat, allowing the passage of effluent through the valve into the sewer. In surcharge conditions, the flow (which has been reversed) cannot flow back because the valve closes against its seat so that adverse flow is impossible.

Where a reflux valve is to be installed, it must be located as follows:

- Where the connection type is direct and the drain has an inspection shaft, the reflux valve must be installed downstream of the shaft.
- Where the connection type is boundary trap, the reflux valve must be installed immediately downstream from the boundary trap.
- Elsewhere in the property, sewerage drain to protect against surcharge from fixtures that are not protected by an ORG.

All reflux valves must be accessible for inspection and maintenance.

Where cast-iron reflux valves are installed above ground, they must be accessible so that the flap mechanism can easily be removed, and the top of the valve can be used as an access point to the drain to clear blockages. Where cast-iron reflux valves are

installed below ground, they must be installed in an inspection chamber (see Figure 4.22).

Where a DN 100 PVC reflux valve is to be installed below ground, it may be located in an inspection chamber, or below ground with a DN 150 PVC shaft extended to surface level. Where a DN 100 PVC reflux valve is authorised for installation without an inspection chamber, the flap in the valve is attached to a DN 25 mm class 12 PVC pipe, which is solvent welded and extended through the centre of the shaft to the underside of a 150 mm screwed airtight cap near surface level (see Figure 4.23). This enables the flap mechanism to be removed for service from ground level. The benefit is that it is not required to be installed in a chamber. This is a significant cost saving.

Note that some sewerage authorities require that a reflux valve be installed 80 mm higher than the downstream connection to improve flow away from the valve mechanism (see Figure 4.23).

While reflux valves can restrict the airflow in a drainage system, the normal venting requirements apply. However, there can be problems in positive pressure areas, such as near boundary traps and stacks in multi-storey buildings, and in these instances, additional venting may be required.

Before a reflux valve can be installed, some sewerage authorities require a signed indemnity form. The property owner or agent may be required to complete and sign the indemnity form. You should never install a reflux valve until you have met the requirements of the sewerage authority with regard to the indemnity form.

The purpose of the indemnity form is to place responsibility for maintenance of the reflux valve with the property owner.

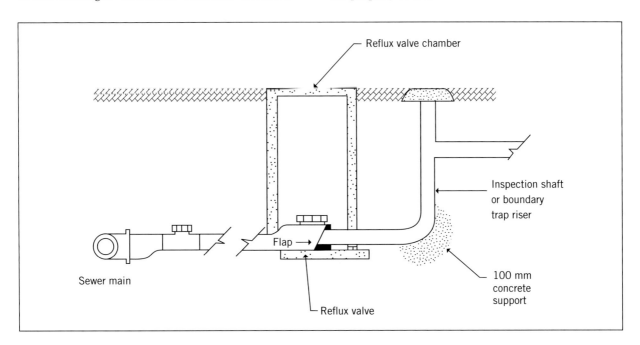

FIGURE 4.22 Cast-iron reflux valve installed in an inspection chamber

Source: Gary Cook.

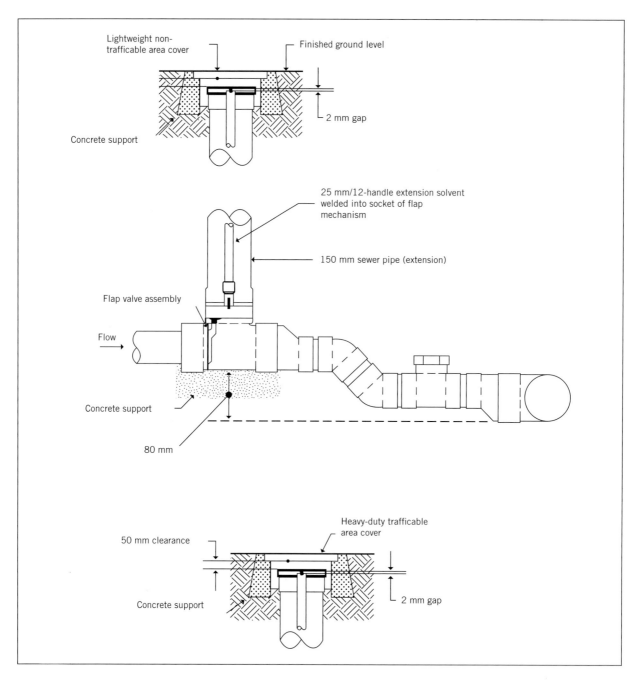

Lightweight non-trafficable area cover

Finished ground level

2 mm gap

Concrete support

25 mm/12-handle extension solvent welded into socket of flap mechanism

150 mm sewer pipe (extension)

Flap valve assembly

Flow

Concrete support

80 mm

50 mm clearance

Heavy-duty trafficable area cover

Concrete support

2 mm gap

FIGURE 4.23 PVC reflux valve installed below ground accessible from ground level

Source: Gary Cook.

Some sewerage authorities require a reflux valve maintenance sign to be installed on the reflux valve access cover, as shown in **Figure 4.24**.

Stacks

When planning the layout for a sanitary drainage system, you will need to know the location of stacks to connect first-floor fixtures and above. This is because there are requirements for bends at the base of stacks and restrictions on the connection of fixtures in close proximity to the base of a stack. (See Unit 5 for more detail.)

A stack is a vertical pipe extending through more than one floor level and is always provided with a vent or air admittance valve. Two-storey residential buildings are becoming increasingly popular in high-density housing, and stacks are commonly used in domestic plumbing installations.

For stacks up to two floor levels, no connection should be made to the drain within 500 mm upstream

SEWER REFLUX VALVE MAINTENANCE REQUIRED

FIGURE 4.24 Example reflux valve maintenance sign

and 500 mm downstream of the stack connection. Also, no connection may be made within 600 mm above the bend at the base of the stack.

Where a stack extends through no more than two floor levels, a single 88° bend may be used at the base of the stack. For most residential sanitary drainage installations, a 100 mm × 88° bend would be provided to connect a stack.

Where the number of floor levels is three or more, the requirements are greater. (Read AS/NZS 3500.2.)

Junctions

When planning the layout for a sanitary drainage system, it is important to be aware of the junction requirements to construct branches. When junctions are installed on a gradient, they must be 45°. This is commonly referred to as an 'oblique junction'. Where a junction is installed on its back to connect a closet pan, it must be a 45° junction.

Special fittings for filled or unstable ground

AS 2870 Residential slabs and footings has special design requirements for drains associated with residential slab and footing systems on moderately, highly or extremely reactive soils.

Always check the building plans or consult with the site supervisor regarding the site classification. If the sanitary drain is to be laid on a Class H1, H2, E or P site, then provisions for movement will need to be designed to ensure that the drain will suit the ground conditions. Such provisions might include swivel and telescopic joints, and you would need to calculate the relevant quantities and plan to install them. Examples are provided in Unit 10.

If there are no building plans or other information regarding the site classification, you can make arrangements for a geotechnical engineer to carry out investigation and provide a report.

Inspection openings

Inspection openings are required in sanitary drains to allow access for inspection, testing and maintenance. Therefore, they need to be placed so that access can readily be achieved on both main drains and branch drains. Inspection openings must be installed:

- at the connection to the sewerage authority authorised point of discharge

- immediately at, or upstream of, the bend at the top of a jump-up
- at every change in horizontal direction of greater than 45°
- on the main drain at the end of each straight section at intervals of not more than 30 m
- on branch drain connections to closet pans, not more than 2.5 m from the building
- on the downstream end of any drain where it passes from underneath a building, except where waste fixtures only are connected to that drain
- where any new section of drain is connected to an existing drain
- at every change in gradient greater than 45°
- at other locations as required by the regulatory authority.

Inspection openings in underground drains should be:

- the same size as the drain for those DN 150 and smaller
- not less than DN 150 for drains larger than DN 150. Inspection openings may be in the form of:
- inspection branches, square junctions and side access bends
- inspection chambers
- reflux valves.

It is good practice to extend inspection openings that are beneath concrete, paving or floor surfaces to the finished surface level, allowing ease of access. (Read AS/NZS 3500.2 and local regulatory authority requirements.)

Cut-ins

A branch insertion into an existing drain is commonly referred to as a 'cut-in'. If your work will require a cut-in to an existing drain, an inspection opening must be installed.

The inspection opening can provide access to serve three purposes, namely:

1 to smooth render any internal gaps, as is occasionally necessary when cutting into a vitrified clay drain
2 to insert a testing plug to test the new section of drain
3 to clear blockages of the drain (see Unit 7).

Disconnection of sanitary drains

Occasionally, properties are redeveloped and the sanitary drainage system must be disconnected from the sewer main at the authorised point of discharge. The following requirements apply to disconnection of sanitary drains:

- Disconnection is to be made at the authorised point of discharge in an approved manner.
- No water, soil, rock or other substances are permitted to enter the authorised point of discharge.

It may be necessary to carry out dewatering of the excavation, particularly if the ground is water charged. This is necessary to prevent water entering the authorised point of discharge during disconnection and sealing works.

When disconnecting a branch drain, the same requirements apply, except the drain must be disconnected as close as possible to the drain remaining in use.

The method of sealing must be suitable to the materials remaining in use, and must be watertight to prevent infiltration of water. (Read AS/NZS 3500.2.)

✔✔ AS/NZS 3500.2 SANITARY PLUMBING AND DRAINAGE

Plan drawing and recording

In this section, we will develop our example project into a drawing to enable us to determine the quantities of materials and fittings to complete the job.

First, there are many sewerage authorities with different procedures and practices for drawings. You will most likely find that the symbols used in the example appear unfamiliar, and in this case your teacher or trainer can assist you in developing the drawing using your local procedures and practices. The abbreviations and symbols shown in **Figure 4.25** will be used.

Development of example project: Now, following the procedure described in this unit, we will summarise the relevant requirements and apply them to our example project as follows:

1 Fixtures and fixture unit rating

Using the fixture unit ratings as listed in AS/NZS 3500.2, we can calculate the fixture unit rating of individual fixtures, or groups of fixtures, as follows:

Bathroom group	6 fixture units
WC	4 fixture units
Laundry	5 fixture units
Sink	3 fixture units
Powder room	5 fixture units
Ensuite group	6 fixture units
TOTAL:	29 fixture units

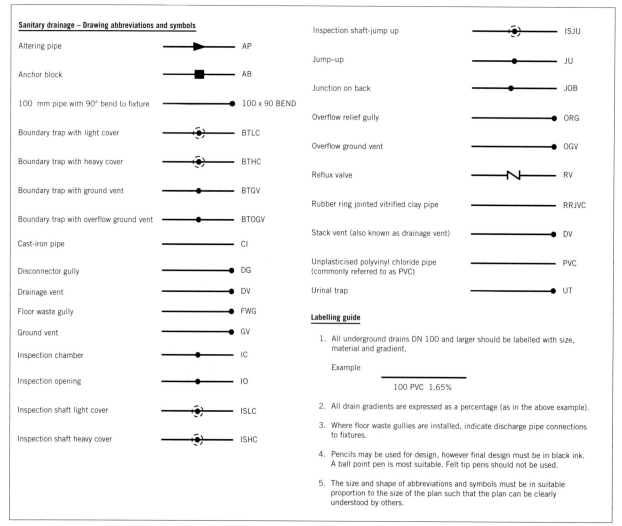

FIGURE 4.25 Example drawing abbreviations and symbols

GET IT RIGHT

LOCATION OF DRAINS

FIGURE 4.26 Drain layout

Planning the layout of drains includes locating other services and obstructions prior to commencing works. Figure 4.26 shows a drain being laid near a building foundation where there are electrical services in the path of the proposed drain alignment.

1   What separation is required between a sanitary drain and an electrical conduit if the conduit location is indicated with orange marker tape and is mechanically protected?

_____

_____

_____

2   What separation is required between a sanitary drain and an electrical conduit if the conduit location is neither indicated with orange marker tape nor mechanically protected?

_____

_____

_____

3   If you cannot lower the drain to achieve the required separation, or raise the drain due to the location of existing slab drains, what should you do?

_____

_____

_____

2 Location and size of approved point of discharge

We found from the property sewerage plan for our example project that the location and size of the approved point of discharge is:
- 2.24 m depth of cover over the soffit
- 4.2 m measured along the property boundary (or fence) line between the boundary corner and centreline of approved point of discharge
- 100 mm size of drain connection at approved point of discharge.

3 Types of connection to approved point of discharge

We found from the property sewerage plan that the type of connection is direct; therefore, an inspection shaft is required.

4 Depth of cover

We assumed a level site and no vehicular traffic at the rear of the garage. The minimum cover over the drain at the furthermost distance from the approved point of discharge must be 300 mm.

5 Line of drains

In the example project, we found for the branch drain to the bathroom that the edge of the trench must be at least 85 mm from the footing. The drain can be laid between the building and property boundary fence without encroaching on the restricted zone beneath the building footing.

6 Drain gradient

We found that the available fall is 1.94 m, and the required fall is 0.619 m; therefore, we have fall well in excess of that required for our 100 mm drain at 1.65%.

7 Jump-up

As we have more than enough fall, the inspection shaft can be constructed as an inspection shaft jump-up.

8 Proximity to other services

It is necessary to obtain information on the location of existing and proposed services. For the example project, we will assume that there are no other services or obstructions.

9 Size of drains

The size of our main drain will be DN 100 between the approved point of discharge and stack to the first-floor ensuite.

10 Maximum distances in unvented branch drains

The length of branch drain from the main drain to the ground-floor bathroom fixtures exceeds 10 m; therefore, it must be vented. The ground-floor closet pan adjacent to the bathroom has a fixture outlet size DN 100; therefore, the branch drain should be 100 mm.

11 Ventilation of drains

The stack vent will provide for upstream ventilation of the main drain. The length of branch drain from the main drain to the ground-floor bathroom fixtures exceeds 10 m; therefore, it must be vented, and we will assume that an air admittance valve will be installed.

12 Surcharge protection

An overflow relief gully is required as surcharge protection. There are options to install the ORG in several locations, but in this case we will plan to install the ORG to connect the ground-floor bathroom fixtures.

13 Stacks

The bend at the base of our stack will be a 100 × 88° bend.

14 Junctions

Junctions in our below-ground sanitary drains will be 45° oblique type.

15 Inspection openings

Inspection openings will be installed as specified in AS/NZS 3500.2.

Taking the above information, we will plan the layout of our residential external sanitary drainage system by transferring it to our plan. We will also add a plumbing fixture legend (see Figure 4.27).

When your project is completed, your plan might change slightly – for example, if there are obstructions encountered during excavation that alter the planned position. In this case, the plan will need to be altered. This will then become the 'as-constructed' plan.

As mentioned earlier, sewerage authorities have different requirements for recording and submitting 'as-constructed' plans. This might change in the future; however, in the interim, you will need to record and submit 'as-constructed' sewerage plans in a format determined by the relevant sewerage authority, using their abbreviations and symbols.

Regardless of whether sewerage authorities require 'as-constructed' plans, the benefits in maintaining your own records should not be underestimated. Professional plumbing contractors maintain their own records relating to the drainage work they have completed.

Materials and quantities

Once you have planned the layout for the external sanitary drainage system, you can then determine the materials, and quantities of pipes and fittings, that are required. The most common material used for sanitary drainage systems in Australia is PVC-U (unplasticised polyvinyl chloride; commonly referred to as 'PVC'). Some images of different PVC fittings can be found in Unit 6. Here we will look briefly at a variety of materials and their benefits.

- PVC-U: This is not recommended where there may be extreme ground movement, or where there are certain types of chemical contamination of the soil (e.g. petroleum products).
- Cast iron: Used for applications where vehicular damage or ground movement may occur. Where there is ground movement, an engineer must be engaged to provide a suitable design for drain support.

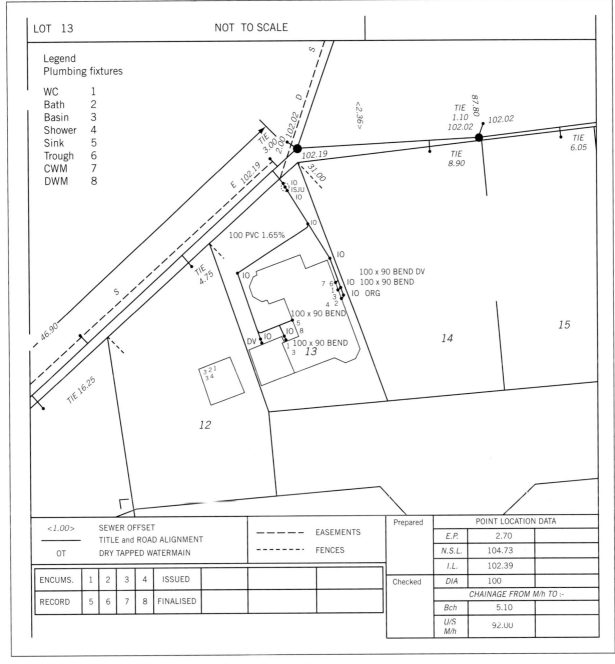

FIGURE 4.27 Plan of residential external sanitary drainage system

- Vitrified clay: Used for corrosive wastes, and where a more rigid wall pipe is required. Pipes may fracture where there is ground movement or differential settlement within a pipe length.
- High-density polyethylene: Used where there are reactive soils and for a wide range of corrosive wastes.
- Copper: Used mainly in above-ground installations, but may also be used below ground. Susceptible to corrosion from urinal wastes.
- Copper alloy (brass): Used in similar situations to copper, but is less susceptible to corrosion. Must not be used for waterless urinal wastes.

Quantities of pipes and fittings

The quantities of pipes and fittings are determined from the plan you have developed. The quantities should be interpreted as the minimum required.

When drain laying, you might find minor variations in the angles at which pipes meet; therefore, you should always have spare 5° bends. Regardless of some minor changes in horizontal alignment, the vertical alignment must always be true – that is, the vertical pipe on all inspection shafts, boundary trap shafts and jump-ups must always be checked with a spirit level and installed vertical.

Development of example project: Based upon our plan of the external sanitary drainage system, we must establish the quantities of pipes and fittings. When this list is complete, it can be used to obtain competitive prices from suppliers. Table 4.5 is an example of the quantities of pipes and fittings and other materials obtained from the plan developed for the example project.

Clean-up

Work area

When planning the layout for a sanitary drainage system, there is little required for clean-up. Unit 10 will deal more specifically with the clean-up requirements after installation of the sanitary drainage system. However, at this planning stage, you are required to make provisions for clean-up. Clean-up occurs during the progress of works and at completion.

You must allow time to:

■ ensure the worksite is safe for other tradespeople during the progress of work
■ clear the worksite of excess materials and trip hazards

■ remove excavated material that is unsuitable for backfill
■ clean up food and drink waste, wrappers and containers
■ store materials and fittings that may be used later.

Tools and equipment

At the planning stages for the job, you should allow time to determine that you have the correct tools and equipment in accordance with the type of work. You should maintain tools and equipment in a safe and serviceable condition.

Personal protective equipment (PPE) should be available on-site during the progress of work. Planning includes making sure that the PPE on-site is suitable for drainage excavation work.

Plan for time to collect, clean and maintain all tools and equipment. Also, allow time to return hire equipment in good condition in accordance with the hire agreement.

 COMPLETE WORKSHEET 2

TABLE 4.5 Materials list developed from the plan of the example project

Item	Description	No. required	Unit	Price	Totals
1	100 mm PVC sewer pipe	11	6m		
2	100 × 90° PVC sewer bend	6	ea		
3	100 × 45° PVC sewer junction	5	ea		
4	100 × 45° PVC sewer bend	5	ea		
5	100 × 15° PVC sewer bend	2	ea		
6	100 × 5° PVC sewer bend	6	ea		
7	100 PVC sewer sweep junction	1	ea		
8	100 PVC IO	11	ea		
9	100 PVC DG	1	ea		
10	100 PVC finishing collar and domed grate	1	ea		
11	100 PVC socket and screwed cap	1	ea		
12	100 inspection shaft light cover	1	ea		
13	75 × 50 hardwood timber	2	m		
14	Concrete mix (20 MPa) 20 kg bag	7	ea		
15	PVC priming fluid (red) 500 mL	1	ea		
16	PVC solvent cement (blue) 500 mL	1	ea		
17	Pipe bedding material (not included) depends on ground conditions (see Unit 10)				
				Total excluding GST	
				GST	
				Total including GST	

SUMMARY

- When preparing to lay a residential sanitary drainage system, it is necessary to gather information, including plans and specifications that outline the location of buildings and fixtures, existing services and easements. Also, familiarise yourself with local regulations and inspection requirements. Before the job starts, you will need to compile a list of materials such as pipes, fittings, concrete mix, priming fluid, solvent cement, bedding and other consumables that will be required to complete the job. It is good practice to plan to have additional fittings such as 5° bends as a contingency for minor variations in drain alignment.

- Planning the system layout involves inspecting the site to view the position of the point of discharge, existing obstructions and services and available fall and cover. Before commencing work, you must be prepared to comply with the local regulatory requirements and AS/NZS 3500.2 which specify all of the design requirements including pipe location, support, materials, fittings, sizes, gradients and depth of cover, ventilation and overflow relief.

- You must make provision for clean-up, ensuring the safety of the worksite and the proper maintenance and storage of tools.

WORKSHEET 1

Student name: _____

Enrolment year: _____

Class code: _____

Competency name/Number: _____

Task: Review the section 'Preparing for work', then complete the following.

1   Using AS/NZS 3500.0, research and define the term 'main drain'.

_____

_____

_____

_____

2   Using AS/NZS 3500.0, research and define the term 'branch drain'.

_____

_____

_____

_____

3   Research your local sewerage authority requirements, and answer the following:

   a   What is the name of your local sewerage authority?

_____

   b   Does your local authority keep records of property sewerage drains?

_____

   c   Does your local authority require 'as-constructed' sewerage plans when property sewerage drains are altered/extended, and if so within how many days of completion?

_____

   d   What are the plumbing permit requirements of your local authority – for example, number of days before commencement, types of work? Explain your answer.

_____

4   a   What minimum depth of cover is required for a PVC sanitary drain in a domestic driveway without brick or concrete cover?

_____

   b   What minimum depth of cover is required over a PVC sanitary drain in a domestic driveway where the thickness of driveway concrete is 75 mm?

_____

5   When planning the layout and installation of a sanitary drain, why is it good practice to first locate the point of discharge?

_____

_____

_____

6   Research the soffit level requirements of your local authority. What is the minimum required height of the overflow gully grate or lowest fixture above the soffit of the sewer main connection?

_____

7   What code and scheme determines whether a plumbing and drainage product must have WaterMark?

_____

_____

8   Is it necessary for copper waste fittings to have WaterMark?

_____

9   Research your local authority requirements for inspection of sanitary drains.

    a   Are inspections required for sanitary drains? YES/NO

_____

    b   Are inspections required for sanitary drains under buildings? YES/NO

_____

    c   Is it a requirement that sanitary drains be tested before backfilling? YES/NO

_____

 **WORKSHEET 2**

Student name: _____

Enrolment year: _____

Class code: _____

Competency name/Number: _____

Task: Review the section 'Planning the system layout', then complete the following.

1   What are three factors that must be determined from site inspection?

   a _____

   b _____

   c _____

2   What is the minimum size of the main line of sanitary drain?

   _____

   _____

3   What is the minimum size of a branch drain?

   _____

   _____

4   What are the fixture abbreviations for the following?

   a Basin _____   c Shower _____

   b Bath _____   d Sink _____

5   Research AS/NZS 3500.0 and define the term 'fixture unit'.

   _____

   _____

   _____

6   What are two types of connection to the approved point of discharge?

   a _____

   b _____

7   Why are boundary traps specified for property drains?

   _____

   _____

8　Where a boundary trap is located in a building, can a ground vent be installed at the top of the boundary trap shaft?

_____

_____

9　When a boundary trap is installed, what two types of ventilation are required?

_____

_____

10　Explain where an inspection shaft must be located.

_____

_____

_____

11　The following diagram shows a trench in sand adjacent to a concrete footing.

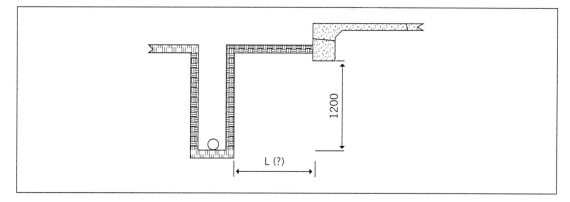

a　What is the minimum distance for 'L'?

_____

b　If the distance 'L' was 1 m, is the trench permitted in this location? Explain your answer.

_____

_____

_____

_____

c　How would you determine the depth of the concrete footing?

_____

_____

12　What is the total fixture unit rating for a basin, water closet, bath and shower in the same room?

_____

13 What is the minimum gradient for a DN 65 branch drain having a rating of 21 fixture units?

_____

_____

14 What is the nominal minimum gradient for the following drains?

a 65 mm _____

b 80 mm _____

c 100 mm _____

15 Why are jump-ups used in sanitary drains?

_____

_____

16 Explain when and why anchor blocks are used in sanitary drains.

_____

_____

_____

_____

17 Provide four reasons for providing ventilation in sanitary drains.

a _____

b _____

c _____

d _____

18 Research AS/NZS 3500.0 and define 'air admittance valve'.

_____

_____

_____

19 Why must air admittance valves not be used for ventilation of the main sanitary drain?

_____

_____

_____

20 Size the external sanitary drainage system shown below.

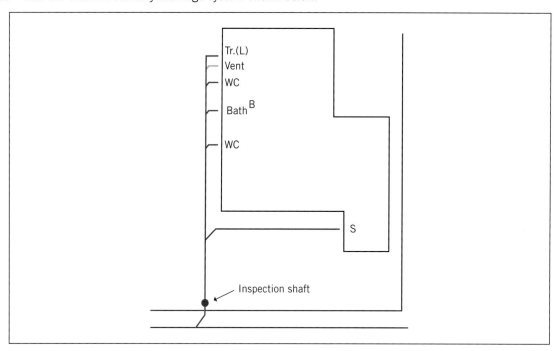

21 What is the maximum allowable length for an unventilated branch drain with discharge pipe, and why?

_____

_____

_____

22 Research AS/NZS 3500.0 and define the term 'surcharge'.

_____

_____

23 What is usually used to prevent surcharge and overflow of sewage into a building?

_____

_____

_____

24 What is the minimum vertical distance that the overflow relief gully grate must be beneath the outlet of a water closet pan?

_____

25 What is the minimum height that an overflow relief gully grate must be above the following?

a   Natural surface level

_____

b   Paved surface level graded away from the grate

_____

26 What types of junctions are used to join drains on a gradient?

_____

_____

_____

27 What types of joints might be required to be installed in filled or unstable ground?

_____

_____

28 Each of the following two diagrams represents a two-storey building connected to a sanitary drainage system. On the diagrams, insert the following:

- either boundary trap or inspection shaft
- drainage vents
- minimum size of main drain and all branch drains
- overflow relief gully
- inspection openings.

a Non-boundary trap area

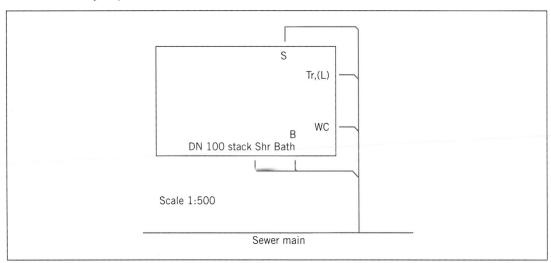

b Boundary trap area

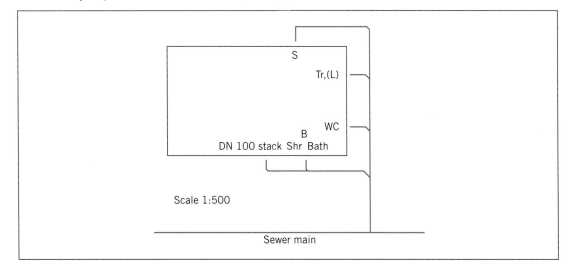

5

PLANNING THE LAYOUT FOR A RESIDENTIAL SANITARY PLUMBING SYSTEM

Learning objectives

This unit provides the basic principles and knowledge for planning the layout of an above-ground residential sanitary plumbing system. Areas addressed in this unit include:

- principles of sanitary plumbing, and basic causes of trap seal loss
- an introduction to the following plumbing systems:
  - elevated pipework using drainage principles
  - fully vented system
  - fully vented modified system
  - reduced-velocity aerator stack system
  - single-stack system
  - single-stack modified system
- determining suitable materials
- building constraints influencing plumbing system selection
- general requirements such as venting, discharge pipes and restricted zones of connection
- basic requirements of different plumbing systems
- selection of plumbing systems
- determining material quantities from an example project.

Introduction

Sanitary plumbing systems are constructed above ground to collect and convey the discharges from sanitary fixtures to the below-ground sanitary drainage system. The sanitary plumbing system is an assembly of pipes, fittings, fixtures and appliances, which must be connected in such a way as to ensure that:

- noise is minimised
- blockage and leakage are unlikely
- foul air and gas escape to the building are unlikely
- there is access to the system for maintenance and clearance of blockages
- damage to the system is unlikely
- ingress of stormwater into the system is unlikely
- it enables the use of water-efficient sanitary fixtures
- it contains adequate ventilation for correct operation.

To achieve these objectives, plumbers are required to have a basic understanding of the principles of sanitary plumbing and relevant standards in order to effectively plan the layout of a sanitary plumbing system before it is installed.

Sanitary plumbing systems are constructed at all types of properties. It could be new work, or it could involve the renovation, extension or maintenance of an existing system.

Units 4 and 10 cover the requirements for the layout and installation of below-ground sanitary drainage systems. In this unit, we focus on planning the layout of residential above-ground sanitary plumbing that would normally connect to a below-ground system.

While studying this subject, you will be required to study and complete exercises referring to AS/NZS 3500.2 Sanitary plumbing and drainage.

AS/NZS 3500.2 SANITARY PLUMBING AND DRAINAGE

Preparing for work

Plans and specifications

Before you begin to plan the layout of a sanitary plumbing system, you will need to obtain plans and specifications that will enable you to determine options for the positioning of pipes. Plans may be available from your project manager or job supervisor. You may also need to notify your network utility operator and/or plumbing regulatory authority by obtaining and submitting permits and/or work notices. Planning requires you to gather all relevant information before commencing work.

Table 5.1 lists sources of information and the type of information each source can provide.

TABLE 5.1 Sources and types of information

Source of information	Information provided
Network utility operator	Location of existing services, easements, positions and depths of connection points, type and conditions of connection, regulations
Plumbing regulatory authority	Plumbing regulations and standards
State and local regulations	Regulatory requirements
Site or block plan	Position of building
Building plan	Location of fixtures, ducts and wall cavities
Plans and specifications	Layout, fixture type and colour
WHS regulations	Safe work regulations – e.g. confined space entry and working at heights
AS/NZS 3500.2	Pipe sizing, clipping
Site inspection	Work conditions, slopes of land, obstacles, access
Manufacturer catalogues	Fixture types, colours and dimensions, fixture options

Plans and specifications are necessary for planning the position of fixtures, fixture discharge pipes, and stacks to connect to the drain. Work notices must be lodged with state, territory or local authorities, and may need to be submitted at certain stages of the job.

Plans and specifications may provide details of the work, including:

- sewerage plans indicating the type of connection, position and depth of the connection point and any easements
- location and type of other services
- type and size of materials to be used
- location of pipes (e.g. under-floor, within walls or ducts)
- type of fixing devices to be used
- work health and safety (WHS) requirements in accordance with state or territory legislation.

Principles of sanitary plumbing

Fixture traps

A fundamental performance requirement of a sanitary plumbing system is the prevention of foul air and gas escape into the building. This has been the main focus in the design of sanitary plumbing, drainage and ventilation systems since the 1850s. One way to achieve this is at individual fixtures, where a trap resembling a 'U' bend is used to retain a depth of water that acts as a barrier to the passage of air through the trap.

Fixture traps must retain a residual water seal of not less than 25 mm under normal operating conditions. This water seal is measured between the weir and dip of the fixture trap (see **Figure 5.1**).

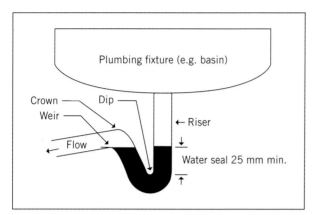

FIGURE 5.1 Trap components and retention of water seal

Plumbing fixtures may either be fitted with traps during installation or manufactured with integral traps; regardless, every fixture must have a trap, and the trap must be accessible and located in the same room as the fixture or appliance that it serves. Where approved, a self-sealing device may be used instead of a trap, and this performs the same function as the fixture trap.

After the sanitary plumbing has been installed, all traps should be tested. If the water seal is less than 25 mm, the trap must be vented or other alterations made to ensure retention of a water seal of at least 25 mm. (Read the section 'Testing of sanitary plumbing and sanitary drainage installations' in AS/NZS 3500.2.)

AS/NZS 3500.2 SANITARY PLUMBING AND DRAINAGE

Failure to install and maintain trap water seals may have contributed to the severe acute respiratory syndrome (SARS) epidemic in Hong Kong in 2003. It is therefore an important component in the plumbing system.

Trap seal loss
There are six main causes of trap seal loss (see **Figure 5.2**):
- spontaneous syphonage (sometimes referred to as 'self-syphonage')
- induced syphonage
- capillary attraction
- evaporation
- momentum
- wind pressures.

Syphonage is the most common cause of trap seal loss.

Spontaneous syphonage
Cause: Occurs within the trap and waste pipe of a fixture when the fixture is operated. It is generally caused by one or a combination of:

- small bore of discharge pipe
- long discharge pipe
- lack of ventilation.

It commonly occurs in waste pipes of small diameter, such as hand basins.

Example: The discharge through a pipe fills it completely, causing a full bore of water travelling by gravity to pull air behind it, causing a reduction in pressure between the full bore of water and the trap seal. The atmospheric pressure acting on top of the water in the trap will then force the water seal out of the trap and down the waste pipe.

Solutions: Improving ventilation in the fixture discharge pipe by installation of a vent or air admittance valve. Installation of a flow-restricting grate.

Induced syphonage
Cause: Occurs within the trap and waste pipe of a fixture as a result of the operation of one or more other fixtures. Waste from one fixture is discharged into a common discharge pipe or stack. Rushing past the discharge opening of a lower discharge pipe connection, it causes a reduction in pressure in the lower discharge pipe.

Solutions: Proper design of the system by correct junction location, discharge pipe size and length, and ventilation will assist in balancing air pressures on both sides of the trap seals to prevent syphonage. Improving ventilation in the lower fixture discharge pipe by installation of a vent or air admittance valve can assist in resolving induced syphonage.

Capillary attraction
Cause: This can occur when material such as lint, hair or string extends over the weir of the trap, and water follows the material over the weir, thereby draining the trap of water.

Solution: Use of smooth bore trap materials and seamless joints. Ensure that grates are fitted to fixture outlets to capture materials.

Evaporation
Cause: When the trap seal is not regularly replenished by operating the fixture, the water seal evaporates, usually because fixtures are not used for many weeks.

Solution: Regularly operate the fixture, or fit an automatic trap priming device.

Momentum
Cause: If the trap riser is excessively long, the discharging water can gain sufficient momentum to carry away the trap seal.

Solution: Place traps as near as possible to the fixture outlet. The standard maximum recommended length between the fixture outlet and water level in trap seals is 600 mm. Avoid the use of 'S' traps.

Wind pressures
Cause: Where vent terminals are exposed to high-velocity wind, the sanitary plumbing system can be

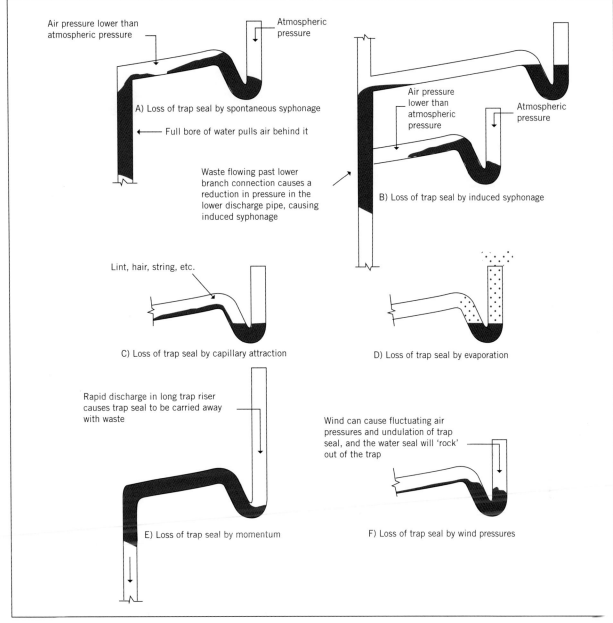

Air pressure lower than atmospheric pressure

Atmospheric pressure

A) Loss of trap seal by spontaneous syphonage

Full bore of water pulls air behind it

Waste flowing past lower branch connection causes a reduction in pressure in the lower discharge pipe, causing induced syphonage

Air pressure lower than atmospheric pressure

Atmospheric pressure

B) Loss of trap seal by induced syphonage

Lint, hair, string, etc.

C) Loss of trap seal by capillary attraction

D) Loss of trap seal by evaporation

Rapid discharge in long trap riser causes trap seal to be carried away with waste

Wind can cause fluctuating air pressures and undulation of trap seal, and the water seal will 'rock' out of the trap

E) Loss of trap seal by momentum

F) Loss of trap seal by wind pressures

FIGURE 5.2 Six main causes of trap seal loss

exposed to fluctuating air pressures, which results in the undulation of water in the trap. The fluctuating air pressures effectively 'rock' the water seal out of the trap.

Solution: Fit wind-deflecting-type cowls. While drainage vents must remain open, other open trap vents can be replaced with air admittance valves.

Venting

Ventilation is necessary in sanitary plumbing systems in order to:

■ provide for the escape of foul air and gases that can accumulate in the sewerage system

■ dry the internal surface of piping, which assists in preventing material adhering to the pipe wall and the formation of blockages

■ assist in balancing the air pressures between the inlet and outlet of trap seals to prevent syphonage and loss of trap seals.

Ventilation occurs naturally due to the difference in temperature between the inside and outside of the plumbing, movement of waste, and wind pressure on open vent terminals.

In order for ventilation to be effective, pipes must be correctly sized and installed at a self-cleansing gradient. Where there are transitions in pipe size, the tops of the pipes should be in continuous alignment. See **Figure 5.3**.

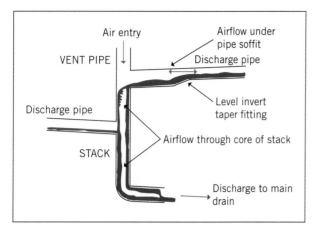

FIGURE 5.3 Elevation section: ventilation in sanitary plumbing

Length of discharge pipes and measurement

Later in this unit we discuss the maximum lengths of discharge pipes for different systems of plumbing. As we have just discussed ventilation, you will be aware that vents are necessary, and this generally means that, for all types of sanitary plumbing design, there is a maximum length of any unvented discharge pipe. If a discharge pipe exceeds a length specified in AS/NZS 3500.2 without a vent, there is greater potential for blockages and loss of trap seals to occur.

The maximum length is measured along the pipe from the weir of the fixture trap to the connection to a graded discharge pipe, stack or other drainage trap (see Figure 5.4). (Read AS/NZS 3500.2.)

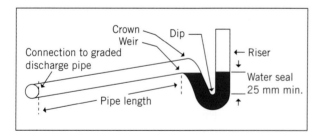

FIGURE 5.4 Measuring the maximum length of a discharge pipe

Pipe gradient

All discharge pipes must be installed at the minimum gradient, which generally depends upon the size of pipe. This ensures proper drainage to assist in preventing blockages. For example, the minimum gradient of a 50 mm discharge pipe is 2.5%. Gradients are generally expressed as a percentage. Gradients are also sometimes

expressed as a ratio, and for 2.5%, the equivalent ratio is 1 in 40. The best way to understand the relationship between these two figures is rise/run or 1/40, which is the same as 2.5/100. (Read AS/NZS 3500.2.)

Hydraulic loading

When a sanitary fixture is operated, there is discharge to the sanitary plumbing installation, and when multiple fixtures are operated, there is a combined discharge and hydraulic loading imposed on the sanitary plumbing installation. There are different types of fixtures, and each has a different rate of discharge, frequency of use and likely time of operation.

The 'fixture unit' (FU) is a unit of measurement that has been developed to express the hydraulic load imposed by a sanitary fixture on the sanitary plumbing installation. Each type of sanitary fixture has been assigned a 'fixture unit rating'.

The systems of sanitary plumbing and drainage in AS/NZS 3500.2 are sized using fixture unit ratings, whereby each fixture is given a rating – for example, bath = 4 FU. (Research the term 'fixture unit' in AS/NZS 3500.0 Plumbing and drainage and AS/NZS 3500.2.) These ratings have been calculated based on tests that show how much water is normally discharged through different fixture outlets. An example of this would be comparing a water closet with a basin. A water closet can discharge as much as 9 litres of water in approximately four to five seconds, whereby a basin may discharge only 2 litres of water in seven seconds.

AS/NZS 3500.2 SANITARY PLUMBING AND DRAINAGE

The six different systems of plumbing discussed in the following sections must be sized on, among other things, the basis of the total number of fixture units.

Elevated pipework using drainage principles

In Units 4 and 10 we discuss the requirements for planning the layout and installing a below-ground sanitary drainage system. This includes consideration of the gradient, fixture unit loadings and ventilation of drains. The same considerations can be used when planning the layout of elevated pipework for up to four floors above the invert of the connection point to the boundary trap or inspection shaft. The main restriction is that no graded discharge pipe or branch, except a stack, is permitted to connect to any vertical section of pipework. This system is commonly used in residential plumbing. (Read AS/NZS 3500.2.)

Fully vented system

A fully vented system is a system of sanitary plumbing where every fixture trap is vented, except for traps discharging to a floor waste gully. A relief vent is also normally installed. This system is rarely used, due to the high cost of vent piping. (Read AS/NZS 3500.2.)

Fully vented modified system

A fully vented modified system differs from the fully vented system of sanitary plumbing in that each connection to the stack is vented, but some trap vents are omitted. This system is commonly used in high-rise plumbing. (Read AS/NZS 3500.2.)

Reduced-velocity aerator stack system

This system has been introduced in recent years with the development of the aerator junction fitting (a proprietary junction fitting). An aerator junction is installed in the stack to connect discharge pipes at each floor level, and a de-aerator is installed at the base of the stack. Airflow requirements of the system are provided through the stack vent. (Read AS/NZS 3500.2.)

Single-stack system

Single-stack systems are designed on the principle that the air within the discharge pipes, stack and vent provides sufficient ventilation considering the number and type of fixtures, and no trap vents are required. The restrictions on numbers of fixtures and lengths of discharge pipes are the main limiting factors. This type of system is commonly used in, and is well suited to, high-rise plumbing, where groups of fixtures are in close proximity to the stack through multiple floor levels. (Read AS/NZS 3500.2.)

AS/NZS 3500.2 SANITARY PLUMBING AND DRAINAGE

Single-stack modified system

Single-stack modified systems differ from single-stack systems in that the stacks can receive a greater discharge loading due to the addition of a relief vent and cross-vents. This type of system is sometimes used where there are a greater number of floor levels than would be acceptable to the design of a single-stack system. (Read AS/NZS 3500.2.)

AS/NZS 3500.2 SANITARY PLUMBING AND DRAINAGE

Safety and environmental requirements

When planning the layout for a residential sanitary plumbing system, you will need to consider the safety and environmental requirements of the job. Each state/territory has WHS regulations that require employers and self-employed people to identify hazards and assess and control risks at the workplace in consultation with their workers.

While planning the layout, you will observe issues that attract your attention, and you should note any special requirements so that you have all of the necessary WHS issues covered before the job commences. Planning the layout of a sanitary plumbing system is not only about installing piping; it is also about planning how long the job will take to complete. For example, if scaffolding is required to safely carry out the work, you will need to allow time for the erection and disassembly of the scaffolding. If you can foresee wet conditions, it is best practice to maintain flexibility to rearrange the work schedule. For example, power leads and tools must only be used in dry conditions due to the possibility of electrocution.

You will also need to consider any environmental requirements. When connecting to the sanitary drain, you will need to be prepared to provide protection to the connection points near ground level to prevent the entry of any stormwater sand, silt and rubbish. Stormwater must not be allowed to enter the sanitary drainage system, and *must only be discharged to an approved point of discharge for stormwater.*

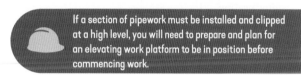
If a section of pipework must be installed and clipped at a high level, you will need to prepare and plan for an elevating work platform to be in position before commencing work.

Quality assurance

In planning the layout of a sanitary plumbing system, there may be requirements for quality assurance. In most areas, a regulatory authority will require an inspection of all, or a percentage of, your plumbing work, and this is a form of quality assurance of your work for the sewerage authority and property owner.

On larger building sites, it is common to see additional requirements for quality assurance of your work, such as sign-off on portions of your work for progress payment, and additional tests and inspections.

All pipes and fittings used in a sanitary plumbing system must comply with *the National Construction Code (NCC) Volume 3, Plumbing Code of Australia* and bear WaterMark and Australian Standards certification. Not all sanitary fixtures are required to bear WaterMark.

Preparation for quality assurance requirements may involve 'hold points', where your work must stop in order to allow time for the necessary tests and inspections (see Table 5.2).

TABLE 5.2 Examples of tests and inspections

Examples of tests	Examples of inspections
Hydrostatic or air testing	Pipe clipping and support
Testing of sanitary fixtures by subjecting them to normal use. Fixture traps must retain a water seal of at least 25 mm	Provision for expansion
	Provision for ventilation
	Pipe gradient

Later in this unit, we will discuss the planning requirements for inspection and testing of sanitary plumbing.

You need to be aware of your company's particular quality assurance requirements. The following are some quality assurance issues to be considered when planning the layout for a residential sanitary plumbing system:

- conforming with company operating procedures
- ensuring that any necessary applications are lodged with authorities
- ordering materials and products in accordance with job plans and specifications
- receiving and inspecting materials and products for conformance with orders
- storage of materials and products to prevent deterioration
- operation and maintenance of tools and equipment
- compliance with contract work schedules
- maintaining records of job variations.

LEARNING TASK 5.1

Undertake research on the internet to find the Australian Building Codes Board (ABCB):
- WaterMark Certification Scheme Schedule of Products
- WaterMark Certification Scheme Schedule of Excluded Products.

Are the following products required to have WaterMark?
1 Sewerage reflux valves
2 Basins
3 Water closet pans

Sequencing of tasks

In planning the layout for a residential sanitary plumbing system, it is important that tasks are sequenced so that the building project will run smoothly. Each trade on a building site has a job to do, and understanding the project schedule and communicating with others is fundamental to the success of the plumbing business. For example, as previously mentioned, when preparing for work you might note that scaffolding is required to install high-

level pipework. In that case, you will need to know the date of commencement of plasterwork that might cover the piping, so that you will have time to install, test and, if required, have the piping inspected before that date.

Another example might be that tiling is required for shower bases, so you will need to know the date when the tiling is to commence to enable you to have time to complete and test the waste pipe connection.

Proper planning and sequencing of work will assist you in understanding the dates and times of work on one or more projects, so that you can utilise your time in the most efficient way for the successful operation of the plumbing business.

Tools and equipment

To plan the layout of a sanitary plumbing system, tools and equipment are necessary for site inspection, measuring, drawing and estimating material quantities.

Where possible, it is best practice first to visit the site for a visual inspection and measurement with the project drawings and specifications. Examples of useful tools and equipment may include:

- 30 m wind-up measuring tape
- compass
- torch
- note pad and pencil
- digital camera
- personal protective equipment (PPE), such as sunhat and sunscreen.

A site visit with the project supervisor may be of benefit in determining where piping, materials and your tools can be securely stored for the duration of the work.

Second, off-site measurement and inspection of plans and specifications is carried out to determine the layout of the sanitary plumbing system. The plans and specifications should provide information on allowances for piping within the building construction in order to plan the layout and positions for the piping and to estimate material quantities. Useful equipment may include:

- drawing board
- scale ruler/s, with multiple scales that include 1:10, 1:20, 1:50, 1:100, 1:200 and 1:500
- pencils for drawing the design layout
- note pad
- eraser
- calculator
- computer and printer.

 COMPLETE WORKSHEET 1

Planning the system layout

Site inspection

As discussed, it is best practice to visit the site for a visual inspection and measurement with the project drawings and specifications. A site meeting with the project supervisor may be of benefit in clarifying project drawings and specifications where the level of detail is inadequate. The level of detail in project drawings is important in determining allowances for piping within the building construction. Also, the project supervisor should be able to allocate secure storage space for piping, materials and your tools for the duration of the work.

In the following sections, we discuss the relevance of the allowances for piping within the building construction to determining the layout of the sanitary plumbing system.

Quantity, location and type of fixtures

The project drawings and specifications provide details of the quantity, location and type of plumbing fixtures required.

We have previously discussed fixture unit ratings and fixture unit loading. When the fixture unit ratings of all fixtures or groups of fixtures are added, we arrive at the total fixture unit loading and options for use of the six different types of plumbing systems mentioned earlier under 'Principles of sanitary plumbing'.

The location of fixtures will also determine what options are available for the type of plumbing system to be used, as there may be restrictions within the building construction that require minimum pipe sizes and venting.

The type of fixtures will determine the minimum pipe sizes, because generally the piping downstream of a fixture must be at least equivalent to the size of the fixture outlet. Also, the type of fixture influences the type of plumbing system, as some systems have limitations on the numbers of the same type of fixture that may be connected.

Determining materials

Factors affecting selection and use of materials

Materials used for discharge pipes and pipe supports must be compatible and comply with AS/NZS 3500.2. Materials for a given application can be selected directly from the Australian Standard. Brackets, clips and hangers can be selected from manufacturers' catalogues and must be installed in accordance with AS/NZS 3500.2, the job specifications and the manufacturers' instructions.

AS/NZS 3500.2 SANITARY PLUMBING AND DRAINAGE

A number of factors will determine options for use of materials and how they should be installed. These factors include:

- properties of liquid discharges
- properties of solid discharges
- soil type and condition
- environmental factors
- temperature of discharge
- characteristics of common materials and products.

We will now examine some of the above factors.

Properties of liquid discharges

A variety of liquids can travel through discharge pipes.

- Household wastes from kitchen, bathroom, laundry, toilet, etc. are generally considered acceptable wastes for all pipe materials.
- Wastes within some commercial and institutional properties may require further consideration. For example, copper and brass pipe materials must not be used in contact with acidic urinal discharges. It is recommended that no metallic pipe materials be used in contact with undiluted waterless urinal discharges. *Note:* Where a waterless urinal discharge pipe is connected to a sanitary plumbing system, the common discharge pipe or drain should ideally be flushed by fixture/s of at least two upstream fixture units. This will assist in reducing the build-up of uric scale in the pipework.
- Discharges such as acids, grease, fats, dirt and oil are usually considered to be 'trade' wastes. The network utility operator will require that these wastes be intercepted or treated before discharging to the authority sewerage system. The piping used to collect trade waste discharges must be of suitable materials that resist degradation. Different pipe materials have different chemical resistance to trade wastes. For example, copper and brass pipes are not recommended for acidic discharges. Also, PVC (polyvinyl chloride) should not be used for some petroleum trade waste discharges. For further information, see pipe manufacturer chemical resistance charts or AS 2032 Installation of PVC pipe systems.

AS 2032 INSTALLATION OF PVC PIPE SYSTEMS

Properties of solid discharges

Discharge pipes are used to convey many different types of wastes containing solids, and the suitability of pipe materials depends upon the type of solid. Consideration of pipe material suitability for conveying solids generally relates to trade wastes when solids are unacceptable for discharge to the authority sewerage system. For example, waste from a wash-down area may contain sand and silt. Some pipe materials, such as PVC, may be easily worn away and scoured by sand and silt, and so a more durable material such as cast iron should be considered.

Soil type and condition

If a discharge pipe is to be laid in the ground, it is worth considering the type of soil in which the pipe will be placed. Some pipe materials and types of soil are not compatible. The selection of the correct pipe material to be used will determine whether the pipe is to last a long time or a short time, due to the effects of corrosion. Some soils may be too acidic or alkaline, or contain too much salt for certain types of pipes. For further information, see pipe manufacturer chemical resistance charts, or for PVC, see AS 2032.

AS 2032 INSTALLATION OF PVC PIPE SYSTEMS

Environmental factors

When choosing a suitable pipe material for installation, the cost, ease of installation and durability of the material must be considered. There are also environmental considerations, such as the minimisation of sound from wastes travelling in discharge pipes and stacks. Some pipes that are more expensive than PVC are made from materials that have been specifically created to minimise sound transmission.

Other environmental factors that should be considered relate to the location of pipes in positions where they are:

- exposed to sunlight or excessive heat
- exposed on a car park wall, with no protection from being damaged by a vehicle
- exposed in a factory, where they could be damaged by impact from a forklift
- exposed to corrosive chemicals such as acids.

GREEN TIP

Many projects require 'sound attenuation' in sanitary plumbing systems, where it may be necessary to lag PVC pipes or install another pipe material to achieve plumbing that produces *an acceptable level of noise*. In these situations, alternative pipe systems are available, including push fit polypropylene systems that *do not require lagging* and are therefore quicker to install.

Temperature

The temperature of the liquid discharge can adversely affect discharge pipes. For example, the discharges from glass or dish washing machines may be as high as 90°C. The pipes carrying these high-temperature discharges may require additional support along their length, and allowances may also need to be made for

the higher expansion rate of the pipe. It is generally recommended that PVC pipes not be exposed to continuous discharges of temperatures greater than 60°C. Also, the resistance of PVC to some chemicals is less satisfactory at higher temperatures.

Characteristics of common materials and products

Materials used for discharge pipes include:

- unplasticised polyvinyl chloride (PVC-U), commonly referred to as 'PVC'
- high-density polyethylene (HDPE)
- polypropylene
- copper and copper alloy (brass)
- galvanised steel
- glass
- glass-filament-reinforced thermosetting plastic (GRP)
- fibre-reinforced concrete (FRC).

PVC-U has a high expansion rate, which has to be considered when using it with high-temperature discharges. With a wide range of fittings available, the material is ideal for the vast majority of sanitary plumbing systems.

HDPE can be used for trade waste installations with temperatures up to 100°C for short periods and 80°C for constant flow. Always check manufacturers' specifications to ensure that HDPE pipe materials are compatible with the proposed waste discharges.

Copper can be used in many applications where high durability is required. It is both fire and impact resistant. Copper pipe can be used to fabricate junctions for use in sanitary plumbing systems, although a large range of fittings is available. Copper pipe is available in different wall thicknesses. Type A copper has the thickest wall and Type D the thinnest. AS/NZS 3500.2 sets out the limitations on the use of copper pipes and fittings. Copper pipe is unsuitable for the discharge of urinals.

Copper alloy (brass) is similar to copper pipe in that it can be used to fabricate junctions, or, similarly, fittings are available ready-made. Refer to AS/NZS 3500.2 for information on the limitations of use. It is not suitable for waterless urinal waste discharges.

Galvanised steel pipe is tough and highly resistant to damage by impact; however, it is rarely used due to its susceptibility to corrosion and blockages. Threads are made on pipes to join pipes and fittings. Refer to AS/NZS 3500.2 for information on the limitations of use.

Glass-filament-reinforced thermosetting plastic (GRP), polypropylene and fibre-reinforced concrete (FRC) installation requirements vary between manufacturers. Refer to AS/NZS 3500.2 and manufacturers' instructions.

AS/NZS 3500.2 SANITARY PLUMBING AND DRAINAGE

Plumbing fittings chart

Unplasticised polyvinyl chloride (PVC-U), usually referred to as 'PVC', is the most common material used in Australia because it is easy to install, lightweight, low in cost and durable. The chart in **Figure 5.5** provides a small sample of the range of PVC fittings.

a) 100 mm 15° F & F Plain Bend

b) 100 mm 45° F & F Plain Bend

c) 100 mm 88° F & F Plain Bend

d) 100 mm 88°F & F Side Access Bend

e) 100 mm 5° M & F Plain Bend

f) 100 mm 15° M & F Plain Bend

g) 100 mm 45° F & F Side Access Bend

h) 100 mm 45° F & F Left Hand Access IO Junction

i) 100 mm 88° F & F Plain Junction

j) 100 mm 88° M & F Rear Access IO Junction

k) 100 mm Square IO Junction

l) 150 × 100 mm 45° F & F Unequal Junction

m) 100 mm 45° F & F Plain Junction

n) 100 mm VC to PVC Bush

o) 80 × 65 mm Floorwaste Gully

FIGURE 5.5 Vinidex PVC Fittings Chart © 2009. For more information on PVC fittings, visit https://www.vinidex.com.au.

p) 80 x 80 mm Floorwaste Gully

q) 80 x 50 x 40 x 40 mm 88° M & F Floorwaste Junction

r) 100 x 50 mm Flat Reducer

s) 100 mm Slab Repair Coupling

t) 100 Screwed Access Cap

u) 40 mm Copper to PVC Coupling

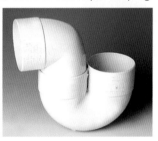

v) 100 mm Disconnector Trap

w) 100 mm M & F Concentric Pan Adaptor

x) 100 M & F Adjustable Eccentric Pan Adaptor

FIGURE 5.5 (*Continued*)

Fittings for flexible location of fixtures

Some manufacturers have developed plumbing fittings that permit flexible location of fixtures. The Iplex Smartpan (see Figure 5.6) allows flexible positioning of toilets, baths and shower trays. It eliminates the requirement to adjust a toilet pan to suit the trap riser. Pan, bath or shower base outlets can be positioned anywhere within the 250 mm diameter lid, giving approximately 65 mm flexibility in any direction from the centre, or 85 mm when used in conjunction with an offset pan collar (Based on text at: https://www.iplex.co.nz/products/smartpan/).

Fittings for plumbing where there are space constraints

Some manufacturers have developed plumbing fittings that fit within tight spaces when compared with an assembly of conventional fittings. The Iplex Smartrap, for example (see Figure 5.7), can significantly reduce underfloor trap depth. With an overall height of 176 mm when installed, it can deliver up to a 75% saving in the space required.

FIGURE 5.6 Iplex Smartpan. For more information, visit http://www.iplex.com.au.

Source: Iplex Pipelines: http://www.iplex.com.au/iplex.php?page=lib&lib=18&sec=121

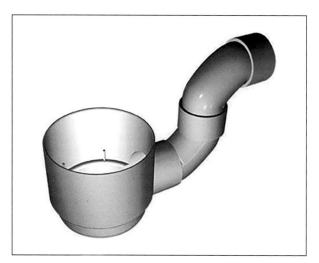

FIGURE 5.7 Iplex Smartrap. For more information, visit http://www.iplex.com.au.

Building constraints and factors influencing plumbing system selection

Gradient

Buildings must be able to accommodate sanitary plumbing systems. One of the main challenges facing plumbers is the lack of consideration given by architects and builders to the plumbing requirements.

The requirements for accommodating sanitary plumbing with building structures are significant, due not to the pipe sizes, but to the gradient of discharge pipes and vents (see Figure 5.8). For example, a DN 100 PVC discharge pipe on a gradient of 1.65% over a distance of 6 m theoretically requires a space depth of 210 mm. However, if the size of the pipe is reduced to DN 50, the gradient must be increased to 2.5%, and over the distance of 6 m the 50 mm pipe theoretically requires a space depth of 206 mm. So, it can be seen that reducing the pipe size may achieve very little.

In the past, plumbers have often installed pipes at less than minimum gradients within the constraints of the building structure. However, more blockages are now occurring with the introduction of more water-

efficient plumbing fixtures, even with pipes installed at the correct minimum gradient. Installing pipes to the correct gradient is critical to proper drainage.

FROM EXPERIENCE

The performance of sanitary plumbing and fixtures depends upon the pipe gradient. Therefore, your ability to plan the location of piping to achieve the correct gradient is a critical skill.

Cutting building members

When planning to install sanitary plumbing, it is best practice to avoid cutting any building members such as floor and ceiling joists, lintels, bearers and beams. Plumbing fixtures or building members should be positioned to avoid cutting any load-bearing members.

If it is necessary to cut or hole saw load-bearing members, then the adequacy of the members must be verified, and designed and approved by a building practitioner.

It is normal practice to cut plates and noggings for pipe penetrations. However, cuts must be to the minimum depth. Plates and noggings must be of sufficient strength to retain their position and prevent movement.

Provision for expansion and movement

Different pipe materials have different rates of expansion, and this must be considered generally when installing sanitary plumbing, but particularly when planning to install piping within building constraints. Piping must not be bent around building members, and should be free to expand.

Considerations for copper and copper alloy sanitary plumbing systems

Students should read AS/NZS 3500.2 to gain an understanding of the requirements for expansion joints in copper and copper alloy (brass) sanitary plumbing systems.

AS/NZS 3500.2 SANITARY PLUMBING AND DRAINAGE

General considerations for expansion in PVC sanitary plumbing systems

Students should read AS 2032, noting that:
- expansion joints should be installed in cold pipes at maximum intervals of 6 m, and in hot pipes at maximum intervals of 4 m (a hot discharge has a temperature of 45°C or higher)
- the maximum length of pipe between fixed points without an expansion joint or provision for expansion is 2 m for cold pipes, and 1 m for hot pipes.

Expansion joints in vertical PVC discharge pipes

Expansion joints (see Figure 5.9) should be located:

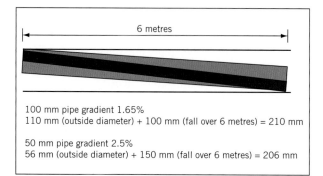

6 metres

100 mm pipe gradient 1.65%
110 mm (outside diameter) + 100 mm (fall over 6 metres) = 210 mm

50 mm pipe gradient 2.5%
56 mm (outside diameter) + 150 mm (fall over 6 metres) = 206 mm

FIGURE 5.8 Allowances for PVC pipe installation in the building structure

FIGURE 5.9 100 mm PVC DWV F&F expansion joint

Source: Reproduced with permission of Storm Plastics Pty Ltd., www.stormplastics.com.au

- at the base of a stack, or at the end of a drain connection for a discharge pipe
- on each floor at which fixtures are connected, above the highest branch connection
- at maximum intervals in accordance with the above general considerations (see Figure 5.10).

Expansion joints in graded PVC discharge pipes

Expansion joints must be provided in graded discharge pipes as follows:

- upstream of the entry of a graded discharge pipe to a vertical stack in accordance with the above general considerations
- upstream of each change of direction in graded discharge pipes in accordance with the above general considerations
- at maximum intervals subject to the above general considerations.

Expansion joints must be securely supported at the socket, with the clip attached to a fixed support to prevent movement. Where the expansion joint is installed in a graded discharge pipe and clip shanks are long, the expansion joint should be fixed and prevented from moving by attaching two clip shanks, each 45° apart and fastened to the building structure.

Providing for expansion without expansion joints

Expansion joints may be omitted in a graded discharge pipe where expansion in the pipe can be accommodated by thermal movement of an offset leg that does not alter the gradient of the pipe (see Figure 5.11). The requirements are as follows:

1 For pipe sizes 40 mm and 50 mm:

Maximum pipe length (L)	Maximum length of offset leg (l) (not clipped)
2.0 m	0.5 m
3.0 m	0.6 m
4.0 m	0.8 m
6.0 m	1.0 m *Example*

2 For pipe sizes 65 mm, 80 mm and 100 mm:

Maximum pipe length (L)	Maximum length of offset leg (l) (not clipped)
2.0 m	0.75 m
3.0 m	1.0 m
4.0 m	1.1 m
6.0 m	1.2 m

This is commonly referred to by plumbers as the 'Big L, little l' rule, where pipe clips are positioned to permit thermal expansion of PVC pipe. For example, referring to the above figures and the *italicised example*, it can be seen that for a 50 mm PVC pipe 6 m long, a minimum length of 1 m is required in the offset leg to accommodate the expansion of the 6 m-long pipe. In this example, the offset leg must not be clipped within 1 m of the offset bend.

Expansion joints may also be omitted in a graded discharge pipe where expansion movement of the pipe can be accommodated at a trap of plastic material, provided that the length of the pipe does not exceed 6 m for cold pipes or 4 m for hot pipes, and the trap is in the same vertical alignment as the discharge pipe. In other words, when the graded discharge pipe expands, there must be no twisting of the joints in the trap that might loosen or wear the trap sealing rings, resulting in leaks.

In summary, consideration must be given to the expansion of PVC pipe in sanitary plumbing systems. As a guide, every 1 m of PVC pipe requires an allowance of 3 mm for linear expansion. Therefore, over a 6 m length of PVC pipe, one might see a change in length of approximately 18 mm, and this must be allowed for in the building construction.

Piping systems of other materials may have different rates of expansion. For information on provisions for expansion in these systems, refer to manufacturers' installation instructions.

Clearance from other services

The building construction must be able to accommodate the diameter of discharge pipes, and the pipes must be protected from damage. Discharge

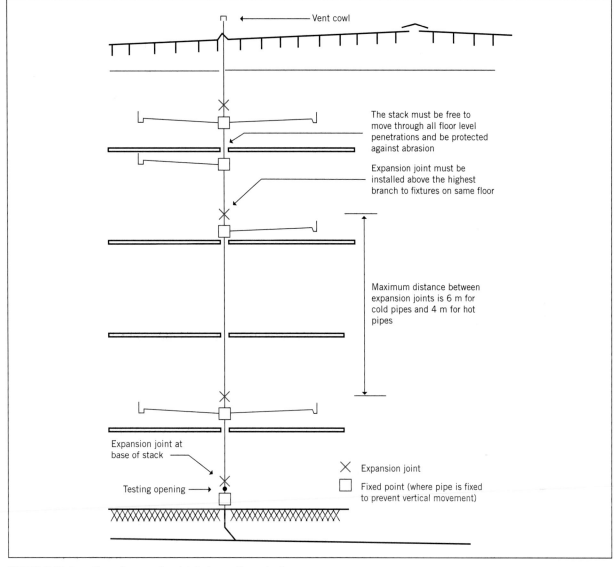

FIGURE 5.10 Location of expansion joints in sanitary stacks

pipes should be located at least 100 mm clear of electrical, gas and cold water services. PVC and other plastic discharge pipes must be located at least 75 mm from insulated heated water pipes and 150 mm from uninsulated heated water pipes.

Pipe diameter

Pipe diameters are determined by the:

- size of the sanitary fixture outlet
- fixture unit loading on the pipe
- system of plumbing used, and the rules applicable to that system.

Primary factors influencing system selection

In review, the primary factors influencing selection of the plumbing system to be installed are:

- gradient
- cutting building members
- provision for expansion and movement
- pipe diameters and clearance from other services.

Secondary factors influencing system selection

Earlier in this unit we discussed the basic principles of sanitary plumbing for the following six systems:

1. elevated pipework using drainage principles
2. fully vented system
3. fully vented modified system
4. reduced-velocity aerator stack system
5. single-stack system
6. single-stack modified system.

There are specific rules and limitations for each of the above systems, and these are regarded as the secondary factors influencing system selection. For example, when installing the system 'elevated pipework using drainage principles', it is not permitted to connect a graded discharge pipe or branch, except a stack, to any vertical section of pipework. Therefore, if there is no choice but to connect to a vertical section, it may be necessary to choose a different system of plumbing such as the 'single-stack system'.

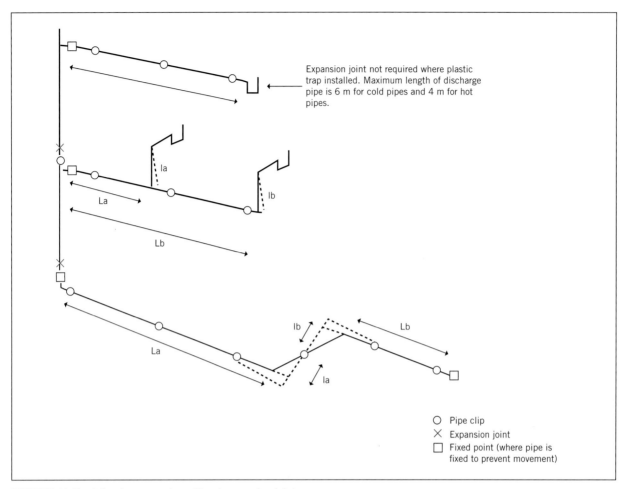

Expansion joint not required where plastic trap installed. Maximum length of discharge pipe is 6 m for cold pipes and 4 m for hot pipes.

Ia

Ib

La

Lb

Ib

La

Lb

Ia

○ Pipe clip
✕ Expansion joint
☐ Fixed point (where pipe is fixed to prevent movement)

FIGURE 5.11 Providing for expansion without expansion joints

Later in this unit, detail is given on the different systems of plumbing and specific rules applicable to each, but first you should study the general requirements for sanitary plumbing systems.

General system requirements

Fixture unit ratings

Earlier in this unit, we discussed hydraulic loading. This relates to the number of fixture units, which is an expression of the hydraulic load imposed by fixtures on various parts of the sanitary plumbing installation.

AS/NZS 3500.2 includes a table setting out fixture unit ratings, and these ratings are used to size drains, stacks and discharge pipes. The rules for different systems of plumbing determine the allowable fixture unit loadings.

Grates

With the exception of water closet pans, slop hoppers, bedpan washers and bedpan sterilisers, the outlets of all sanitary fixtures must be fitted with a grate. Therefore, in residential plumbing, it is likely that all fixture outlets will have grates, with the exception of water closet pans.

It is best practice to fit outlets with removable grates. Where fixture traps cannot be accessed, grates must be removable.

Fixture traps

Earlier in this unit, we studied the various components of fixture traps and the retention of trap seals. It is also necessary to study the limitations on the location of traps in relation to the fixtures. The discharge from all sanitary fixtures must pass through an accessible trap or self-sealing device before entering the sanitary plumbing system. The trap must be fitted as close as practicable to the outlet, but not more than 600 mm for fixtures, with the following exceptions:

■ fixtures permitted to discharge through a floor waste gully, provided the fixture discharge pipe does not exceed 1.2 m untrapped (basins and drinking fountains are not permitted to discharge untrapped through a floor waste gully)

■ when a pair of fixtures is connected through a common trap (see **Figure 5.13**).

GET IT RIGHT

PROVIDE FOR EXPANSION

FIGURE 5.12 PVC junction fitting with cracks

Failure to provide for thermal movement can result in pipe or fitting fracture and leaks. In Figure 5.12, a sweep junction to a vent in elevated pipework has fractured and leaked, leading to a building dispute and causing an accommodation building to be vacated.

1   How many expansion joints must be installed in a PVC pipeline 10 m long with fixed ends where the discharge temperature is 44°C?

_____

_____

_____

2   How many expansion joints must be installed in a PVC pipeline 10 m long with fixed ends where the discharge temperature is 48°C?

_____

_____

_____

3   How would you securely fasten the expansion joints so that they do not move with the pipe?

_____

_____

_____

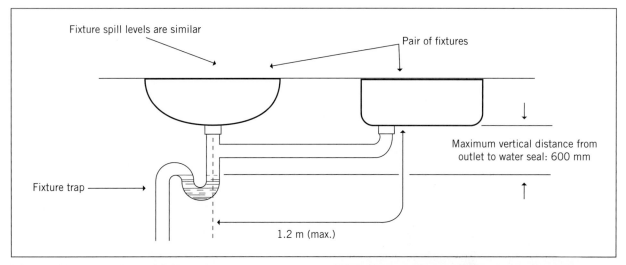

FIGURE 5.13 Fixture pair connected to same trap

Fixture pairs

A pair of waste fixtures may discharge to a common fixture trap, subject to the following conditions:

1 The fixtures have a similar spill level.
2 Only the following fixtures are permitted:
   – basins
   – sinks
   – showers
   – laundry troughs
   – ablution troughs.
3 Both fixtures must be installed in the same room.
4 The fixture trap for a pair of showers may be a floor waste.
5 The fixture pair trap size is set out in AS/NZS 3500.2.

Note: Fixture trap size for basins is a minimum of DN 40 diameter.

The untrapped fixture discharge pipe must be:
- as short as possible
- connected into the trap above the trap seal
- a maximum of 1.2 m in length.

Gradients of discharge pipes

Listed below are the minimum gradients of discharge pipes commonly used in residential sanitary plumbing. The simplest way to remember them is to associate the gradient with a range of sizes. The minimum gradients are:
- 40 to 65 mm discharge pipe: minimum gradient 2.5%
- 80 to 100 mm discharge pipe: minimum gradient 1.65%.
AS/NZS 3500.2 specifies the minimum gradients of all discharge pipes.

Length of discharge pipes

If a discharge pipe exceeds a length specified in AS/NZS 3500.2 without a vent, there is greater potential for blockages and loss of trap seals.

The maximum length is measured along the pipe from the weir of the fixture trap to the connection to a graded discharge pipe, stack or other drainage trap (see Figure 5.14).

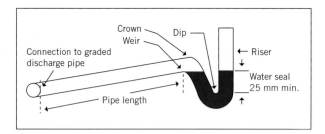

FIGURE 5.14 Measuring the maximum length of a discharge pipe

The maximum length of discharge pipes depends upon the type of plumbing system, and we will study six different plumbing systems later in this unit. If the length of discharge pipe exceeds that specified in AS/NZS 3500.2, it still may be connected, provided that a vent or air admittance valve (AAV) is installed.

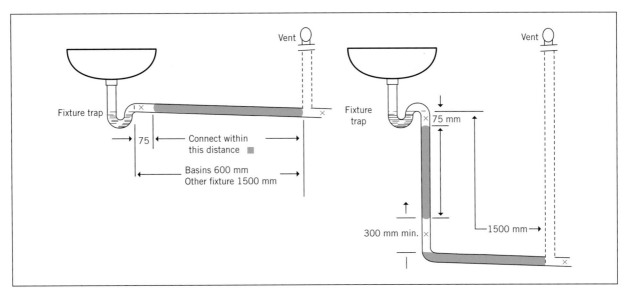

FIGURE 5.15 Connection of trap vents to fixture discharge pipes

Trap venting

If a fixture discharge pipe exceeds the permitted length, a vent or AAV must be installed. The vent or AAV must be connected to the discharge pipe and constructed in accordance with AS/NZS 3500.2. Figure 5.15 illustrates the requirements for connecting trap vents to fixture discharge pipes.

Installation of trap vents and air admittance valves

Trap vents and AAVs can be installed in one of four ways. For further details, read AS/NZS 3500.2.

Gradient of trap vents

All vents should be installed at a minimum grade of 1.25%, allowing any condensation or other liquids that form in or enter the vent to drain to the sanitary plumbing/drainage system, regardless of the size of the vent.

Size of trap vents

For fixture discharge pipes, the size of the vent depends upon the size of the fixture trap. The following are the minimum sizes of trap vents. The simplest way to remember them is to associate the size of the trap vent with a range of fixture trap sizes:

- 40 mm fixture trap: 32 mm trap vent
- 50 to 100 mm fixture trap: 40 mm trap vent.

Termination of trap vents

AS/NZS 3500.2 specifies the general requirements for termination of vents.

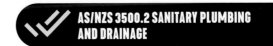

Connections to vertical stacks

The connection of discharge pipes to a vertical stack may be by either manufactured junction fittings or site-formed junctions, depending on the type of material used.

The types of junctions allowed are 45° junctions, sweep junctions, entry at grade with throat radius, and entry at grade without throat radius. The type of junction that may be used depends on the size of the stack and the size of the branch. When connecting branches DN 50 or less to stacks DN 65 or less, 45° entry, sweep junctions or entry at grade with throat radius may be used. When connecting branches DN 65 to a DN 65 stack, a 45° entry or sweep junction may be used.

When connecting branches DN 65 or less to a stack DN 80 or larger, any of the junction types mentioned above may be used.

Restrictions

A fixture discharge pipe 500 mm or less in length and connected to a vertical stack with a square junction may only be used provided:

- an 'S' trap is fitted to the fixture, and the fixture discharge pipe has a vertical drop between the outlet of the fixture trap and stack connection
- a 'P' trap is fitted to the fixture, and the fixture discharge pipe is installed at a gradient of at least 6.65%, or
- a self-sealing device is fitted to the fixture.

Restricted zones

There are restricted zones in the areas of the stack immediately opposite and a distance vertically down from another junction in the stack.

Opposed connections at the same level must be made using double 'Y' junctions with an included branch angle of 90° (see **Figure 5.16**).

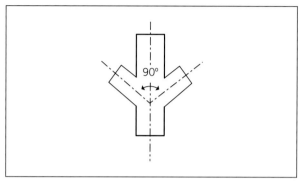

FIGURE 5.16 Double 'Y' junction

Generally, when using PVC fittings, you will have no problems complying with restricted zone requirements due to the dimensions of fittings. However, you should familiarise yourself with the section, 'Junctions in stacks' in AS/NZS 3500.2.

AS/NZS 3500.2 SANITARY PLUMBING AND DRAINAGE

Connections to single-stack graded offsets

Single stacks of a maximum 10 floor levels, and not more than five floor levels above the upper offset bend, may have one graded offset provided the following guidelines are followed:

■ The minimum horizontal distance between the centrelines of the vertical stack is 2 m.

■ No connection can be made closer than 2.5 m downstream or 900 mm upstream of the upper offset bend. The exception is a closet pan, which may be connected 600 mm above the upper offset bend, but it must be provided with a DN 40 trap vent.

■ No connection may be made within 450 mm above or 600 mm below the lower offset bend.

For other single-stack offset restrictions, see AS/NZS 3500.2. See also **Figure 5.17**.

AS/NZS 3500.2 SANITARY PLUMBING AND DRAINAGE

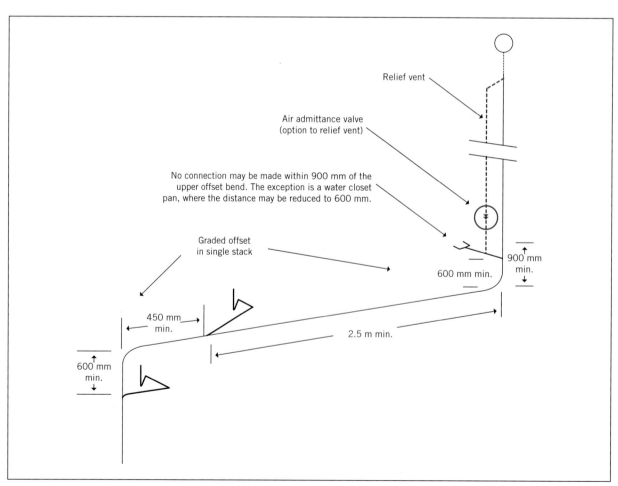

FIGURE 5.17 Connection within a graded offset in single stack

Connections near the base of stacks

There are restricted zones near the base of stacks where discharge pipes must not be connected. This is due to the potentially adverse effects of air pressure and foaming on fixtures. Figure 5.18 illustrates the restricted zone dimensions for stacks to drains or graded pipes, and above the base of stacks of up to two and three floor levels. For information on stacks of more than three floor levels, read AS/NZS 3500.2.

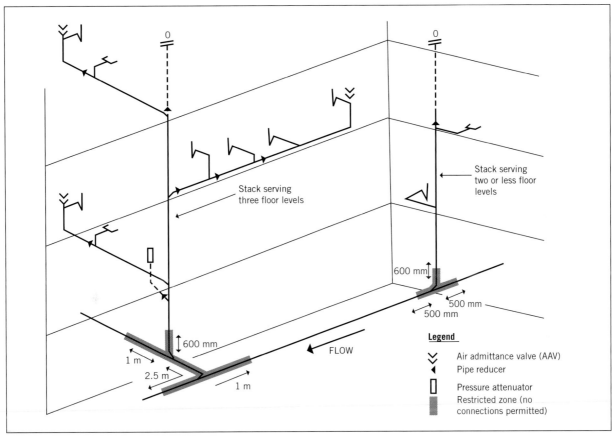

FIGURE 5.18 Restricted zones near the base of stacks

Connection of stacks to drains or graded pipes

Where a stack extends through no more than two floor levels, one 88° bend may be used at the base of the stack. For stacks connecting to junctions, or extending through *more than* two floor levels, refer to AS/NZS 3500.2.

Floor waste gullies

A floor waste gully is a trap installed in a building with a floor grate. Fixture discharge pipes can be connected to the riser.

The requirements for installation of floor waste gullies are not specified in AS/NZS 3500.2, but appear in building codes and regulations. One notable exception is where a wall-hung urinal is installed; in this case, the floor surface must be graded to a floor waste gully.

Generally, floor waste gullies are installed in wet areas such as bathrooms and laundries to intercept the accidental spillage of waste water.

Installation

Where required by local building regulations, floor waste gullies must be installed at the lowest point in the floor of a room or compartment containing a bath, basin, shower, water closet, bidet, urinal, trough or slop hopper.

Floor waste gullies may be omitted where:

- a room contains a water closet, or a water closet and basin only, and the cistern has an internal overflow, or the cistern overflow discharges to the outside atmosphere and the basin has an internal overflow
- the floor is graded to a urinal channel or shower
- the floor is graded to an external doorway above ground level
- a dry floor waste is permitted.

The water seal of every floor waste gully must be maintained by a waste fixture, charge pipe or hose tap in the same room, in accordance with the methods outlined in AS/NZS 3500.2.

AS/NZS 3500.2 SANITARY PLUMBING AND DRAINAGE

There are limitations on the types of fixtures that may be connected to a floor waste gully, and these are:

- *fixtures that may be connected:* basins, baths, bidets, bar sinks, clothes-washing machines, showers and laundry troughs
- *fixtures that are prohibited:* sinks and soil fixtures such as urinals and water closets.

It is common practice to discharge clothes-washing machine waste through a laundry trough fixture discharge pipe. However, clothes-washing machine discharge to floor waste gullies has often caused foaming problems. Therefore, where a floor waste gully is required it may be primed by connection of a laundry trough, However, it is preferable to separately connect the clothes-washing machine discharge pipe to the sanitary plumbing and drainage system.

Fixtures that are permitted to connect to floor waste gullies must, except for tundish discharges, be located in the same room as the gully, for the following two reasons:

1 In the event of blockage in a floor waste gully, the overflow will be readily visible.
2 Sound transmission between adjoining rooms is avoided.

All permitted waste fixtures, except basins and drinking fountains, may be connected untrapped to a floor waste gully, providing that the length of waste pipe does not exceed 1.2 m. All permitted waste fixtures may be connected trapped to a floor waste gully, providing the length of waste pipe does not exceed 2.5 m. See **Figure 5.19**.

With the exception of fixture pairs, each fixture shall be connected to a floor waste gully riser by an individual waste pipe at a gradient of not less than 2.5%.

The height of a floor waste gully depends on the size of the gully riser and the angle of entry of the fixtures discharging pipe. The minimum height of a floor waste gully riser is specified in AS/NZS 3500.2. The maximum length of a floor waste gully riser is 600 mm from the water seal to the level of the floor grate.

AS/NZS 3500.2 SANITARY PLUMBING AND DRAINAGE

Removable grate

Floor waste gullies must be installed with an accessible removable grate and a riser not less than DN 80 at floor surface level. Where the only function of the floor waste is to drain the floor of spilled water and wash-down water (no fixtures are connected), a DN 50 outlet grate and riser may be used.

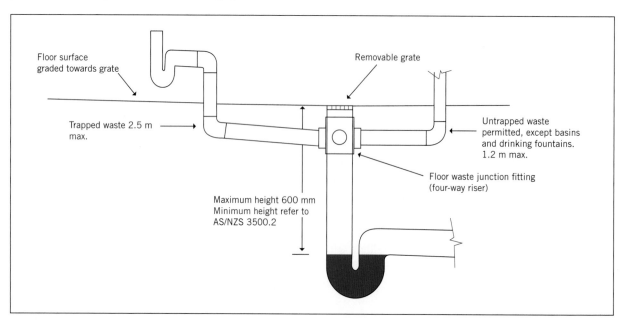

FIGURE 5.19 Floor waste gully connections

The outlet size of a floor waste gully trap determines the maximum fixture unit loading discharging into the trap. For example, a floor waste gully with a DN 65 outlet may only receive a maximum of 10 fixture units, including not more than one bath. (Read AS/NZS 3500.2.)

AS/NZS 3500.2 SANITARY PLUMBING AND DRAINAGE

Dry floor wastes

Floors in residential buildings may be drained to the outside of the building by the installation of a dry floor waste (see **Figure 5.20**). Dry floor wastes must:

- be a minimum size of DN 40 and fitted with a removable grate
- discharge where they will not cause water damage or create a slip hazard
- terminate between 25 mm and 100 mm above ground level
- be fitted with an air break over a tundish if more than 1.8 m above ground level.

In multi-storey residential buildings, dry floor wastes may discharge to a common pipe, provided the common pipe is:

- a minimum size of DN 50
- fitted externally
- connected to floor waste with air breaks and tundishes.

Commercial dry floor wastes must be a minimum of DN 50.

Note: Dry floor wastes cannot be fitted in rooms containing a urinal.

Testing and inspection openings

All discharge pipes in above-ground sanitary plumbing must have inspection openings as follows:

- in any stack or common discharge pipe where necessary for inspection or testing
- on every common discharge pipe that connects to a stack
- no more than 30 m apart in any pipe on grade
- at the base of every stack.

All 100 mm pipes must have inspection openings 100 mm in diameter. AS/NZS 3500.2 specifies the

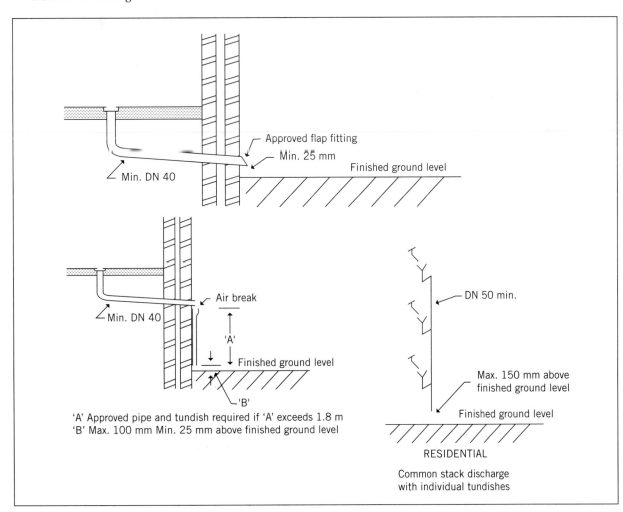

FIGURE 5.20 Dry floor waste layouts

Source: Department of Education and Training (http://www.training.gov.au) © 2013 Commonwealth of Australia.

required sizes of testing and inspection openings relative to each pipe size.

Inspection openings must be accessible for maintenance and clearance of blockages.

Elevated pipework using drainage principles

Elevated pipework is the most common system of sanitary plumbing used in residential applications up to three storeys (see Figure 5.21).

In Units 4 and 10, we study the requirements for planning the layout and installing a below-ground sanitary drainage system. This includes consideration of the gradient, fixture unit loadings and ventilation of drains. The same principles can be applied to the layout of elevated pipework up to four floors above the invert of the boundary trap or inspection shaft. The main restriction is that no fixture discharge pipe is permitted to be connected to any vertical section. (Read AS/NZS 3500.2.)

Fully vented system

A fully vented system is rarely used, due to the significant cost of providing vents to every fixture trap. A relief vent is normally installed. The stack and relief vent is sized in accordance with the maximum fixture unit loadings as outlined in AS/NZS 3500.2.

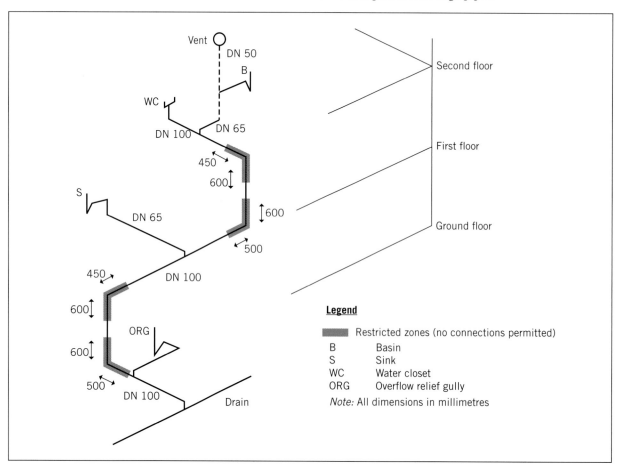

The following section includes an example for sizing vented graded discharge pipes for both fully vented and fully vented modified systems.

Size of vented graded discharge pipes

While the size of the fixture outlet determines the minimum size of the trap and fixture discharge pipe, the fixture unit rating is not considered in sizing the fixture discharge pipe. The fixture unit ratings are used when sizing the common discharge pipe, which connects more than one fixture.

When the fixture units of each fixture connected to the same discharge pipe are added, we arrive at the total fixture unit loading on that discharge pipe.

Example 5.1 assists in explaining the method for calculating fixture unit loadings and the minimum size of vented graded discharge pipes.

FIGURE 5.21 Elevated pipework (above ground) using drainage principles

EXAMPLE 5.1

Refer to the tables in AS/NZS 3500.2, 'Fixture unit ratings' and 'Maximum fixture unit loadings for graded discharge pipes'.

First, the minimum size of trap outlet and fixture discharge pipe is obtained from the table titled 'Fixture unit ratings'.

Fixture 1   Laundry trough =   5 fixture units, outlet size 40 mm

Fixture 2   Bath =   4 fixture units, outlet size 40 mm

Fixture 3   Shower =   2 fixture units, outlet size 40 or 50 mm

Fixture 4   Basin =   1 fixture unit, outlet size 40 mm

What is the total number of fixture units when these fixtures are connected to the same vented graded discharge pipe?

$$5 + 4 + 2 + 1 = 12 \text{ fixture units}$$

The size of graded discharge pipe serving these four fixtures must be at least the size of the largest fixture outlet. From the table, 'Maximum fixture unit loadings for graded discharge pipes', we can see that for a 50 mm discharge pipe at gradient 2.5%, the maximum number of fixture units is eight; however, we have 12 fixture units. Therefore, the downstream section of our graded discharge pipe will need to be increased in size to 65 mm, because this will allow a maximum fixture unit loading of 21 fixture units. Each section of pipe is sized in accordance with the total of fixture units discharging through that pipe (see Figure 5.22).

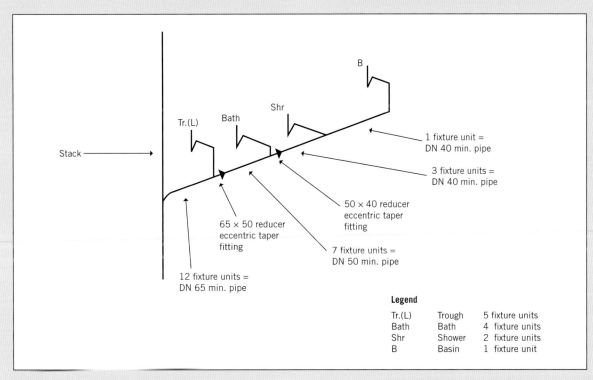

FIGURE 5.22 Sizing vented graded discharge pipes

Fully vented modified system

The fully vented modified system is commonly used because some individual trap vents can be omitted. Groups of fixtures can be group vented.

The vented graded discharge pipe is sized as per Example 5.1. The following section includes an example for venting a graded discharge pipe in the fully vented modified system.

Venting a graded discharge pipe

When connecting a vent or AAV to a graded or common discharge pipe, it must meet the requirements of AS/NZS 3500.2.

AS/NZS 3500.2 SANITARY PLUMBING AND DRAINAGE

In the case of basins and bidets, the vent or AAV should be connected no closer than 75 mm and no further than 600 mm from the crown of the fixture trap, provided no change of direction occurs between the trap and the vent/AAV.

In the case of fixtures other than basins and bidets, the vent/AAV should be connected between 75 mm and 1.5 m, provided that where an 'S' trap or a bend is fitted downstream of a 'P' trap, the vent or AAV connected on the vertical discharge pipe is at least 300 mm from any bend at the base of the vertical section.

In Example 5.2, there are three main options for providing group ventilation:

1 Installation of an AAV, sized in accordance with the fixture unit loading on the graded discharge pipe. (See the table in AS/NZS 3500.2, 'Minimum determined airflow capacity of air admittance valves when used as a trap, group vent, or branch drain vent'.)

2 Group vent extended to open air, sized in accordance with the fixture unit loading on the graded discharge pipe. (See the table in AS/NZS 3500.2, 'Size of group vents'.)

3 Group vent connected to the relief vent, sized in accordance with the fixture unit loading on the graded discharge pipe. (See the table in AS/NZS 3500.2, 'Size of group vents'.)

AS/NZS 3500.2 SANITARY PLUMBING AND DRAINAGE

EXAMPLE 5.2

We now follow Example 5.1, and add ventilation to the graded discharge pipe. In the fully vented modified system of plumbing, group vents are provided for each 10 fixtures or part thereof. In our example, following the previous section, there are only four fixtures, and so only one group vent is required. Refer to the table in AS/NZS 3500.2, 'Size of group vents'.

First, we must determine the largest section of graded discharge pipe to be vented. In the example, the largest section is DN 65.

From the table 'Size of group vents', it can be seen that for a DN 65 common discharge pipe, the size of group vent required is DN 40.

See Figure 5.23 for the sizing example and venting options.

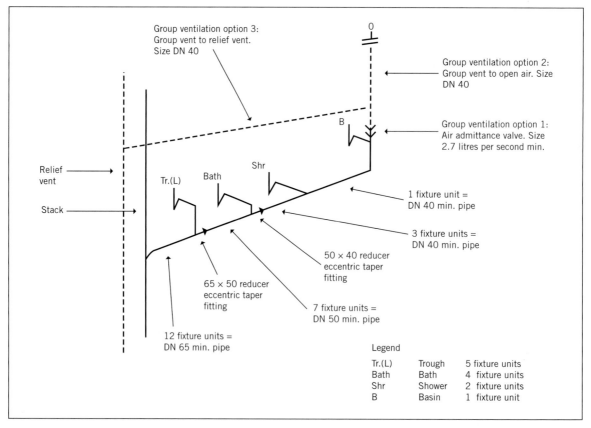

FIGURE 5.23 Group ventilation of graded discharge pipes

Determining stack and relief vent requirements and sizing

After the graded discharge pipes and group vents have been sized, the stack and relief vent requirements and sizing must be determined (see Example 5.3). The following sequence can be followed to complete the sizing of the fully vented modified system:

1 Determine whether a relief vent is necessary. If there are only two consecutive floors of fixtures, no relief vent is required. In all other cases, a relief vent must be installed.

2 Add the fixture unit loadings for the entire stack and for each floor, and then size the stack in accordance with AS/NZS 3500.2. As a guide, when there are three or fewer floor levels, the fixture unit loading from each floor must not exceed one third of the stack capacity.

3 Size the relief and stack vent in accordance with AS/NZS 3500.2. This will depend on the size of the stack, total number of fixture units and developed length of the vents. The developed length is the total length of relief vent and stack vent from the lower stack connection to the vent terminal. Pressure attenuators may be used as an alternative to the piped relief vent (see Figure 5.18 earlier in the unit).

AS/NZS 3500.2 SANITARY PLUMBING AND DRAINAGE

Where there are fewer than eight floors, one pressure attenuator is required at the base of the stack. The connection of relief vents or pressure attenuators must be clear of prohibited zones, as described earlier in this unit.

Reduced-velocity aerator stack system

The reduced-velocity aerator stack system (REVASS) is a more recent development that has been used in large, high-rise residential projects. The REVASS slows the downward flow of waste, prevents trap syphonage and reduces the need for vents. Airflow in the system is provided through the stack vent.

This system was introduced with the development of the aerator junction fitting (a proprietary junction fitting). An aerator junction is installed in the stack to connect discharge pipes at each floor level, and a de-aerator is installed at the base of the stack.

Stacks are sized in a similar way to the fully vented and fully vented modified systems.

For further information on particular products, and system specifications and features, research 'Geberit Sovent System' online and read AS/NZS 3500.2. Section 11.

AS/NZS 3500.2 SANITARY PLUMBING AND DRAINAGE

Single-stack system

The single-stack system is commonly used in residential sanitary plumbing, due to the minimal venting requirements and low cost. Single-stack systems are designed on the principle that the air within the fixture discharge pipes, the stack and the stack vent provides adequate ventilation to enable the permitted numbers and types of fixtures to be connected without the need for individual trap and group fixture ventilation.

AS/NZS 3500.2 contains illustrations of two basic single-stack systems, namely:

1 *single-stack systems, domestic or residential buildings*, where fixtures are connected individually or through floor waste gullies

2 *single-stack systems, commercial or industrial buildings*, where fixtures are connected individually, through floor waste gullies, or in ranges of the same type of fixture.

AS/NZS 3500.2 SANITARY PLUMBING AND DRAINAGE

A 45° stack junction may be used to connect two waste fixtures

In addition to the standard methods of connections mentioned above, two waste fixture discharge pipes, including discharge pipes from floor waste gullies, may be connected at the same level within 1 m of the stack by means of a 45° junction, in accordance with Figure 5.24 and AS/NZS 3500.2.

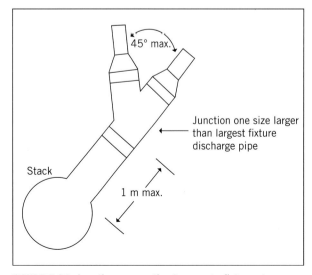

FIGURE 5.24 Junction connecting two waste fixtures to a single stack

AS/NZS 3500.2 SANITARY PLUMBING AND DRAINAGE

Sizing stacks – residential buildings

The size of the stack is determined by adding the individual fixture unit ratings. The maximum number of floor levels through which the stack may pass is specified in AS/NZS 3500.2. If the stack connects to the drain more than 2.4 m below the lowest floor level, this vertical distance must be counted as one or more floor levels. For example, if the stack connects to the drain at 4.8 m below the lowest floor level, then an additional two floor levels must be added.

AS/NZS 3500.2 SANITARY PLUMBING AND DRAINAGE

Fixture discharge pipes

Generally, the lengths of unvented fixture discharge pipes are as follows:

- fixture outlet DN 100: 6.0 m maximum
- fixture outlets less than DN 100: 2.5 m maximum.

If the length of fixture discharge pipes exceeds either of these, then a trap vent or air admittance valve must be fitted.

Other considerations, apart from size and length, must also be given to the installation of fixture discharge pipes. There are also restrictions on the number of bends and the gradient.

Bends in fixture discharge pipes

Basins and bidets must have no more than two bends in the horizontal plane and two bends in the vertical plane. Other fixtures must have not more than two bends in the horizontal plane and three bends in the vertical plane.

Bends of 45° or less are not considered a change of direction.

If more bends are required, the fixture discharge pipe would need to be vented, using either a trap vent or an AAV.

Venting of stacks

Stack vents in single stacks must be the same size as the stack. An exception is where stacks extend not more than three floor levels with a maximum loading of 30 fixture units and the top section is on grade. In this case, the stack vent may be reduced to DN 50.

Offsets in single stacks

An offset may be steep, at more than 45° to the horizontal, or be graded at less than 45° to the horizontal. The minimum gradient of offsets is 2.5% for waste stacks up to DN 80 and 1.65% for DN 100 or larger stacks.

Steep offsets

DN 100 stacks may have a steep offset where the height of the stack does not exceed 10 consecutive floor levels, and there are restrictions on the connections of laundry troughs. There are also restrictions on the connection of fixtures near upper and lower offset bends.

Where a DN 100 stack has a steep offset below the lowest connection:

- the height of the stack above the offset must not exceed three floor levels
- the maximum permitted loading is 30 fixture units through the offset
- laundry troughs are not permitted
- the minimum distance between the lowest connection and upper offset bend is 100 mm (see Figure 5.25).

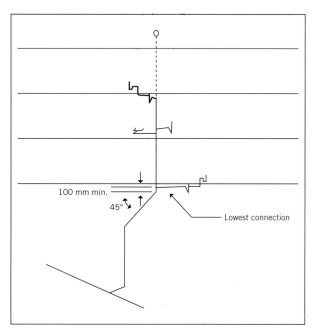

FIGURE 5.25 Single-stack steep offset below lowest connection

Graded offsets

Single stacks of a maximum of 10 floor levels and not more than five floor levels above the upper offset bend may have one graded offset, providing that:

1 The minimum horizontal distance between the centrelines of the vertical stack is 2 m.
2 No connection can be made closer than 2.5 m downstream or 900 mm upstream of the upper offset bend. The exception is a closet pan, which may be connected 600 mm above the upper offset bend, but it must be provided with a DN 40 trap vent.
3 No connection may be made within 450 mm above or 600 mm below the lower offset bend (see Figure 5.26).

For other single-stack offset restrictions and variations, see AS/NZS 3500.2.

AS/NZS 3500.2 SANITARY PLUMBING AND DRAINAGE

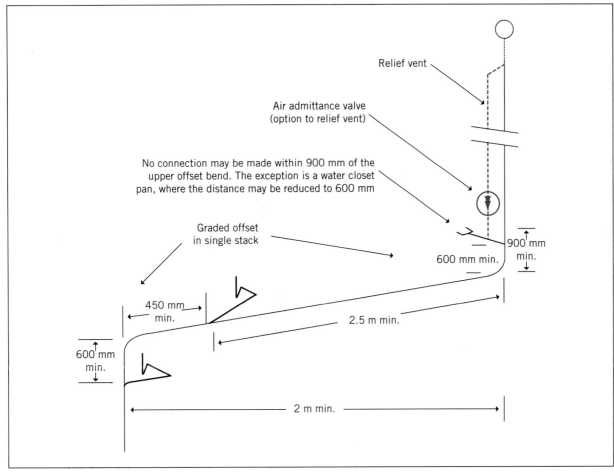

FIGURE 5.26 Single-stack graded offset

EXAMPLE 5.3

Following Example 5.2, look at Figure 5.27. We can now complete the sizing of the stack and relief vent. The following steps are used:

1   Add the fixture units at each floor. The first floor has five fixture units, the second floor has 12 fixture units, and the third floor has five fixture units. Therefore, the total fixture unit loading on the stack is 22 fixture units.

2   In AS/NZS 3500.2, find the table 'Maximum loadings on stacks in fixture units'. Always remember to look in the correct section of the standard.

3   In our example, we will assume that the water closet discharge pipes will be DN 100; therefore, the stack must be at least DN 100. (Note that a maximum of two water closet pans may be connected to a DN 80 stack; however, it is generally more cost-effective to install DN 100 discharge pipes and stacks.) Where asked for minimum pipe sizing, a DN 80 stack must be used to connect a maximum of two closet pans.

4   Now look at the table 'Maximum loadings on stacks in fixture units'. You will see that for three or fewer floor levels and a DN 100 stack, we are allowed a maximum of 65 fixture units per floor level and a maximum of 195 fixture units per stack. Since the maximum per floor

level in our example is only 12 fixture units, and our total fixture unit loading is 22 fixture units, our stack size DN 100 is satisfactory.

5   Now we will size our relief vent. Remember that a relief vent is required because we have more than one floor level separating the highest and lowest branch connections. We must measure our relief vent between the lower stack connection and stack vent terminal. For the purposes of our example, imagine that the measured *developed* length of our relief vent is 12 m.

6   Using AS/NZS 3500.2, find the table 'Size of relief vents and stack vents'. You will see that for a DN 100 stack, a maximum fixture unit loading of 150 fixture units and a maximum developed relief vent and stack vent of 25 m, a DN 65 relief vent and stack vent is required. Note that a DN 50 relief vent and stack vent would have been adequate if the developed length of our relief vent and stack vent had been 9 m or less.

7   The graded discharge pipes for each floor level must be sized correctly, remembering that no graded common discharge pipe should be less than the size of group vent required. In the example (see Figure 5.27), you will see that for the DN 100 graded discharge pipe to the water closet, the pipe upstream to the basin is DN 50. This is

because a DN 50 group vent is required for a DN 100 graded discharge pipe. The size of the AAV may be smaller than DN 50; however, this is not relevant to the airflow required in the graded discharge pipe.

8   Branches in stacks to relief vents and pressure attenuators must be the correct angle. The lower stack connection

is at 45° to the stack clear of the restricted zone, and the upper stack connection of the relief vent to the stack vent is made on an ascending gradient of 1.25%.

9   Testing and inspection openings must be installed in accordance with AS/NZS 3500.2, and as mentioned earlier in this unit.

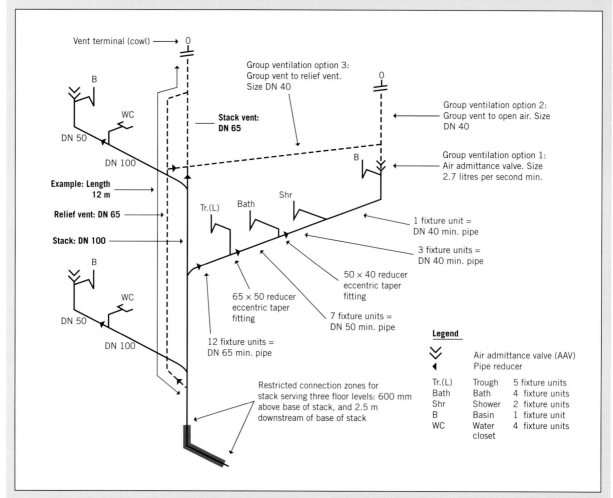

FIGURE 5.27 Stack and relief vent sizing

Waste stacks

Waste stacks may be used to connect only waste fixtures. A DN 65 waste stack may be used to connect kitchen sinks and laundry troughs, provided that:

■   the stack does not exceed two floor levels, and

■   only two sinks, or one sink and a trough, are separately connected from each floor.

A DN 50 waste stack may be used to connect three waste fixtures to the top of a DN 50 vertical stack vent where the total stack loading will not exceed 30 fixture units; however, this applies only to basins, showers and kitchen sinks.

Aside from the above conditions for DN 65 and DN 50 waste stacks, all other waste stacks must be sized, so the fixture unit loading from any floor level must not exceed a quarter of the maximum loading (see Example 5.4).

EXAMPLE 5.4

Find the table 'Size of waste stack' in AS/NZS 3500.2. Imagine that we propose to install a DN 65 waste stack. The maximum fixture unit loading is 15. Therefore, the maximum loading to a DN 65 waste stack from any floor level is a quarter of 15 (i.e. three fixture units).

Variations to single-stack systems

AS/NZS 3500.2 lists the following variations to the requirements for single-stack systems:

■   DN 80 stacks serving up to three floors with a maximum loading of 30 fixture units in domestic or residential buildings

- DN 80 stacks serving up to two floors with the top section graded in domestic or residential buildings
- DN 100 stacks serving up to three floors with the top section graded, with a maximum loading of 30 fixture units in domestic or residential buildings
- DN 100 stacks serving one first floor with the top section graded, with a maximum loading of 90 fixture units in domestic or residential buildings. (Read AS/NZS 3500.2.)

AS/NZS 3500.2 SANITARY PLUMBING AND DRAINAGE

Single-stack modified system

The single-stack modified system is rarely used in residential sanitary plumbing in low-rise buildings, as the design is more suited to large fixture unit loadings in high-rise buildings.

Single-stack modified systems are designed on the same principle as single-stack systems in that the air within the fixture discharge pipes, the stack and the stack vent provides ventilation to enable the permitted numbers and types of fixtures to be connected without the need for individual trap and group fixture ventilation. A higher maximum discharge loading from a higher number of floor levels is achieved by introducing a relief vent and cross-vents, and yet no increase in nominal size is necessary.

AS/NZS 3500.2 contains illustrations of two basic single-stack modified systems, namely:

1 *single-stack modified systems, domestic or residential buildings*, where fixtures are connected individually or through floor waste gullies, and a relief vent and cross-vent/s are installed
2 *single-stack modified systems, commercial or industrial buildings*, where fixtures are connected individually, or in ranges of the same type of fixture, and a relief vent and cross-vent/s are installed.

For more information on the application of single-stack modified systems, read AS/NZS 3500.2.

AS/NZS 3500.2 SANITARY PLUMBING AND DRAINAGE

Selecting plumbing systems

In Unit 4, you will have studied plan reading, symbols and material quantities. When a plumber is planning the layout for a sanitary plumbing system, it is necessary to be able to read and interpret various types of building drawings to determine the building constraints and then consider what factors will influence plumbing system selection. This was discussed earlier in this unit.

As part of the planning process, you will need to do the following:

- Determine the quantity, location and type of plumbing fixtures from the design drawings, plans and specifications. Floor wastes may also be required.
- Determine options and the best position of fixture discharge pipes, considering the maximum permitted length of fixture discharge pipes and the number of bends permitted in both the vertical and horizontal planes.
- Determine options for the location of junctions, considering restricted zones and requirements for the orientation of junctions.
- Determine options for the location of graded discharge pipes and stacks (e.g. in pipe ducts, ceiling cavities, wall cavities and under floors).
- Determine system venting requirements and the location of vent terminals. *Note*: AAVs must be accessible.
- Consider the effects of the location of pipes. For example, the pipes should ideally be placed where noise of waste moving through the pipes is acceptable.
- Consider options and features of different pipe materials. (*Note:* There is a growing emphasis on the use of sustainable/recyclable materials.)

Location of pipes and fixtures

Pipes and fixtures need to be located:
- in accordance with plans and specifications
- as determined by building constraints
- so as not to cause damage or interference to surrounding structures (e.g. doors and windows).

Fixture dimensions

Fixture specifications will give the plumber the exact fixture dimensions to determine the location of fixture outlets – for example, for closet pans, whether it be a vertical height measurement for a 'P' trap WC or a horizontal distance from a wall for an 'S' trap WC. Remember to allow for finished surfaces, such as wall and floor tiles.

Specifications in relation to concealed pipework

Specifications may require discharge pipes and stacks to be positioned internally in concealed positions. In this case, traps and testing/inspection openings need to be positioned in an accessible location with access panels as required by AS/NZS 3500.2. *Note*: Any trap that will not be accessible must have a removable grate. All AAVs and all self-sealing devices must be permanently accessible. Some IOs must be permanently accessible.

AS/NZS 3500.2 SANITARY PLUMBING AND DRAINAGE

Selecting a system of plumbing

A system or systems of plumbing must be selected, ensuring that the final configuration of pipework will comply with AS/NZS 3500.2 and relevant state/territory requirements.

AS/NZS 3500.2 SANITARY PLUMBING AND DRAINAGE

While we have described six main systems of sanitary plumbing in this unit, the three systems commonly used for residential sanitary plumbing up to three storeys are:

1 elevated pipework using drainage principles
2 fully vented modified system
3 single-stack system.

Determining material quantities

Once the layout of the sanitary plumbing system has been determined, a drawing can be prepared. It is recommended, particularly when learning to plan the layout for sanitary plumbing systems, that learners prepare isometric drawings. Isometric drawings give learners a pictorial representation of the proposed piping layout, which is useful in developing an understanding of plumbing systems and relevant material quantities.

Example project

Figure 5.28 is an isometric drawing of a sanitary plumbing system for a three-storey building. The system 'elevated pipework using drainage principles' has been selected, and the piping has been sized accordingly.

A materials list for the sanitary plumbing system is in Table 5.3. The plumbing fittings chart earlier in this unit (see Figure 5.5) can assist in identifying some of the PVC plumbing fittings.

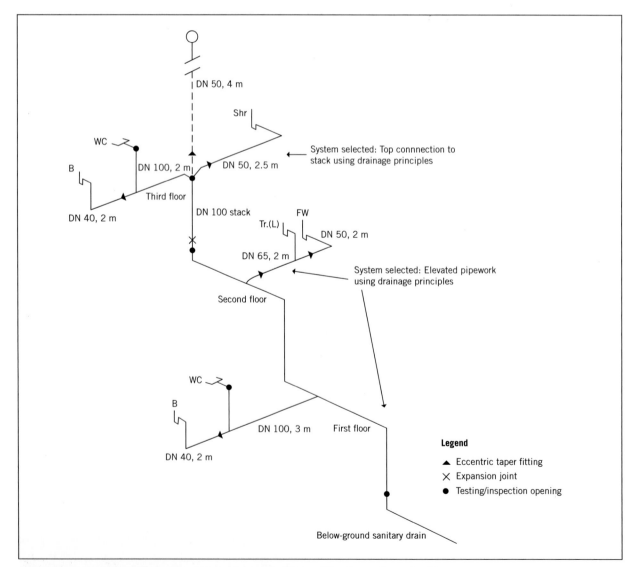

FIGURE 5.28 Example project, isometric drawing: material quantities for sanitary plumbing

TABLE 5.3 Materials list for example project

Item	Description	Quantity
1	50 PVC DWV plain vent cowl	1
2	50 PVC DWV pipe	10 m
3	100 × 50 PVC flat reducer	2
4	100 double 'Y' IO PVC junction	1
5	100 × 50 PVC eccentric taper fitting (LIT)	1
6	50 PVC 'P' trap	3
7	50 PVC plug and washer	2
8	100 PVC DWV pipe	22 m
9	100 × 88° F&F IO PVC bend	2
10	100 M&F PVC concentric pan adaptor	2
11	100 × 40 PVC eccentric taper fitting (LIT)	2
12	40 × 85° F&F PVC plain bend	4
13	40 PVC 'P' trap	2
14	40 PVC plug and washer	2
15	40 PVC DWV pipe	4 m
16	100 × 45° PVC F&F plain junction	4
17	100 × 45° PVC F&F plain bend	4
18	100 PVC expansion joint	1
19	100 PVC square IO junction	2
20	100 × 88° F&F PVC plain bend	4
21	100 × 65 PVC eccentric taper fitting (LIT)	1
22	65 PVC DWV pipe	2 m
23	65 × 88° PVC plain junction	1
24	65 × 50 PVC flat reducer	1
25	65 × 50 PVC eccentric taper fitting (LIT)	1
26	50 CP (chrome-plated) floor-waste grate	1
27	40 plastic coated clip heads	4
28	50 plastic coated clip heads	8
29	65 plastic coated clip heads	2
30	100 plastic coated clip heads	14
31	Clip shanks	28
32	PVC priming fluid (red)	1 litre
33	PVC solvent cement (blue)	1 litre

Estimates for material quantities vary. For example, some plumbers will count pipe lengths rather than in metres of pipe. PVC pipe is normally sold by the 6 m length. Depending on the building structure, other materials will also need to be considered and added to the list shown in **Table 5.3**, such as self-drilling screws, masonry anchors, wall sleeves and roof penetration flashing. Also, you will need to consider consumables such as drills and cutting blades.

Preparation of plans

In the example project, an isometric drawing assisted us to determine the quantities of materials and fittings to complete the job. Isometric and plan design drawings are often prepared by hydraulic designers for large projects, and 'as-constructed' drawings are prepared when works are complete.

Sewerage and regulatory authorities have different procedures and requirements for drawings. You may find that the symbols used in the example isometric drawing in **Figure 5.28** are different in your area; however, the principles are the same.

Your teacher or trainer can assist you in developing drawings in accordance with your local procedures and requirements. It is also important to research your local authority requirements for recording 'as-constructed' details of sanitary plumbing.

COMPLETE WORKSHEET 2

Clean-up

Work area

When planning the layout for a sanitary plumbing system, there is little required for clean-up. Unit 8 will deal more specifically with the clean-up requirements after installation of the sanitary plumbing system. However, at this planning stage, you are required to make provisions for clean-up. Clean-up occurs during the progress of works and at completion.

You must allow time to:
- ensure the worksite is safe for other tradespeople during the progress of work
- clear the worksite of excess materials and trip hazards
- remove material that is hazardous
- clean up food and drink waste, wrappers and containers
- store materials and fittings that may be used later.

Tools and equipment

At the planning stages for the job, you should allow time to determine that you have the correct tools and equipment in accordance with the type of work to be undertaken. You should maintain tools and equipment in a safe and serviceable condition.

Personal protective equipment must be available on-site during the progress of work. Planning includes making sure that elevating work platforms, for example, are on-site and suitable for any work involving installation of piping at high levels.

Plan for time to collect, clean and maintain all tools and equipment. Also, allow time to return hire equipment in good condition in accordance with the hire agreement.

SUMMARY

- When preparing to install a residential sanitary plumbing system, it is necessary to gather information, including plans and specifications that outline the location of fixtures, dimensions of fixtures and floor and wall finishes. Also, familiarise yourself with local regulations and inspection requirements to ensure that your sanitary system will be compliant. Preparation also includes understanding the principles of sanitary plumbing, safety and environmental requirements. Most fixtures, pipes and fittings must be approved for installation. Fixtures and pipework must be installed in coordination with other trades and at the appropriate stages of construction.

- Before commencing work, you must be prepared to comply with the local regulatory requirements and AS/NZS 3500.2, which specify the design requirements, including pipe loadings (fixture units), pipe gradient, provisions for expansion, pipe support and ventilation. You will need to compile a list of materials such as pipes, fittings, traps, pipe clips and brackets, priming fluid, solvent cement, fixings and other consumables that will be required to complete the job. It is good practice to plan to have additional fittings such as 5° bends as a contingency for minor variations in pipe alignment.

- You must make provision for clean-up, ensuring the safety of the worksite and the proper maintenance and storage of tools.

WORKSHEET 1

Student name: _____

Enrolment year: _____

Class code: _____

Competency name/Number: _____

Task: Review the section 'Preparing for work', then complete the following.

1   Identify four sources of information.

   a _____

   b _____

   c _____

   d _____

2   What is the minimum residual water seal for a fixture trap?

   _____

3   What are the six main causes of trap seal loss?

   a _____

   b _____

   c _____

   d _____

   e _____

   f _____

4   What are two ways of preventing trap seal loss caused by wind pressures?

   a _____

   b _____

5   When installing a Level Invert Taper fitting to a graded discharge pipe, why is it necessary to install the tops of the pipes in common alignment?

   _____

   _____

   _____

   _____

WORKSHEETS

5

6   The following gradients are expressed as a ratio. Convert them to percentages.

   a   1:40 _____

   b   1:60 _____

   c   1:100 _____

7   The following gradients are expressed as a percentage. Convert them to ratios.

   a   1% _____

   b   2% _____

   c   3.35% _____

   d   2.5% _____

8   Is it necessary for offset pan connectors to have WaterMark?

   _____

   _____

9   What is the total fixture unit loading for the following:

   a   One water closet and one basin?

   _____

   b   One trough?

   _____

   c   One trough with a clothes-washing machine connected to the trough trap?

   _____

   d   One bathroom group, including basin, bath, shower and water closet?

   _____

10 What safety equipment is required to fix piping at a high level?

   _____

   _____

   _____

11 Explain why proper planning and sequencing of work is important.

   _____

   _____

   _____

   _____

12 What are two ways of solving the problem of spontaneous syphoning of a fixture trap?

_____

_____

_____

_____

13 Research and state the minimum gradient for an 80 mm discharge pipe.

_____

14 State three reasons why venting is necessary in a sanitary plumbing system.

a _____

b _____

c _____

15 Between what points is a discharge pipe measured?

_____

_____

 **WORKSHEET 2**

Student name: _____

Enrolment year: _____

Class code: _____

Competency name/Number: _____

Task

Review the section 'Planning the system layout', then complete the following.

1 Why is it beneficial to have a site meeting with a project supervisor?

_____

_____

_____

_____

2 What pipe materials should not be used for waterless urinal waste pipes?

_____

_____

3 What is the maximum temperature for continuous discharges through:

a HDPE pipe?

_____

b PVC pipe?

_____

4 Using AS/NZS 3500.2, research eccentric taper fittings, then draw an eccentric taper fitting and indicate the direction of flow.

5 Imagine that you have the option to install either a DN 65 or DN 100 pipe over a distance of 4 m. Research the minimum gradient for each and state the amount of fall that is required for each pipe size:

a DN 65 _____

b DN 100 _____

6 Explain why it is important to install discharge pipes at the correct minimum gradient.

_____

_____

_____

_____

7 State two positions where expansion joints are required to be installed in vertical PVC stacks.

a _____

b _____

8 Imagine that you have a DN 50 PVC pipe 4 m long. What is the maximum length of offset leg that must not be clipped when providing for expansion without installing an expansion joint?

_____

_____

9 Following your answer to question 8, draw an isometric diagram to explain your answer.

10 What is the minimum clearance between discharge pipes and an insulated heated water pipe?

_____

_____

11 What are the four primary factors that must be considered when planning the layout and selecting a system of sanitary plumbing?

a _____

b _____

c _____

d _____

12 Under what circumstances must grates in fixture outlets be removable?

_____

_____

13 What is the minimum gradient for a DN 65 fixture discharge pipe?

_____

14 When installing a basin discharge pipe with a vent, where must the vent be connected? Explain your answer by drawing a diagram.

15 Research termination of vents in AS/NZS 3500.2, and state six restrictions on the location of an open vent terminal.

a _____

b _____

c _____

d _____

e _____

f _____

16 Name the fitting that may be used instead of an open trap vent.

_____

_____

17 What type of bend may be used at the base of a stack extending through no more than two floor levels?

_____

18 Research the requirements for bends and junctions at the base of DN 100 stacks through more than two floor levels, and draw one bend and one junction indicating the dimensions of each.

19 Research and state the restricted zone dimension for connections above the bend at the base of a stack through four floor levels.

_____

_____

20 What are two residential plumbing fixtures that must not be connected to a floor waste gully?

a _____

b _____

21 Give two reasons why fixtures discharging to a floor waste gully must be located in the same room as the floor waste gully.

a _____

b _____

22 Research and draw a floor waste gully connecting one bath, one basin and one shower. Show each fixture with a trap. Indicate all minimum and maximum dimensions.

23 The following diagram is a plan view of a common discharge pipe connecting two closet pans, a floor waste gully and a basin.

a Indicate the minimum size and gradient of the common discharge pipe on the diagram.

_____

b What would be the size of the group vent?

_____

c What would be the maximum distance between the floor waste gully and connection of the group vent?

_____

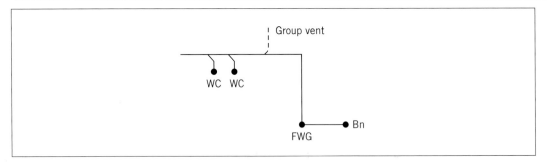

Source: Department of Education and Training (http://www.training.gov.au) © 2013 Commonwealth of Australia.

24 If a graded offset is required in a stack that is to be installed under the rules of the single-stack system of plumbing, what would be the minimum centre-to-centre measurements of the offset?

_____

25 The single-stack modified system of sanitary plumbing requires the installation of a relief vent and cross-vents. What advantage is gained by adding a relief vent and cross-vents to the single-stack system?

_____

_____

26 Where a relief vent is required to be installed in the fully vented modified system, where should it connect to the stack at its base, and at what angle?

_____

_____

27 If the installation of fixtures cannot comply with the requirements of the single-stack system of plumbing, it might be necessary to select the fully vented modified system. How does this system vary from the single-stack system?

_____

_____

_____

28 When planning the layout of a sanitary plumbing system using the system 'elevated pipework using drainage principles', is it permitted to connect a discharge pipe to a vertical riser?

_____

29 Optional exercise: On the following page you are required to complete an isometric drawing to indicate the layout of a sanitary plumbing system to connect plumbing fixtures in a three-storey residence. The fixtures are as follows:

- Third floor: Bathroom group with bath, basin, shower, water closet and floor waste.
- Second floor: Ensuite with basin, shower and water closet, and floor waste.
- Ground floor: Sink with dishwasher, and trough with clothes-washing machine connected to an overflow relief gully.

The drawing must include traps as required, all pipe sizes and gradients, pipe reducers, vents, and inspection and testing openings.

The system of plumbing to be used is the single-stack system.

Note: Do not copy from any other drawing.

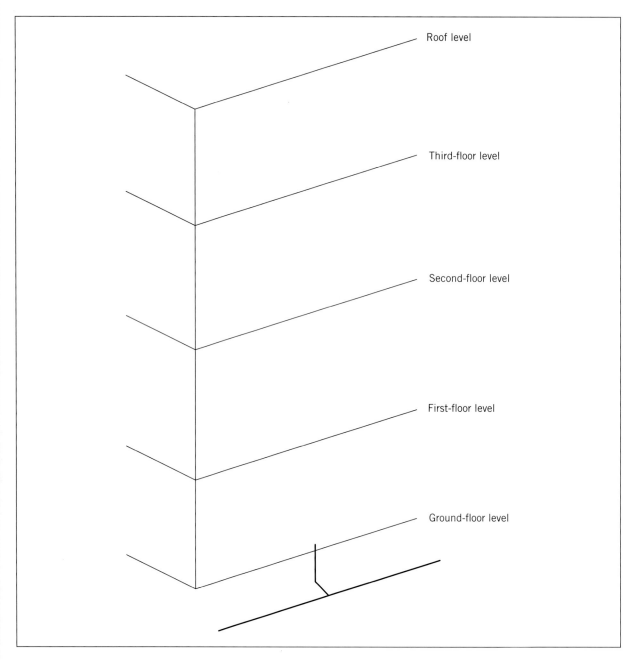

Roof level

Third-floor level

Second-floor level

First-floor level

Ground-floor level

Source: Department of Education and Training (http://www.training.gov.au) © 2013 Commonwealth of Australia.

WELD POLYETHYLENE AND POLYPROPYLENE PIPES USING FUSION METHOD

6

Learning objectives

This unit provides practical guidance on welding polyethylene and polypropylene low-pressure pipes for stormwater, sanitary, waste and vent pipes. Areas addressed in this unit include:

- preparing for the work, and planning the welding process
- establishing appropriate work health and safety (WHS) and environmental conditions
- materials and welding parameters
- tools and equipment
- preparing the work area
- identifying welding requirements
- assembling and checking welding equipment
- welding, pressure testing and inspection
- cleaning up the work area, maintaining equipment and documentation.

Introduction

Plumbers and drainers are required to weld polyethylene and polypropylene pipes to convey stormwater, sanitary and trade wastes to an authority's approved point of discharge or a pre-treatment system.

In this unit, we focus specifically on preparation and welding techniques, and ensuring welded joints are properly made and inspected and tested for defects. The relevant considerations relating to pipe sizing, pipe laying and gradients are described in other units.

This type of welding work is carried out on all properties, but usually where there are ground conditions or waste discharges that are unsuitable for PVC (polyvinyl chloride), such as trade wastes and high temperature discharges. The work is carried out on new properties and where there are alterations or additions to existing systems.

Polyethylene and polypropylene pipes can be laid in the ground and also installed and clipped above ground.

While studying this subject, you will be required to respond to questions and complete exercises referring to this unit and AS/NZS 3500.2.

> **AS/NZS 3500.2 SANITARY PLUMBING AND DRAINAGE**

Before studying this subject, you should have completed training in WHS as outlined in Chapter 3 of *Basic Plumbing Services Skills*.

Preparing for work

Plans and specifications

Before work can commence on-site, you must have obtained all the plans and specifications. The following checklist reviews the requirements prior to welding:

- plans and specifications of proposed building/ building alterations
- approval from local government/sewerage authority
- materials approved for the application
- standards relating to materials and installation
- identified responsibility and requirements for inspection and testing of welds
- pipe manufacturers' specifications and installation instructions
- regulatory authority requirements for inspection and testing of installed plumbing
- materials suited to fluid being conveyed
- materials suited to ground and soil conditions
- depth of cover requirements
- environment required for welding
- pipe manufacturers' specified service and calibration on welding equipment.

Safety and environmental requirements

Preparing for work involves obtaining all of the necessary equipment to comply with WHS and environmental regulations in your state/territory. Examples also include:

- personal protective equipment (PPE) specified in the pipe manufacturer material safety data sheet (MSDS)
- environmental considerations specified in the pipe manufacturer MSDS for safely disposing of materials.

The welding processes must only be carried out in conditions that are safe and controlled for protection of workers and the general public, and also for predictable and reliable welding.

Each state and territory has WHS regulations that require employers and self-employed people to identify hazards and assess and control risks at the workplace in consultation with their workers. *It is everyone's job to identify hazards and control risks.*

> **GREEN TIP**
>
> If you are unsure about how to safely use, store or dispose of chemicals used in the plumbing industry, refer to the MSDS.

Quality assurance

You must be aware of the project requirements for quality assurance, such as the necessary inspections and tests required. Inspection and testing will be discussed in greater detail later in this unit.

For some projects, specifications include that the contractor provides a quality assurance plan, which demonstrates that the contractor has the resources, capacity and sufficient controls to satisfy the requirements of the project.

A typical quality assurance plan would contain the following:

- document control and distribution
- a list of quality management systems (e.g. ISO 9001)
- a list of personnel with responsibilities, training and experience
- welding procedures (typically referring to manufacturers' specifications)
- sample weld records (i.e. blank pro forma sheet)
- an inspection and test plan (to suit the specification).

AS/NZS 2033: 2008 Installation of polyethylene pipe systems specifies that the acceptable methods for fusion jointing are electro-fusion, butt-fusion and socket-fusion.

The following fusion jointing guides can be accessed via https://pipa.com.au/technical/pop-guidelines/:

1 For electro-fusion refer to PIPA POP-001.
2 For butt welding parameters refer to PIPA POP-003.

Only trained and certified operators are permitted to carry out fusion jointing. Access further information via https://pipa.com.au/welder-training/.

GET IT RIGHT

HDPE SHORT RADIUS BENDS

FIGURE 6.1 Incorrect bend configuration at base of stack

Figure 6.1 shows an HDPE pipe installed for a stack that will pass through 10 floor levels. The change of direction at the base of the stack is made up of two M & F 45° bends joined by a short electrofusion coupling. AS/NZS 3500.2 specifies that two bends are required, separated by a straight pipe of length not less than twice the bore of the pipe. Therefore, this arrangement is substandard. Note that most HDPE 88° bends have a very short centreline radius and therefore single bends cannot be used at the base of stacks that extend through more than two floor levels.

1   What is the section in AS/NZS 3500 that specifies the requirements for bends at the base of stacks?

_____

_____

_____

2   What is the minimum centreline radius for a bend at the base of a DN 100 stack through more than two floor levels?

_____

_____

_____

3   Why do you think a single 88° bend might be a problem in the situation in this photograph?

_____

_____

_____

Sequencing of tasks

Before commencing work on-site, it is best practice to arrange for a meeting with the project supervisor and trade supervisors. A good site meeting should provide, among other things, the proposed construction schedule and the contact details of trade supervisors.

Some building projects have a fairly loose construction schedule, and this is when communication with other trades is even more important, as significant problems can arise when there is a clash of trades or work is not carried out in the correct sequence and at the right time.

As well as the sequencing of work in coordination with other trades, you will need to plan and carry out the welding and installation of the pipework in the correct sequence. The best approach is first to establish and reveal the unknown factors, such as services and obstructions. Make sure that the work area in which the piping is to be installed has been prepared, and always ensure that conditions will be a) clean and b) dry before commencing welding.

One consideration that is particularly important is the effect of sunlight on high-density polyethylene (HDPE) pipes. The pipes are typically black; they will expand and contract considerably with changes in temperature and can quickly bow in trenches. Therefore, it is good practice to plan to quickly lay and backfill HDPE pipes. In some conditions, this may be challenging, because the weld and cool times demand that new joints are not disturbed for specified periods of time.

As discussed in Unit 1, before work commences the worksite must be prepared to ensure that the piping can be installed in an efficient and safe manner.

AS/NZS 3500.2 SANITARY PLUMBING AND DRAINAGE

Safety equipment

Safety equipment specific to welding operations and installation of piping includes:

■ respiratory protective device with P2 filter – recommended if dust is generated
■ safety goggles
■ overalls
■ hard hat
■ ear muffs
■ sunscreen
■ gloves
■ safety boots
■ trench support (see Unit 2).

Tools

A range of hand and power tools and mechanical equipment is required. These may include:

■ butt welding machine
■ welding plate
■ metal ruler and tape measure

■ electrofusion machine
■ pipe scraper
■ cutting tools, including hacksaw, mitre box and panel saw, files
■ chamfering tool.

Special equipment – selection and operation

Welding machines are supplied by manufacturers of polyethylene and polypropylene pipe systems. The welding procedures explained in this unit are carried out with machines supplied by Geberit for welding HDPE. Machines must always be operated in accordance with manufacturers' instructions.

Electrofusion machine ESG-3

Electrofusion is particularly suited to installations where access is difficult, such as in trenches or where there are fittings in close proximity. Attributes of the ESG-3:

■ useful welding tool for all pipe dimensions between [diameter] ø 40 and ø 315 mm
■ simultaneous welding of up to three electrofusion couplings saves time (see Figure 6.2)
■ remote control facilitates installation work
■ compact and robust design for everyday building site work
■ built-in overvoltage protection enabling operation with a generator.

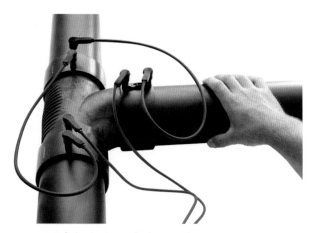

FIGURE 6.2 Geberit electrofusion welding

Butt welding machine Universal d50-315

Attributes of the Universal d50-315 (Figure 6.3) are:

■ robust base element for welding accessories up to ø 315 mm
■ retainer for welding plate support and plane
■ welding plate support swivelling
■ swivel support
■ movable clamping carriage
■ hand wheel.

 Whenever using electric welding machines, ensure that the environment is dry and power leads have been tested and tagged.

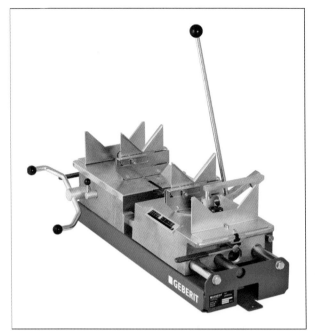

FIGURE 6.3 Geberit butt welding machine Universal d50-315

COMPLETE WORKSHEET 1

Identifying welding requirements

Plans and specifications

The project plans and specifications will usually provide information on the intended purpose of the pipeline, including the size and material required. All pipes and fittings used in sanitary plumbing and drainage systems (including trade waste pipelines) must comply with *the National Construction Code (NCC) Volume 3, Plumbing Code of Australia* and bear WaterMark and Australian Standards certification. After the piping has been welded and installed, it will need to be subjected to hydrostatic or air testing to comply with AS/NZS 3500.2. The welding procedure must be carried out correctly at all stages in accordance with the manufacturer's instructions.

AS/NZS 3500.2 SANITARY PLUMBING AND DRAINAGE

Welding equipment

The welding equipment must be assembled and checked for correct operation according to the manufacturer's instructions. Ensure that the welding equipment is assembled and checked for correct operation by carrying out test welding and destructive testing of samples in accordance with the procedures outlined later in this unit.

At this point, your teacher or instructor should introduce you to welding equipment, pipes and fittings.

COMPLETE WORKSHEET 2

Weld and pressure test pipes

Types of welding using Geberit procedures and equipment

The following sections are based on the *Geberit Piping Catalogue*, 2018, pp. 119–26 (Published by Geberit Pty Ltd, Macquarie Park, NSW. https://www.geberit.com.au.).

Welded joints can be created by:
- electrofusion sleeve coupling
- butt welding.

General information about welding

Piping up to 75 mm can be butt welded by hand. From 90 mm, pipe can be welded using the Geberit Universal or Media welding machines.

When welding Geberit HDPE pipes and fittings, the quality of the weld is primarily dependent on:
- pipe and joint preparation
- clean and dry conditions
- weld heat and cool times
- material characteristics
- manufacturing specifications
- tolerances.

This applies particularly to electrofusion welding, where the electrofusion sleeve couplings and pipes and fittings must be matched to the automatic welding control settings on the electrofusion machine.

Note: The Geberit electrofusion machine, electrofusion sleeve couplings and fittings are a self-contained manufacturer-specific system, which cannot be replaced with other supplier products. Geberit can therefore only guarantee the suitability of pipes, fittings and electrofusion sleeve couplings for welding if Geberit products are exclusively welded to other Geberit products.

Note: The welding bead should be about half as thick as the pipe wall thickness (Figure 6.4).

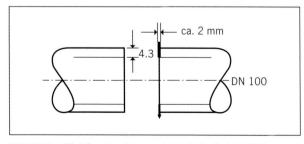

FIGURE 6.4 Welding bead approximately half pipe thickness

Creating a butt welded joint

A welding plate is required as a processing tool (Figure 6.5).

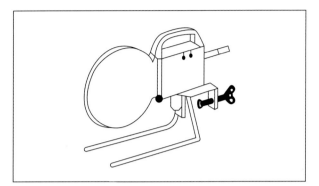

FIGURE 6.5 A welding plate

The reference values required for creating a butt welded joint are outlined in Table 6.1.

Butt welding manually

The manual butt welding procedure in Figure 6.6 should be used with the reference values in Table 6.1.

1 Cut the pipes to size at right angles to the pipe axis and if necessary clean them.
2 Heat the pipe ends.
3 Press the pipe ends lightly on the plate.
4 Only hold the pipe ends so that heat can flow evenly.
5 Immediately push the pipe ends together after welding beads form.
6 Increase the welding pressure slowly to the reference value. [See Table 6.1.]

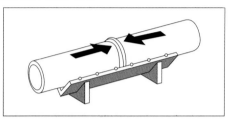

ⓘ **Do not accelerate the cooling process by applying cold items or water.**

7 Examine the butt welding.

Result

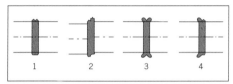

1 Correct
2 Incorrect, off the axis
3 Incorrect, welding pressure too high at the beginning of welding
4 Incorrect, uneven welding temperature

FIGURE 6.6 Butt welding manually

TABLE 6.1 Reference values for welding Geberit HDPE

dø [mm]	Welding allowance per weld seam [cm]	Heating-up time [min]	Time until full pressure build-up [s]	Welding and cooling time [min]	Welding pressure [N]
40	0.3	0,40	4	3	60
50	0.3	0,40	4	3	70
56	0.3	0,40	4	3	80
63	0.3	0,40	4	3	90
75	0.3	0,40	4	4	100
90	0.4	0,50	5	5	150
110	0.5	1,00	5	5	220
125	0.5	1,10	5	5	280
160	0.7	1,30	5	5	450
200	0.7	1,50	5	5	570
250	0.8	2,00	5	5	900
315	1.0	2,30	6	6	1400

Prerequisite
* Ambient temperature: −10 °C to +40 °C
* Clean welding plate surface
* Welding plate temperature: 220 °C signal lamp green
* Up to ø 75 mm the welding can be done by hand. From ø 90 mm the Geberit welding machines Universal or Media must be used.

Butt welding by machine

The machine butt welding procedure in **Figure 6.7** should be used with the reference values in **Table 6.1**.

1 Align and clamp the fittings or pipe ends which have been cut at right angles and deburred in the welding machine.

2 Plane the ends to the required dimensions.

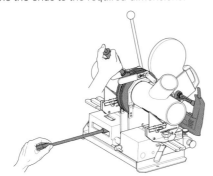

3 Press the pipe ends lightly on the plate.

4 Only hold the pipe ends so that heat can flow evenly.

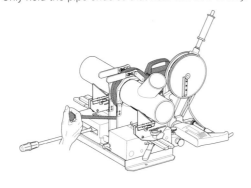

5 Remove the welding plate after the welding bead has formed.

6 Immediately push the pipe ends together.

7 Increase the welding pressure slowly to the reference value.

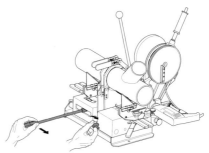

ⓘ **Do not accelerate the cooling process by applying cold items or water.**

8 Allow the pipe ends to cool.

9 Unclamp the pipe assembly after the welding and cooling time.

10 Examine the butt welding.

Result

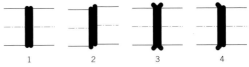

1 Correct

2 Incorrect, off the axis

3 Incorrect, welding pressure too high at the beginning of welding

4 Incorrect, uneven welding temperature

FIGURE 6.7 Butt welding by machine

LEARNING TASK 6.1

Research the instructions for butt welding using the Geberit butt welding machine and find out:

1 the heating-up time for 110 mm pipe

2 the welding and cooling time for 110 mm pipe

3 the total theoretical time to butt weld 110 mm pipe.

Welding using an electrofusion sleeve coupling

The process for welding using an electrofusion sleeve coupling is shown in **Figure 6.8**.

- Geberit electrofusion sleeve couplings with integrated thermal fuses are required.

Geberit electrofusion sleeve coupling (height 60 mm)

- The Geberit electrofusion welding machine ESG 3 is required as a processing tool.

Geberit electrofusion welding machine ESG 3

FIGURE 6.8 Operation of the electrofusion machine

»

>> The Geberit electrofusion welding machine is only designated for welding Geberit HDPE pipes and fittings with electrofusion sleeve couplings ø 40–160 mm. It can do up to 3 connections at the same time ø 40–110 mm and one ø 125–315 mm.

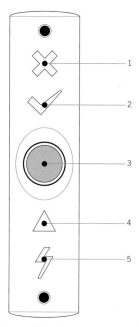

Operating interfaces of the electrofusion machine
1. General malfunction display
2. Welding complete display
3. Start display and actuation
4. Ready for welding display
5. Mains connection display

Prerequisite

- Permissible ambient temperature: –10 °C to +40 °C
- Mains voltage: 185–265 V / 50–60 Hz, power consumption max. 1100 W
- Fuse: Electronic overflow protection. The machine is equipped with a mechanism which prevents a double weld when a sleeve connection cable is connected
- Operation with emergency generator unit: Minimum power 1500 W

FIGURE 6.8 continued

Welding using electrofusion sleeve couplings: 40 to 160 mm

The process for welding using electrofusion sleeve couplings: 40–160 mm is shown in **Figure 6.9**.

 Correctly performed electrofusion welding must only be carried out once.

 All pipes and fittings must be clean and dry during the welding process.

1. Cut the pipes to size at right angles to the pipe axis and roughly clean dirty surfaces.

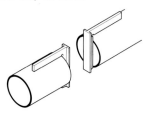

2. Scrape the pipe section/fitting surface in the insertion area of the electrofusion sleeve coupling, Emery cloth must be used to remove the oxidised layer from the pipe. An approved pipe scraper is also acceptable and is recommended for larger pipe sizes from 160 mm to 315 mm.

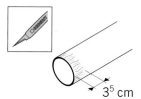
3⁵ cm

 The small amounts of HDPE residue left over from this process will not affect the quality of the connection. No further cleaning procedures are required.

3. Mark clean pipe sections/fittings in the insertion area of the electrofusion sleeve coupling with an insertion depth of 3 cm.

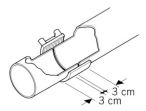
3 cm
3 cm

4. Insert the pipe sections/fittings into the electrofusion sleeve coupling and examine the insertion depth: The axes of the welding ends must match.

 Only connect the sleeve connection cable with the electrofusion sleeve coupling/electrofusion tape after clean, dry Geberit HDPE pipes or fittings have been inserted.

5. Connect the device to the mains voltage. Display ⚡ lights up.
6. Connect the sleeve connection cable with electrofusion sleeve coupling/electrofusion tape. Display ▲ lights up.

⚠ **DANGER**
Risk of burns
▶ Do not touch the pipeline, electrofusion sleeve coupling or electrofusion tape during the welding process or during the cooling-down phase.

FIGURE 6.9 Welding using electrofusion sleeve couplings: 40 to 160 mm

>>

7 Press the start button ●. Display ● lights up and display ▲ goes out. Welding is completed after approximately 80 seconds. The start button display ● goes out and the display ✔ lights up.

 Keep the pipeline in an unstressed position during the entire welding procedure.

Result

Welding has been performed correctly and is finished. Completed welding is indicated by the protruding yellow indicator.

FIGURE 6.9 continued

Welding using electrofusion sleeve couplings: 200 to 315 mm

The process for welding using electrofusion sleeve couplings: 200–315 mm is shown in **Figure 6.10**.

- Geberit electrofusion couplings with integrated thermal fuses and indicators are required

Geberit electrofusion sleeve coupling (height 150 mm) with integrated thermal fuse

- The Geberit electrofusion welding machine ESG T2 or ESG 3 is required as a processing tool

Geberit electrofusion welding machine ESG T2 or ESG 3

The Geberit electrofusion welding machine ESG T2 is solely intended to be used for welding Geberit HDPE pipes and fittings with 200–315 mm electrofusion couplings with integrated thermal fuses.

Prerequisite

- Permissible ambient temperature: –10 °C to +40 °C
- Mains voltage: 220–240 V / 50 Hz
- Power consumption: 2500 W
- Fuse: The electrofusion couplings with integrated thermal fuses have two fuses that switch off the welding current once the corresponding temperature is reached. The same electrofusion coupling with integrated thermal fuse cannot be welded a second time
- Operation with emergency generator unit: Minimum power 2.5 kW. No other devices can be connected during the welding process. The starter switch voltage under load is at least 200 V
- Recommendation: Always mount Geberit pressure ring

 DANGER
Moisture or water-filled pipelines
Fatal danger!

▶ Welding must not be carried out.
▶ Stop the water flow.
▶ Dry pipelines and electrofusion couplings with integrated thermal fuses.

 **An isolation transformer (230 V / 2.5 kW) must be included in the circuit when carrying out welding work in damp areas.**

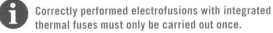 **Correctly performed electrofusions with integrated thermal fuses must only be carried out once.**

1 Attach pressure rings to the pipes. Once the welding time has elapsed, the pressure rings must remain mounted for 15 minutes.

2 Cut the pipes to size at right angles to the pipe axis and roughly clean dirty surfaces.

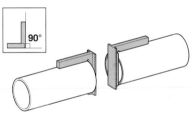

3 Scrape the pipe section/fitting surface in the insertion area of the electrofusion coupling with integrated thermal fuse with the Geberit pipe scraper.

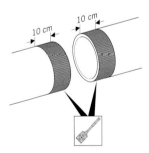

 Remove only the uppermost oxide layer evenly and thinly. There must be no deep cuts/recesses.

4 Remove any burrs and slightly chamfer the pipe ends.

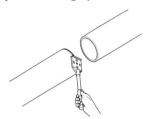

FIGURE 6.10 Welding using electrofusion sleeve couplings: 200 to 315 mm

>>

5 Mark clean pipe sections/fittings in the insertion area of the electrofusion coupling with integrated thermal fuse with an insertion depth of 7.5 cm.

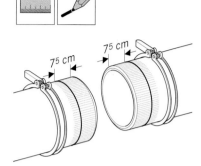

6 Insert the pipe sections/fittings into the electrofusion coupling with integrated thermal fuse and check the insertion depth. The axes of the welding ends must match.

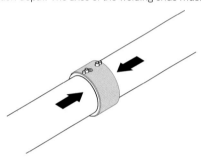

 Do not remove the thermofilm. Only connect the connection cable with the jointing nut after clean, dry Geberit HDPE pipes or fittings have been inserted.

7 Connect device to mains voltage and sleeve connection cable to the electrofusion coupling with integrated thermal fuse.

 CAUTION
Risk of burns
▶ Do not touch the pipeline and the electrofusion coupling with integrated thermal fuse during the welding process and the cooling down phase.

8 Press the start button briefly. The 'weld' signal lamp lights up. Welding current flows through the connected electrofusion coupling with integrated thermal fuse for the next few minutes. The 'weld' signal lamp goes out. The welding process is ended.

 Keep the pipeline in an unstressed position during the entire welding process.

9 Examine the welding: Press the start button briefly. If the lamp goes out when you let go, the welding process has been carried out correctly. If the lamp remains lit up when you let go, the welding time was interrupted and must be repeated once the jointing nut has cooled down.

10 Remove the thermofilm approx. 15 minutes after welding end.

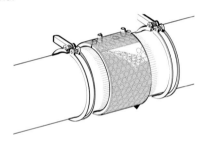

FIGURE 6.10 continued

Creating an electrofusion slip coupling

The process for creating an electrofusion slip coupling is shown in Figure 6.11.

To change the electrofusion sleeve coupling into a slide-over sleeve, remove the central ring. This process can also be used for repairs.

FIGURE 6.11 Creating an electrofusion slip coupling

COMPLETE WORKSHEET 3

Inspection and testing of fusion welds

Testing of butt welds

Visual examination of butt welds

Testing should commence with visual examination of the weld bead. The bead should be uniform and symmetrical in appearance throughout the full circumference of the joint. Figure 6.12 depicts a correct butt welded joint between HDPE pressure pipe and reducer fitting.

The depth of any notches or scores in the pipe should not exceed 10% of the thickness of the pipe wall. If so, the section of pipe should be rejected. For detailed visual assessment of welds refer to *POP014 Industry Guidelines, Assessment of polyethylene welds*, available on the Plastics Industry Pipe Association

FIGURE 6.12 Butt weld must be uniform and symmetrical

of Australia (PIPA) website: https://pipa.com.au/technical/pop-guidelines/

Bead testing

After the weld has cooled, the bead can be tested to ensure that it has properly formed and bonded. This involves removing a section of weld and then bending it in a 'V' shape so that the external pipe surfaces are brought together. The bead should not separate or fracture from the pipe wall.

Destructive testing of butt welds – tensile testing

There are several test methods that can be carried out on weld specimens in a laboratory or in the field. A successful test would see weld material stretching / tearing rather than snapping apart.

Destructive testing of butt welds – bend testing

Bend testing is a simple way of testing the weld. This can be carried out by cutting a section of joint and placing one side of the pipe wall in a bench vice. The weld is then bent backward. A successful test would see that the weld bends without cracking or fracturing. Whilst performing this procedure, wear safety glasses and shield the work area to protect against flying weld fragments.

Testing of electrofusion joints

Visual examination of electrofusion joints

Weld features and acceptance criteria for electrofusion joints are outlined in *POP014 Industry Guidelines, Assessment of polyethylene welds*, which is available on the PIPA website: https://pipa.com.au/technical/pop-guidelines/.

Destructive testing electrofusion welds peel decohesion testing

The acceptance criteria for this testing is defined in AS/NZS 4129: 2008 Fittings for polyethylene (PE) pipes for pressure applications. The brittle failure decohesion should be less that 33.3%. In other words, when the joint is peeled apart, two thirds of the joint should

appear to stretch or tear apart rather than crack or fracture apart.

Strip bend testing

Strip bend testing is a simple way of testing the weld. This can be carried out by cutting a section of electrofusion joint and placing one side of the pipe wall in a bench vice. The weld is then bent sideways. A successful test would see that the weld bends without cracking / fracturing. Whilst performing this procedure, wear safety glasses and shield the work area to protect against flying weld fragments.

FROM EXPERIENCE

When destructive testing, always wear safety goggles (not glasses) and ensure that components are securely anchored – for example, in a bench vice. It is also best practice to place shields to prevent flying material fragments from hitting other people.

Pressure testing and inspection of pipe joints

Both visual and destructive weld testing must be carried out periodically. However, in-field hydrostatic or air testing must always be carried out on new installations in accordance with the procedures outlined in AS/NZS 3500.2.

AS/NZS 3500.2 SANITARY PLUMBING AND DRAINAGE

For example, an air test may be applied to sanitary and trade waste drains. The procedure is as follows:
1 An expandable testing plug (see Figure 6.13) is fitted into one open end through an inspection opening and tightened to expand the plug and seal it against the walls of the pipe. This prevents the passage of air.

FIGURE 6.13 Open end of HDPE drain capped and connected to air pump

2 All other open ends of the drain to be air tested should be plugged (see **Figure 6.14**).

3 An air pump and connection incorporating a pressure gauge is attached to the testing plug and air is pumped into the drain, or section of drain, under test to achieve a test pressure of 15 kPa. (*Note:* Some regulatory authorities may specify a different test pressure.)

4 Normally, at least three minutes is allowed for the temperature in the drain to stabilise.

5 To find any leaks, use soapy water.

FIGURE 6.14 Remaining ends of HDPE drain capped for air testing

Monitoring and recording welding and test results

Successful electrofusion jointing relies upon proper joint preparation. It is just as important to carry out batch testing and keep records. The reason is that if one weld fails, it is possible that many welds fail, as they are likely to have been made under the same conditions.

Batches of welds should be inspected, tested and recorded whenever there is new equipment, a change of equipment, new operator or new project.

Inspection and test records of batches should contain information including:

- the identity of the welder(s) and training records
- pipe and fitting materials, SDR, sizes, WaterMark approval, manufacturer
- environmental conditions, i.e., dry, dust-free and within acceptable range of environmental temperatures
- preparation of pipe ends and fittings and use of welding wipes
- welding heating and cooling periods
- welding equipment used, including clamps
- welding equipment service and calibration records.

Clean-up

The work area must be cleared and materials disposed of, reused or recycled according to legislation, regulations, codes of practice and job specification.

Tools and equipment must be cleaned, checked, maintained and stored according to manufacturers' recommendations and workplace procedures including:

- service work to welding equipment to maintain it within calibration
- inspection, cleaning, maintenance and storage of tools
- disposal of unserviceable tools
- removal and storage of usable pipe off-cuts
- removal, disposal and recycling of waste to correct bins
- cleaning weld equipment
- keeping the work area tidy and clear of trip hazards.

 COMPLETE WORKSHEETS 4 AND 5

SUMMARY

- When preparing to weld polyethylene and polypropylene pipes using a fusion method, you must research the plans, specifications and regulations relating to the proposed work. Always familiarise yourself with the welding equipment and only use equipment specified by the pipe manufacturer. The work area must be clean and dry, and the temperature of the pipe to be welded must not be influenced by contact with cold materials.

- Identify the welding requirements by referring to plans and specifications. Ensure that the welding equipment is assembled and checked for correct operation by carrying out test welding and destructive testing of samples.

- Before welding pipes, ensure that joints are prepared strictly in accordance with the pipe manufacturer's instructions. Follow pipe manufacturer welding and cooling times, and pressure specifications. Periodically inspect and test welds, particularly when there is a change of equipment, a new operator or a new project.

- Monitor and document weld inspection and test results to ensure that all joints have been prepared correctly.

- Always clean up and keep welding equipment within calibration and in a safe serviceable condition.

WORKSHEET 1

Student name: _____

Enrolment year: _____

Class code: _____

Competency name/Number: _____

Task: Review the section 'Preparing for work', then complete the following.

1   How will you find out what PPE must be used when cutting and welding polyethylene and polypropylene pipes?

_____

_____

_____

_____

2   When preparing for work, what are two main conditions in the work area before you commence welding?

_____

_____

_____

_____

3   Why is it good practice to quickly lay and cover HDPE pipes when working in sunlight?

_____

_____

4   What type of respiratory protective device filter might be required when generating dust while cutting polyethylene and polypropylene?

_____

_____

5   Complete the following sentence: Machines must always be operated in accordance with _____

6   When using Geberit welding machine ESG-3

   a   What type of joints are made?

   _____

   b   What is the range of pipe sizes that can be welded?

   _____

 **WORKSHEET 2**

Student name: _____

Enrolment year: _____

Class code: _____

Competency name/Number: _____

Task: Review the sections 'Identifying welding requirements', then complete the following.

1 What approval markings are required on pipes and fittings used in sanitary plumbing and trade waste installations?

_____

_____

_____

2 After a welding machine has been assembled, how would you check it for correct operation?

_____

_____

 **WORKSHEET 3**

Student name: _____

Enrolment year: _____

Class code: _____

Competency name/Number: _____

Task: Review the section 'Weld and pressure test pipes', then complete the following.

1   What are two types of welding using the Geberit equipment and procedures?

    a   _____

    b   _____

2   What are six factors that determine the quality of welds?

    a   _____

    b   _____

    c   _____

    d   _____

    e   _____

    f   _____

3   When butt welding Geberit HDPE pipe, what is the total welding time for 50 mm pipe?

    _____

    _____

4   When electrofusion welding, what is the name of the layer that must be removed?

    _____

    _____

    _____

5   When welding a 200 mm electrofusion sleeve coupling, when must the thermofilm be removed?

    _____

    _____

    _____

6   What is the acceptable depth of a notch or score in the pipe wall?

    _____

    _____

7 Explain how you would strip bend test an electrofusion joint and securely anchor the components?

_____

_____

_____

_____

 **WORKSHEET 4**

Student name: _____

Enrolment year: _____

Class code: _____

Competency name/Number: _____

Task: Review the sections 'Monitoring and recording welding and test results' and 'Clean-up', then complete the following.

1   When should a batch of welds be inspected, tested and recorded?

_____

2   Why is it important to maintain inspection and test records of batches?

_____

_____

_____

_____

3   When recording welding and test results, is it necessary to note environmental conditions during welding? What environmental conditions should be avoided?

_____

4   Research and state the Australian Standard that applies to installation of polyethylene pipe systems.

_____

5   What must be done with unserviceable tools?

_____

 **WORKSHEET 5**

Student name: _____

Enrolment year: _____

Class code: _____

Competency name/Number: _____

Task: Practical exercises

All questions in the previous four worksheets must be completed and checked by your trainer/teacher before you attempt the following practical exercises.

Your trainer/teacher should provide preliminary guidance, conduct an induction and ask underpinning knowledge questions.

Given the plans and specifications, butt fusion weld two joints and electrofusion weld one socket joint up to DN 100 on approved polymer pipes, using appropriate fusion welding processes. Also safely carry out one visual and one destructive test inspection.

7

LOCATING AND CLEARING BLOCKAGES

Learning objectives

This unit provides practical guidance on locating and clearing blockages. Areas addressed in this unit include:

- common causes of blockages
- work health and safety considerations
- drain clearing methods
- description of the use of:
  - plungers
  - hand rods
  - machine-driven cables
- determining the responsibility for blockages in drains
- locating blockages
- gaining access to clear blockages
- preventing blockages.

Introduction

Plumbers are required to locate and clear blockages in the maintenance of sanitary plumbing, sewerage, water and stormwater/roof drainage installations. Some important aspects of this type of work include correctly identifying where blockages are located, and correct selection and safe operation of drain clearing machines, tools and attachments.

By selecting and using appropriate drain cleaning tools and equipment, you will be able to clear blockages effectively, and without causing damage to pipes and fittings.

Blockages are most commonly encountered in sanitary plumbing and drainage systems, although they also occur in stormwater and roof drainage installations. Therefore, this chapter deals primarily with blockages in these systems.

Before studying this subject, you should have completed training in work health and safety (WHS) as outlined in Chapter 3 of *Basic Plumbing Services Skills*.

Common causes of blockages

A blockage is an obstruction to the normal flow in a pipeline, and falls into two categories:
1   total blockage (when there is no flow)
2   partial blockage (when the flow is restricted/slowed).

There are four main causes of obstructions in drainage systems:
1   foreign objects
2   tree roots
3   soil movement
4   sub-standard installation.

Foreign objects

Foreign objects can enter the drainage system while it is being laid or in service. While laying drains and installing piping, you need to cover exposed ends to prevent entry of soil, sticks, stones, bricks, concrete and other objects. While the drainage system is in service, foreign objects such as sanitary napkins, plastic dispensers and children's toys can enter through various openings in the drain, such as water closets and gully surface fittings.

Tree roots

Entry of tree roots into drains is a major problem, especially in older stoneware drains with mortar joints. These pipes can easily fracture with soil movement. Tree roots can enter the drain through the loose inspection openings, pipe joints or fractures in the pipe. Once a small crack has developed, the tree roots can enter and grow, and then completely fill the internal diameter of the pipe. Experience tells us that certain trees cause more problems with roots than others (see Table 7.1). For example, the roots of poplars have been found in drains more than 30 m from the tree.

The pressure of tree roots around the drain can also cause pipe fractures.

TABLE 7.1 Examples of problem tree and plant species

Common name	Botanical name
Athel tree	*Tamarix aphylla*
But but	*Eucalyptus bridgesiana*
Cape Virgilia	*Virgilia oroboides*
Claret ash	*Fraxinus raywoodii*
Desert ash	*Fraxinus angustifolia*
English elm	*Ulmus procera* and related species
False acacia	*Robinia pseudoacacia*
Fig	*Ficus* – all species
Flax-leaf paperbark	*Melaleuca linariifolia* var. *Trichostachya*
Swamp oak	*Casuarina glauca*
Kangaroo paperbark	*Melaleuca halmaturorum*
Lemon-scented gum	*Eucalyptus citriodora*
Lilly pilly	*Eugenia smithii*
Mirror plant	*Coprosma repens*
Moonah	*Melaleuca lanceolata*
Norfolk Island pine	*Araucaria heterophylla*
Pepper tree	*Schinus molle*
Plane tree	*Platanus* – all species
Poplar	*Populus nigra* and related species
Prickly paperbark	*Melaleuca styphelioides*
Pyramid tree	*Lagunaria patersonii*
River red gum	*Eucalyptus camaldulensis*
River she-oak	*Casuarina cunninghamiana*
Salmon gum	*Eucalyptus salmonophloia*
Spotted gum	*Corymbia maculata*
Sugar gum	*Eucalyptus cladocalyx*
Swamp or flat-topped yate	*Eucalyptus occidentalis*
Weeping lilly pilly	*Eugenia ventenatii*
Weeping willow	*Salix babylonica* and related species
Yate	*Eucalyptus cornuta*

Figure 7.1 shows closed circuit television (CCTV) images of common causes of blockages in drains.

Soil movement

Soil movement can damage pipes, regardless of whether they are made of rigid or flexible material

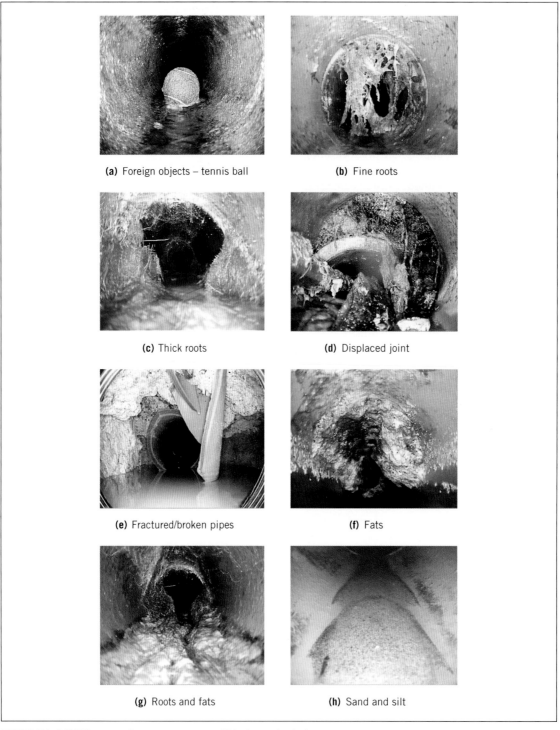

(a) Foreign objects – tennis ball

(b) Fine roots

(c) Thick roots

(d) Displaced joint

(e) Fractured/broken pipes

(f) Fats

(g) Roots and fats

(h) Sand and silt

FIGURE 7.1 CCTV images of common causes of blockages in drains

(e.g. vitrified clay or PVC-U (commonly referred to as PVC). Pipes deform and/or fracture due to ground movement (e.g. landslip, subsidence, filled or unstable ground). Heavy objects moving over the ground may also cause pipes to deform or fracture (e.g. earthmoving machinery, trucks and cars).

Sub-standard installation

If drains/pipes have been incorrectly installed, they can become deformed or fracture in the ground – for example, the use of incorrect fittings installed under tension, drains without proper support/bedding or drains with inadequate provision for movement in reactive soils or unstable ground.

Preparing for work

Each job is different. In order to determine the requirements for a particular job, you will need to begin with some preparation and planning.

Obtaining plans and specifications for the job

You need to plan the job by gathering information that will enable you to find and clear the blockage in the most efficient way. This information can include the following:

- a description of where the blockage was first noticed – this can be obtained from your supervisor, or the person who first noticed the blockage
- any fixtures that have been affected by the blockage – for example, closet pans, basins and sinks
- the owner/occupier's thoughts on what caused the blockage
- the location of the drains/pipework. If the property is connected to a sewerage system, the local sewerage authority/council may have a plan showing the location of the authority connection point, the drain material, fittings and access points (inspection openings)
- a site inspection to locate access points (above- and below-ground inspection openings) and any obstructions to access
- any statutory and authority requirements for opening and resealing the drainage system. If you are unsure, enquire at the relevant authorities as to whether permits are needed for this type of work
- the type of blockage and whether the drain contains any harmful chemicals. Has anyone tried to clear the blockage with chemicals? This might provide valuable information to decide safe and effective clearing methods.

If excavating by hand or with machinery, you will need to know the location of other services – for example, communication and power cables, water, gas, drains, etc. There may be a service in the local area that provides this information, such as Dial Before You Dig (http://www.1100.com.au or phone 1100). If excavating near an authority's sewer main, you may not excavate with machinery within 600 mm of the main.

Work health and safety

Work health and safety (WHS) requirements need to be followed for each particular job in accordance with legislation. If you are unsure about the laws relevant to your work, ask your trainer or supervisor to provide you with this information.

When preparing for work, you may need to have a safe work method statement (SWMS) and to complete a job safety analysis (JSA). The following are some risks that relate specifically to the work of locating and clearing blockages.

Confined spaces

There are risks to your health and safety when working in confined spaces. Your state or territory will have laws that require you to take precautions when working in confined spaces.

Definition

'Confined space' means a space in any vat, tank, pit, pipe, duct, flue, oven, chimney, silo, reaction vessel, container, receptacle, underground sewer or well, or any shaft, trench or tunnel or other similar enclosed or partially enclosed structure, if the space:

- is, or is intended to be, or is likely to be, entered by any person; and
- has a limited or restricted means for entry or exit that makes it physically difficult for a person to enter or exit the space; and
- is, or is intended to be, at normal atmospheric pressure while any person is in the space; and
- contains, or is intended to contain, or is likely to contain:
  - an atmosphere that has a harmful level of any contaminant; or
  - an atmosphere that does not have a safe oxygen level; or
  - any stored substance, except liquids, that could cause engulfment
  - but does not include a shaft, trench or tunnel that is a mine or is part of the workings of a mine.

  Examples of confined spaces include:
- sewerage inspection chambers
- grease arrestors
- petrol and oil arrestors
- stormwater pits and arrestors.

  A training course in Confined Space Entry can be undertaken to help you to understand the risks and procedures associated with working in these areas.

Confined space incidents

The following is the true story of a death related to work in a confined space.

Hazardous atmosphere and oxygen deficiency fatality in sewer

A water board employee was working to clear a blocked sewer. The equipment the employee was using to unblock the sewer became caught and the employee entered the sewer to free the equipment. The clearing of the blockage produced a gush of water and released sewerage gases, and the employee collapsed as he was about to climb out of the access hole. A boy on work experience with the employee attempted to pull him out, but was unsuccessful. The employee fell back into the sewer and the boy went to get help. The employee was unable to be resuscitated after being pulled from the sewer.

Care in using chemicals to unblock drains

There are risks associated with using chemicals to clear blocked drains. Before attempting to clear a blockage, you should be aware of whether any chemicals have been used.

Case study, 2006

The TAFE Commission of NSW pleaded guilty to a charge under the *Work Health and Safety Act 2011* in the Chief Industrial Magistrate's Court regarding an accident where a drain-clearing product caused chemical burns.

The accident involved the use of a product whose active ingredients were caustic soda (sodium hydroxide) and aluminium. Five kilograms of the product – the amount permissible according to the MSDS – was placed in the drain. The drain was then sealed with a hydrostatic plug, and water was injected through a hose in the plug.

The chemical reaction generated a large amount of gas, which blew off the plug and sprayed caustic material on the plumber and his apprentice. The material caused chemical burns to multiple parts of their bodies.

Any drain-cleaning product containing sodium hydroxide or potassium hydroxide and metals (aluminium and zinc) should not be used in a manner in which the product is sealed (by whatever means) in a drain, a pipe or other such enclosed space.

Generally, state and territory WHS legislation requires that anyone carrying or storing chemicals in their work vehicle must:

- have a copy of the MSDS within the work van or storage area
- use the product in accordance with the instructions outlined in the MSDS
- when using the product, undertake a SWMS and a JSA
- wear the required personal protective equipment (PPE), as outlined in the MSDS
- have the required size and type of first-aid kit.

As a general rule and warning, to avoid violent reaction, *always add the drain-cleaning product to water*, and *never* add water to the product first.

Infections

Many pathogenic organisms live in sewers and can easily infect the operator of drain-cleaning equipment through cuts or abrasions in the skin. Some organisms can even penetrate the skin. Avoid touching the corner of your eyes or mouth with unclean hands or gloves. Always use PPE and wash up properly.

Personal protective equipment

Whenever operating drain-clearing equipment, it is important to wear relevant PPE. Wear safety glasses to prevent splashing of water into the eyes. Safety boots should be worn to prevent injury from

falling equipment. To protect your hands from sharp metal, studded mitts should be worn when handling mechanical drain-cleaning rods.

Care in handling machine-driven drain-cleaning equipment

Machine-driven drain-cleaning equipment typically has a cable that is fed into the drain while being driven by an electric motor. Follow the machine operating instructions, and always maintain control of the cable, keeping the length between the machine and the drain as short as possible.

When handling the cable, use metal-studded mitts. When extracting the cable from the drain access point, it is best to turn off the machine. This should eliminate any possibility of the cable or end tool spinning out of control and potentially catching your clothing.

 When working with machine-driven drain-cleaning equipment, only use metal-studded mitts that are in good condition. Also, do not wear loose clothing that could catch onto moving equipment.

Regulations and work notices

When working on sanitary plumbing and drainage systems, you must comply with regulations in accordance with local authority requirements.

Work notices

It may be necessary to notify the local authority – for example, water authorities, councils – before work can commence. Generally, the authorities will allow work to be carried out by a licensed plumber without first obtaining a permit to undertake emergency work such as:

- stopping water wastage and/or damage
- clearing blocked pipes
- protecting public health and safety.

If the authority requires notification and a compliance certificate, this would need to be issued within the time required – for example, within two working days. (Local requirements may vary.)

Quality assurance

Many companies have policies and procedures for assuring quality of services and products to customers. It may be necessary for you to comply with different procedures at different properties. This is particularly relevant at institutional, commercial and industrial premises, and you should ask the customer whether there are any special requirements.

The following are some issues that might relate to your job.

- Quality assurance procedures that relate to your job – for example, at a manufacturing plant, where shutdown of water supply and drainage systems might affect operation of the business.

- Quality procedures that relate to clean-up – for example, at a food-processing plant or at a hospital.
- Procedures that determine how your work is supervised and monitored.
- WHS procedures and reporting incidents and injuries.
- The procedures for recording job details and customer contact.
- Separation of tools and equipment used for water supply and sewerage work. (Some contractors require separate toolkits.)
- Records of your experience, training and qualifications – for example, records of Confined Space Entry training.

Tools and equipment

Safety equipment

In any work situation, you can reduce the risk of injury to yourself and others by using appropriate PPE. The types of PPE that you might need include:

- overalls – to protect yourself when working in a variety of conditions
- safety boots – to protect your feet from injury from falling tools and heavy objects
- safety glasses or goggles – to protect your eyes from dirty water or particles that might be thrown from drain-cleaning rods or hoses
- hat and sunscreen – to protect from ultraviolet light
- face mask – to protect against inhaling airborne particles
- ear plugs or ear muffs – to protect your hearing from excessive noise (depending on the tools and work environment)
- gloves – to protect your hands when working with sanitary snake/rods. Metal-studded gloves are available specifically for drain-clearing tasks
- confined space equipment – to ensure that you can be safely extracted from a confined space by a co-worker and can measure whether the atmosphere is safe to work in.
- liquid hand sanitiser – for protection against infection and disease. *Note:* Select an appropriate sanitiser. The effectiveness of an alcohol-based hand sanitiser will depend upon the alcohol content (e.g. ethanol or isopropyl alcohol). The sanitiser formula should have at least 60% alcohol. Products meeting the World Health Organization (WHO) formula should have 80% ethanol or 75% isopropyl alcohol.

Tools

To locate and clear blockages, you will need both hand tools and special drain-clearing equipment. These items include:

- shovels to expose inspection openings (IOs) in underground drains
- crowbars to expose IOs in underground drains
- levers to remove IO caps
- scale rulers to obtain measurements from plans, and tape measures to measure distances between IOs and fittings
- tool boxes
- pipe wrenches to remove IO caps
- shifting spanners, socket sets and ratchet spanners
- plungers
- manually operated clearing equipment, such as drain-cleaning rods with different attachments to remove blockages
- mechanical drain-clearing machine and attachments
- barricades, signs and guard rails to prevent people from entering the work area, and to protect them from the risk of trip-and-slip injury, excavation and excavated materials
- excavator/backhoe to expose underground drains (if necessary)
- trench shields/shoring equipment to be used in deep excavations.

Special equipment – selection and operation

Drain-clearing methods fall into four basic categories:

1. mechanical (manual or machine)
2. hydraulic
3. chemical
4. bacterial.

It is important to select the correct equipment, and to ensure it is used in accordance with the equipment manufacturer's instructions. One consideration in selecting the correct equipment is that it does not damage pipes and fittings. For example, some metal root-cutting tools can cut into and damage PVC pipe material.

It is best practice to attempt to clear the blockage by the least intrusive methods, such as a plunger, before using other methods.

FROM EXPERIENCE

Selection of appropriate tools to prevent damage to pipes and fittings is a critical skill. There have been instances where plumbers have used metal rods and excessive force in attempting to clear blockages in PVC pipes in high-rise plumbing, resulting in pipe fractures and millions of dollars in property damage.

Mechanical (manual) devices

Plungers

A plunger is the most common device used to clear blockages (see Figure 7.2). It is the easiest to construct and operate, and offers the least risk of damage to pipes and fittings.

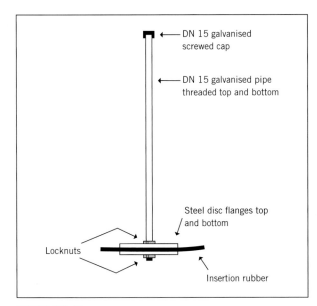

FIGURE 7.2 Plunger

Plungers may be constructed with a steel pipe handle and a thick rubber disc attached to one end. The upper end of the pipe handle must be sealed to prevent effluent spurting out of it. It is operated by inserting it into the vertical section of the drain that is blocked (applying a positive pressure to the column of water) and pulling back the plunger, causing a negative pressure on the column of water. The action is repeated until the blockage is cleared. When plunging a blockage, every effort should be made to ensure that the object causing the blockage, when it is finally dislodged or broken up, is retrieved from the drain. Where practical, attempt to capture the cause of the blockage at the boundary shaft or inspection shaft so as to reduce the likelihood of blockages downstream.

If the blockage cannot be cleared by plunging, other means must be used.

Construction: Plungers can be purchased or made to suit common drain sizes, being 100 mm and 150 mm. The size of the plunger must be suited to the internal diameter of the pipe. The length of the plunger must also suit the task – that is, plunging a deep boundary trap or inspection shaft will require a long handle, whereas a short handle may be suitable when plunging a gully.

Hand rods

Hand-operated drain-clearing rods are used by inserting the rods into the drain and progressively joining rod lengths together until the blockage is found (see Figure 7.3). A clearing tool is attached to the first rod to assist in clearing the blockage.

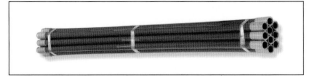

FIGURE 7.3 Hand-operated drain rods

Source: Rothenberger Australia Pty Ltd

The attachment tools vary for different systems, but types can include worm-screw, drop scraper, retrieval, blade, shovel and funnel, and a variety of drill types (see Figure 7.4). The tool type selected depends upon the type of blockage.

FIGURE 7.4 Hand rod attachments: drain brush, worm screw, drop scraper and rubber plunger

Source: Rothenberger Australia Pty Ltd.

Hand rods are operated by inserting them into the drain while rotating them in a clockwise direction towards the blockage, until they either cut through or become entangled in the blockage. Generally, depending on the method of attaching the rods, they should only be rotated in a clockwise direction. This will ensure they do not unscrew and become disconnected in the drain. The rods are then pulled back in a clockwise direction, ideally retrieving the cause of the blockage.

If the blockage cannot be cleared using hand-operated rods, then you can use either machine-driven rods or hydro-jetting methods.

GET IT RIGHT

DAMAGED SANITARY STACK

FIGURE 7.5 View of the inside of a sanitary stack. A screwdriver was inserted into a hole where drain cleaning rods have penetrated the wall of the pipe.

As shown in Figure 7.5, a sanitary stack has numerous fittings and junctions.

1   What method would you use to clear a blockage in this system?

_____

_____

_____

2   Does your method use sharp tools that might damage the pipe?

_____

_____

_____

3   How do you know that your methods are safe enough to prevent damage to piping?

_____

_____

_____

Machine-driven cables

Machine-driven cable (commonly referred to as 'snake') is widely used by plumbers to clear blockages in drains (see **Figure 7.6**). The machines can be purchased or hired.

FIGURE 7.6 Rothenberger R750 sewer clearing machine

Source: Rothenberger Australia Pty Ltd.

Before using this equipment, you must familiarise yourself with the manufacturer's instructions for assembly and operation.

A clearing tool is attached to a cable. The attachment tools vary for different systems, but are similar to those used for hand-operated rods and include hollow and core spirals, drills and cutters. The tool type selected depends upon the type of blockage. A motor rotates the cable while it is fed into the drain. **Figures 7.7** to **7.12** show a range of attachment tools.

The machines typically have a forward and reverse gear, enabling the cables to be driven by power into and out of the drain. Cable can be fed into drains through inspection shafts, IOs, gullies and testing openings.

Most plumbers have these machines. If the blockage cannot be cleared with this equipment, there is the option of calling a specialist with hydro-jetting equipment.

Hydraulic methods

Hydro-jetting uses high-velocity water pressure to blast away accumulated scale, silt, sand, roots and grease build-up on the inside walls of pipes. At around 28 000 kPa, hydro-jetting has the power to dislodge and disperse tough blockages, and at the same time scours the full diameter of the pipe, flushing debris

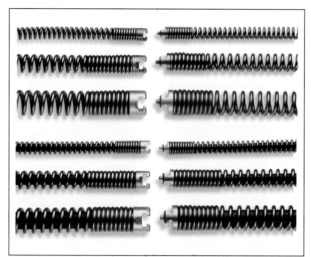

FIGURE 7.7 Hollow and core drain-cleaning spirals

Source: Rothenberger Australia Pty Ltd.

FIGURE 7.8 Drain-cleaning tool – straight drill

Source: Rothenberger Australia Pty Ltd.

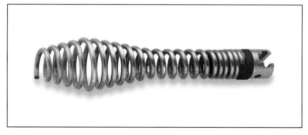

FIGURE 7.9 Drain-cleaning tool – club or bulb head drill

Source: Rothenberger Australia Pty Ltd.

FIGURE 7.10 Drain-cleaning tool – funnel drill

Source: Rothenberger Australia Pty Ltd.

and leaving pipes clear. When carried out correctly, hydro-jetting is a highly efficient way to clear blockages without damaging the pipes.

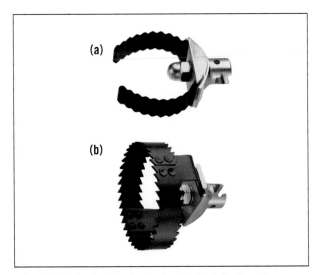

FIGURE 7.11 Drain-cleaning tools: (a) shovel drill, (b) sawtooth spiral, (c) cross-blade drill

Source: Rothenberger Australia Pty Ltd.

FIGURE 7.12 Drain-cleaning tools: (a) shark tooth cutter, (b) two-way root cutter

Source: Rothenberger Australia Pty Ltd.

Hydro-jetting equipment includes heavy-duty hoses with nozzle attachment tools. The nozzles are selected for each job depending on the type of blockage and size of pipe, and are available in a variety of nozzle jet configurations, including root-cutting turbines and spinning chains. Hydro-jetting is a method commonly employed by authorities to maintain their drainage systems, using specialist maintenance contractors. Manufacturers of hydro-jetting equipment provide a wide range of nozzles suited to different drain materials and blockage types, and should be consulted regarding the selection and safe use of their equipment.

When hydro-jetting, maintain control of the high pressure jet hose and keep clear of the drain entry point (see Figure 7.13).

FIGURE 7.13 Hydro-jetting: keep clear of the drain entry point

Chemical methods

Chemicals can be used to assist in removing blockages by breaking up fats and greases that build up in drainage systems. They are not used as a stand-alone method to remove blockages.

Refer to the 'Work health and safety' section of this unit for information on safety precautions, and always follow the manufacturer's MSDS.

Bacterial methods

Bacteria can be used to assist in removing blockages by breaking up fats, oils and greases that build up in drains. This method is not used as a stand-alone method to remove blockages. It uses a technology known as bio-remediation to control the incidence of blockages caused by fats, oils and greases in sewerage systems. Vegetative bacteria microorganisms are injected into the system to feed on fat, sugar and starch wastes. This technology is provided by specialist suppliers and contractors and should only be used by trained persons.

COMPLETE WORKSHEET 1

Planning for work

When you have gathered all of the necessary information, you need to make a plan to complete the job. This should include:

- ensuring all available plans or specifications are on-site
- having appropriate machinery available if required
- having all necessary tools and equipment available to start work
- being prepared for sewer/water spills
- identifying who is responsible for the problem
- informing the customer about the problem, where you will be working, and seeking their permission to proceed with the job
- having access to operate fixtures to test the work after completion.

Hazard control

As a plumber, you must be aware of the hazards you might encounter when locating and clearing a blockage in a plumbing installation. Some of these hazards are:

- drainage or sanitary:
  - sewer and noxious gases
  - confined spaces (e.g. inspection chambers)
  - waste discharges (e.g. when the blockage is dislodged)
  - chemicals
  - pathogenic organisms
  - sharps (e.g. needles)
- stormwater drains:
  - rodents (e.g. mice and rats)
  - stormwater discharges (e.g. when the blockage is dislodged)
  - sharps (e.g. needles)
  - soil, sticks and vegetation.

Basic mechanical and hydraulic principles

Before using drain-clearing machines and manually operated drain-cleaning tools, you need to understand some of the properties of water and how some basic mechanical and hydraulic principles apply to clearing blockages.

All drains are located below the source of the discharge; therefore, when a blockage occurs in the drain, the discharge will build up above the blockage, to a height in the pipe system until it overflows at some point. The head (height) of water in the pipework will steadily build up pressure above the blockage.

You might be clearing the blockage from either an upstream or downstream access point (IO). When clearing a blockage from a downstream access point, you must be particularly aware of this pressure. When the blockage is cleared, the pressure will be released. This release of pressure can force the water

to rush out of the access point at a considerable speed. Make sure you have planned your movements to safely get out of the way before the discharge arrives at your access point, and plan for any potential sewer/water spill.

If it is likely that there will be a sewage spill, you will need to plan the method of collecting the material before you attempt to clear the blockage. Sewage must not be allowed to spill, or to be washed into a stormwater drainage system or onto public property.

The deeper the drain and the higher the connecting pipework, the greater the pressure build-up in the system. Some sewers are very deep, and high pressures can be exerted at lower pipe sections by the build-up of sewage. Care must be taken at all times, particularly in confined spaces, such as access chambers. Also note that there are strict regulations on working in confined spaces.

Remember:

- A blockage can be accessed from upstream or downstream. *Note:* For every 1 m in vertical height, there is around 10 kPa of water pressure in the system.
- Plan to get out of the way safely before the discharge arrives at your access point.
- Plan for any potential sewage/water spillage. Sewage must not be allowed to spill, or to be washed into a stormwater drainage system or onto public property.
- Follow confined space regulations.

Facilities shutdown

If the drain that you are working on serves a number of areas, flats or units, then it may be necessary to shut down some plumbing fixtures or facilities to prevent continuous running water or spillage from the system.

Communication

It is the responsibility of the plumber, when called to a drain blockage, to notify the relevant people if fixtures or facilities need to be shut down, and to address any concerns they may have regarding your work. Notify people in advance, if your work might affect them – for example, where a car needs to be moved to gain access or floor surfaces need to be covered to prevent potential damage.

GREEN TIP

When clearing blockages in buildings where there are numerous occupancy units and fixtures, it is important to notify all persons that are potentially affected. Blockages can be transferred to other sections of plumbing, which can cause overflow from other fixtures.

Locating and clearing blockages

After preparing and planning for the job, you will now be able to locate and clear a blockage safely and effectively.

Property connections – who is responsible?

Generally, the owner of the property is responsible for sewer blockages upstream of the point of connection to the authority's sewer. It is the sewerage authority's responsibility downstream of the point of connection, and you must not attempt to clear a blockage in the authority's system.

First, it is the plumber's responsibility to determine if the blockage is *upstream* or *downstream* of the point of connection to the authority's sewer (see Figures 7.14 and 7.15). The blockage in a sewerage system should be located by a process of elimination. Using the plans of the property sewerage system, locate the boundary trap shaft (BT) or inspection shaft (IS). Usually, a property has either a BT or an IS. Remove the cover from the top of the shaft, and see if the shaft is full of sewage/water.

Remember to be careful and consider the pressure build-up that could be present. There might be a build-up of pressure if the cover is at a low level in the system. If the shaft is not full of sewage/water, the blockage is upstream of the shaft, and it is the owner's responsibility. However, if the shaft is full, then the blockage is either in, or downstream of, the shaft. To determine if the blockage is in the sewer main, inspect the drainage system/fixtures in the adjoining properties.

Referring to Figure 7.16, imagine that you have been called to clear a blockage at Lot 2. If the inspection shaft in Lot 2 is full of effluent, and you find that the inspection shaft in Lot 1 is also full, this would indicate that the blockage is in the sewer main downstream of Lot 1. If the inspection shaft in Lot 1 is not blocked, but the inspection shaft at Lot 3 is blocked, then the blockage would be in the sewer main between the points of connection to Lots 1 and 2.

If the inspection shafts at Lot 1 or Lot 3 are both free draining, then the blockage is downstream of the inspection shaft in Lot 2, but upstream of the Lot 2 main connection. You will then need to find out whether the blockage is upstream of the authority connection point.

To find out the location of the blockage, a sewer cable or hand rods can be used (described and illustrated earlier in this unit); alternatively, you may plunge the shaft. It is possible that plunging might clear the blockage. If these methods are not successful, it may be necessary to excavate to the IO at the authority connection point.

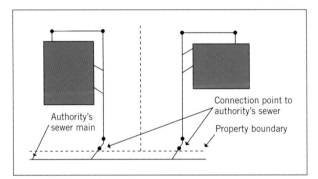

FIGURE 7.14 Sewerage system authority's connection points

Source: Department of Education and Training (http://www.training.gov.au)
© 2013 Commonwealth of Australia.

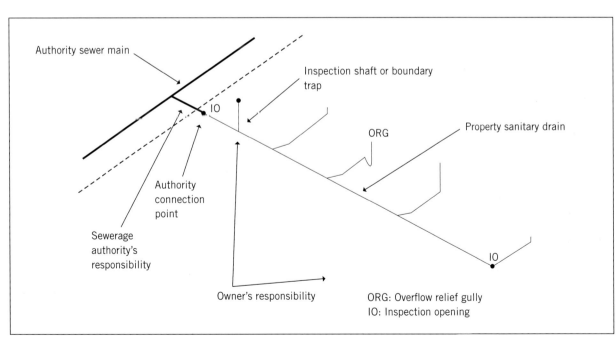

FIGURE 7.15 Determining who is responsible (system layout)

Source: Department of Education and Training (http://www.training.gov.au) © 2013 Commonwealth of Australia.

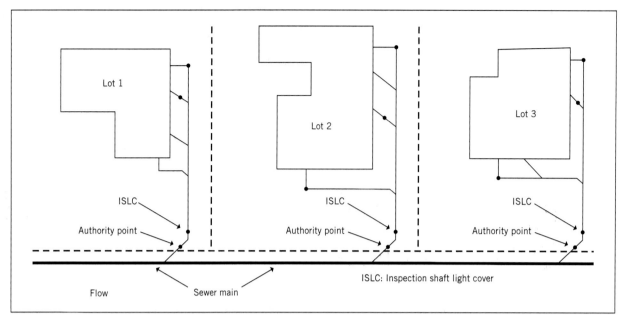

FIGURE 7.16 Locating blockages between adjoining properties

Source: Department of Education and Training (http://www.training.gov.au) © 2013 Commonwealth of Australia.

The location of the blockage is found by first reading the sewerage plan. You might need a scale ruler to determine the distance between the inspection shaft and the authority's connection point. Then, measure the depth of the inspection shaft using a tape measure, and add the two measurements together. For example, if the distance between the authority's connection point and the inspection shaft is 2 m, and the depth of the inspection shaft is 1.5 m, then the approximate distance measured along the pipework from the top of the inspection shaft to the authority's connection point is 3.5 m.

Only *after* you have found this measurement should you attempt to clear the blockage with rods or sewer cable. This is because you must *never insert drain-cleaning rods or a sewer cable past the connection point to the authority's main or into the sewer main*.

In the above example, you could put a mark on the rods/snake at 3.5 m to ensure that while you are attempting to clear the blockage, the mark does not go down past the top of the inspection shaft.

If there is still a blockage when the rods are inserted to the authority's connection point, then retrieve the rods and contact the relevant sewerage authority for advice.

If the blockage is upstream of the authority's connection point, and within the property, then the blockage is the property owner's responsibility.

Locating a blockage upstream of a boundary or inspection shaft

When attempting to clear a blockage in a property drainage system, first inspect all gullies to see if any effluent is flowing out. If it is, then the blockage is downstream of the gully. **Figure 7.17** shows examples of vitrified clay drain sewers where blockages often occur, while the following example relates to **Figure 7.18** and is a method of locating the blockage at the position shown.

Procedure

Refer to **Figure 7.18**, showing the system layout.

1 First you will have determined that the inspection shaft at (A) is not blocked.

2 Check the overflow relief gully (ORG) (B) to see if any effluent is flowing out; if not, move upstream to the next branch (C).

3 The next branch (C) is a direct connection; therefore, you should turn on the tap over the sink and check to see if any water flows down past the inspection shaft (A). If the water can be seen flowing down past the shaft, then this would indicate that the section of drain between the sink (C) and the shaft is clear. Therefore, you will move upstream to the next branch to the trough (D).

4 As in the previous step, branch (D) is a direct connection; therefore, turn on the tap over the trough. In this step, water would not flow down past the shaft, and so you will have determined that the blockage is between the branch to (C) and the trough (D).

5 To find if the blockage is in the main line or in the branch to (D), you must move upstream of branch (D) and flush the water closet (WC) (E), and then see if the discharge flows down past the shaft. If it does, then the blockage would be in the branch drain; however, in this example, it would not flow past the shaft, and you will have determined that the blockage is in the main drain downstream of the branch to the trough, as indicated in **Figure 7.18**.

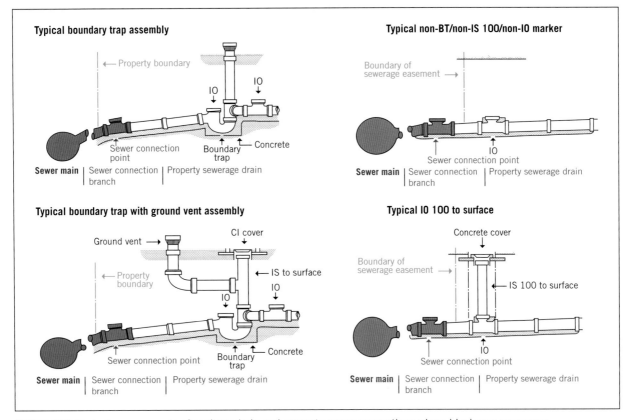

FIGURE 7.17 Typical arrangements of sanitary drain and property sewer connections where blockages may occur

Source: Reproduced with permission South East Water.

Gaining access to clear the blockage

Once the location of the blockage has been identified, you must determine the best place to gain access to the drain.

In the above example, access to the drain can be gained through the inspection opening on the above-ground bend to the WC, through below-ground IOs, or through removal of the closet pan.

Generally, access to the drain can be achieved (see **Figure 7.19**) through:

- inspection shafts and chambers
- above-ground inspection and testing openings
- below-ground IOs
- removal of gully grates
- removal of fixture traps
- removal of closet pans.

If it is necessary to remove a closet pan, be aware of the risk of damage to the pan and floor surfaces. If the pan breaks, there may be very sharp fragments, so always wear appropriate PPE.

The best method for determining the location of IOs is by obtaining a plan for the property from the local authority. Access to plans varies depending on the authority. In some areas – for example, areas converted from septic tank systems – plans are not available.

Some authorities provide property sewerage plans through plumbing merchants, facsimile, email and the internet; however, not all authorities maintain or collect plans of private property sewerage systems. If you are in doubt, ask your supervisor/instructor about the maintenance of these records in your locality.

When attempting to locate IOs, the information on a property sewerage plan can save a lot of time and unnecessary digging.

Where the above options are not available, and the blockage cannot be cleared, a new IO can be 'cut-in' to access the drain. See Unit 10 for a discussion of 'cutting-in' procedures.

LEARNING TASK 7.1

Imagine that you are faced with the task of clearing a blockage in a sanitary drain, and the only means of access is to remove a closet pan that is on a bed of mortar and tiles.

1. Communication with the customer is important. What should you tell the customer about the potential risks to the closet pan and floor surface?
2. What tools should be used to minimise the risk of damage?
3. What PPE should be worn while undertaking this work?

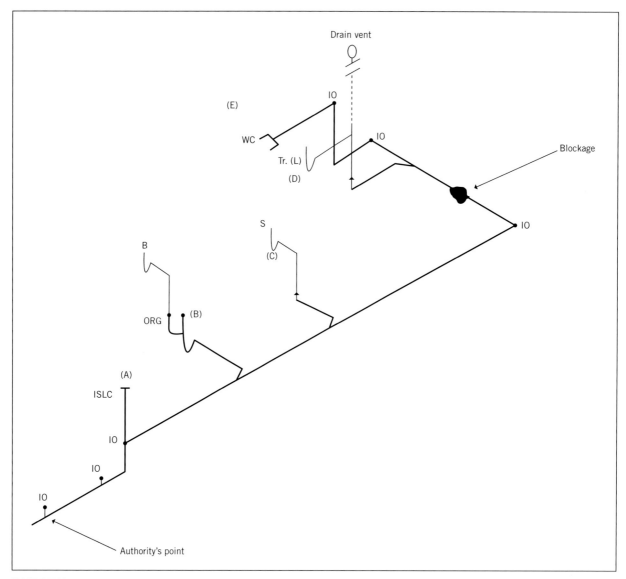

FIGURE 7.18 Locating a blockage in a sanitary drain

Access problems and possible solutions

Table 7.2 indicates some common problems encountered when locating drains and IOs, and suggests possible solutions.

TABLE 7.2 Solutions to accessing blockages

Problems	Solutions
Inspection openings not found in the location on plan	Dig along drain until IO is located, or 'cut-in' an IO.
Depth of access points unknown	Continue to dig deeper until access point is found. (*Note:* The access point is not normally deeper than the inspection shaft or boundary trap.)

Problems	Solutions
No plan; location unknown	Contact a specialist pipe- and cable-locating contractor. If using CCTV equipment, position camera head at the location of blockage and use a pipe locator navigator to locate the camera head (see **Figure 7.20**).
Drain built over	Locate and use a different access method.
Blockage cannot be removed	If mechanical and hydraulic clearing methods are unsuccessful, and the blockage cannot be removed, then a section of drain may need to be replaced.
Blockage is in another property	Consult the relevant authority and the owner of the other property.

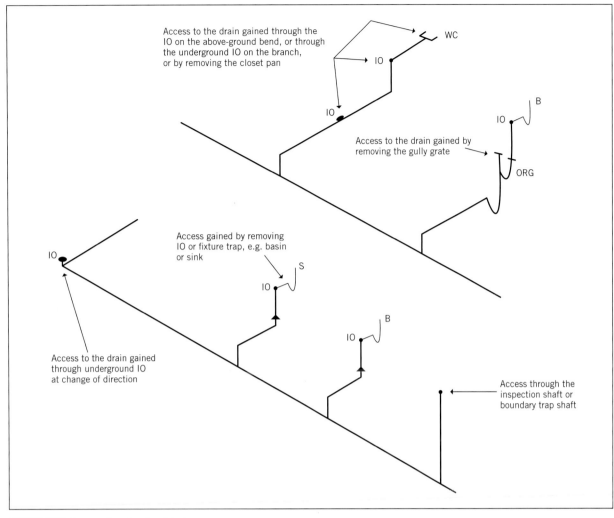

Access to the drain gained through the IO on the above-ground bend, or through the underground IO on the branch, or by removing the closet pan

WC

IO

IO

B

IO

ORG

Access to the drain gained by removing the gully grate

Access gained by removing IO or fixture trap, e.g. basin or sink

IO

S

IO

IO

IO

B

IO

Access to the drain gained through underground IO at change of direction

Access through the inspection shaft or boundary trap shaft

FIGURE 7.19 Access for clearing blockages

FIGURE 7.20 Using a pipe locator to find the position of a camera head in a drain

COMPLETE WORKSHEET 2

Dealing with specific types of roof and stormwater drainage systems

Roof and stormwater drainage systems can be blocked by sand, silt, leaves, sticks and foreign objects. Usually, the result is that stormwater will overflow from ground surface grates or downpipe connections.

These types of blockages should be removed to ensure that water is effectively drained away from buildings, because variations in soil moisture around a building can affect the stability of building foundations.

The best way to locate a blockage in the roof or stormwater drainage system is by a process of elimination similar to the process described earlier in the unit.

Locate the legal point of discharge (LPD) and check to see if the stormwater drain is full of water. Remember to be careful and consider the pressure build-up that could be present. If the drain is not full of water, the blockage is upstream of the LPD. If the

LPD is full, then the blockage is downstream in the council/authority's drainage system, and you will need to contact them.

In the case of roof drainage, locate the downpipes applicable to the overflowing gutter and see if they are full of water. Remember to be careful and consider the pressure build-up that could be present. If the downpipe is not full of water, the blockage is in the roof gutter system. If the downpipe is full, then the blockage is in that downpipe, or downstream in the stormwater drainage system, and will need to be cleared.

Water pipes

Blockages in water pipes are usually caused by damage to pipework or corrosion. Blockages can be indicated by the lack of pressure, or no pressure, through a particular water supply outlet. However, other possible causes of lack of pressure should first be eliminated before investigating the pipes. For example, it is possible that the water supply outlet fitting has a blocked flow restrictor or filter.

Water pipes are occasionally damaged by being kinked or crushed. This can happen where the pipes are in an exposed position during construction. Kinked or crushed pipes must be replaced. Similarly, corroded pipes such as galvanised-steel pipes cannot be unblocked and should be replaced.

Gas pipes

Any work on a gas installation must only be carried out under the supervision of a qualified gasfitter.

Blockages in gas pipes are occasionally caused by damage to pipework or poor installation and commissioning. Blockages can be indicated by the lack of gas supply to appliances, or no pressure. However, other possible causes of lack of pressure should first be eliminated before investigating the pipes. For example, there may be low supply pressure, undersized piping, or a shut-off device may have activated upstream at the gas meter service regulator.

To clear a blockage:

1 all gas meters, gas pressure regulators and gas appliances must be disconnected before any pressure or force is applied

2 no cables or objects should be inserted into pipes; regulated air pressure or inert gas can be used to clear the blockage

3 after the blockage has been cleared, and gas meters, gas pressure regulators and gas appliances are reconnected, the consumer piping must be tested in accordance with the relevant standard. For example, for natural and LP gas refer to AS/NZS 5601.1 Gas installations – General installations.

AS/NZS 5601.1 GAS INSTALLATIONS – GENERAL INSTALLATIONS

Testing after the blockage is cleared

After clearing any blockage, the pipework should be flow tested to ensure that the blockage has been completely cleared. Flow testing is done by individually running/flushing each fixture/appliance to verify that each branch and section of the drain (or pipe) is clear.

Finally, for drainage, verify that water is passing the inspection shaft and the authority's connection point at the downstream end of the drain. If water flows freely, then the blockage has been cleared.

Repairing and resealing of pipework

Once the blockage has been cleared, the system needs to be restored for normal operation. Depending on the type of pipe system and the work undertaken, it may be necessary to:

- replace the IO caps and sealing rings
- replace the traps and sealing rings, and test for leaks by filling with water
- reinstall the water closet pan, and test for leaks
- replace the gully grate
- backfill the excavation, and reinstate the area to its normal condition
- seal any replaced or repaired pipes
- replace any damaged pipes and fittings in accordance with relevant standards and regulations
- replace the inspection chamber cover.

Advising authorities when work is complete

The requirement to submit documentation varies around Australia. If you have altered the drainage system, it may be necessary to lodge an updated plan to the local authority and arrange for an inspection. You may also be required to submit a Certificate of Compliance to the owner/agent and the local sewerage/drainage/regulatory authority.

If you have planned the job correctly, you will have ascertained the requirements of the authorities before commencing the job.

Clean-up and storage

A proper and thorough clean-up is part of the job, and is important not only for WHS, but also as part of your professional obligation.

- Always wash your hands and other areas of your body that have been exposed to spills, etc.
- Inspect the areas of work to ensure they are clean. Cleaning up the areas of work is not just a matter of appearance; it also relates to WHS. The work area may need to be washed down and disinfected, which may include removal of unhygienic waste material.
- Some premises (e.g. food production, hospitals, nursing homes) may have specific clean-up procedures.
- Any waste from the worksite should be disposed of or recycled in accordance with state or territory legislation. If you are unsure of the requirements, consult your supervisor.
- Off-cuts and surplus material should be transported and stored appropriately until required.
- After clearing a blockage, repairing and sealing the pipework, and reinstating the area to its normal condition, backtrack and inspect the whole of the work area.
- Floor surfaces that have been covered must be uncovered and returned to their normal condition.
- Clean up tools and drain-cleaning equipment.
- Clean up PPE.
- Inspect and maintain PPE, tools and equipment on a regular basis according to manufacturers' recommendations. Keep all equipment correctly stored to maintain it in a serviceable condition.
- Clean, maintain and return hire equipment according to the hire agreement.

Preventing blockages

The following may assist in preventing blockages in sanitary drainage:

- Gully grates should be correctly positioned and installed in a loose-fitting condition.
- Inspection shafts and boundary trap risers should be fitted with caps, and the top of the shaft protected with a light or heavy cover.
- Exposed vitrified clay and PVC drainage fittings should be protected with concrete surrounds.
- The amount and type of cover over drains should be in accordance with Australian Standards, and suitable to the type of traffic.
- Drains should be constructed of materials suitable to the ground conditions and nature of discharges.
- Drainpipes and fittings should be jointed in an approved manner in accordance with AS/NZS 3500.2 Sanitary plumbing and drainage.
- Elastomeric drain connectors should be fully supported when installed.

- Drains should be fitted with suitable bedding, side support and overlay materials.

CCTV inspection

CCTV inspection of pipe systems can provide valuable information on the causes of blockages and the basic condition of pipe systems (see **Figure 7.21**). It is important to clear blockages and clean drains and pipes prior to any CCTV inspection.

FIGURE 7.21 Pipe inspection using CCTV equipment

Where CCTV inspection of sanitary plumbing and drainage is required, this work should be carried out by a suitably qualified expert in accordance with AS/NZS 3500.2.

Specialist assessment of the condition of stormwater and sanitary drains can also be done in accordance with CIRCA (Conduit Inspection Reporting Code of Australia). This type of CCTV inspection and assessment must only be carried out by experienced and accredited operators.

SUMMARY

- There are common causes of blockages, and it is important to know what they are and whether repairs are needed to prevent recurring problems.
- Correct preparation ensures that you have as much information as possible before starting the work, and that the work can be carried out safely.
- To clear blockages, tools and equipment must be selected that are suited to the size and material of pipes. Always choose the soft option first by plunging, before using invasive and potentially damaging methods. The blockage may be of tougher material than the wall of the pipe. Using steel rods and heavy-handed methods may cause pipe fractures and significant property damage.
- Planning for work entails having all relevant plans, machinery and tools available, assessing hazards you may encounter and understanding basic mechanical and hydraulic principles. Ensure that all affected property owners/residents are notified and that you have planned for the shutdown of facilities and potential hazards.
- After preparing and planning for the job, you must locate and clear the blockage safely and effectively.

This will mean determining whether the blockage is upstream or downstream of the point of connection to the authority's sewer. Also consider the pressure build-up that could be present, and determine the best place to gain access to the drain once the blockage has been identified.

- Roof and stormwater drainage systems can be blocked by sand, silt, leaves, sticks and foreign objects. The best way to locate a blockage in the roof or stormwater drainage system is by a process of elimination.
- After clearing any blockage, the pipework should be flow tested to ensure that the blockage has been completely cleared.
- Thorough clean-up is an essential part of the job, both as part of WHS responsibilities and as part of your professional obligation.
- Where a blockage cannot be cleared, use CCTV and electronic pipe-locating methods to identify the location of blockage and plan to safely remove any build-up of water above the blockage before replacing the section of affected pipe.

REFERENCES AND FURTHER READING

Biddle, P.G. (1998). *Tree Root Damage to Buildings. Volume 1: 1 Causes, Diagnosis and Remedy*. East Challow, Oxfordshire: Willowmead Publishing.

Safe work Australia (2013). *Guide for managing risks from high pressure water jetting*, December 2013.

Standards Australia (2013). *AS/NZS 4233.1: 2013 High pressure water jetting systems Safe operation and maintenance*.

Water Services Association of Australia (2020). *Conduit Inspection Reporting Code of Australia WSA 05–2020*, 4th Edition, Version 4.1. Sydney: WSAA.

**WORKSHEET 1**

Student name: _____

Enrolment year: _____

Class code: _____

Competency name/Number: _____

Task: Review the sections 'Common causes of blockages', 'Preparing for work' and 'Tools and equipment', then complete the following.

1   Name the four main causes of blockages in drainage systems.

_____

_____

_____

_____

2   Why is it best practice to obtain a plan from the local sewerage authority/council before commencing work?

_____

_____

3   When excavating to locate drains, what other types of services might you find?

_____

_____

4   Give four examples of a confined space in plumbing and drainage systems.

_____

_____

_____

_____

5   What are two things about the atmosphere in a sewerage inspection chamber that define it as a confined space?

_____

_____

6   Research which regulations in your state/territory specify your legal obligations when entering a confined space.

_____

_____

7   What type of gloves should be worn when operating mechanical drain-cleaning equipment?

_____

8   What types of tools should be used carefully in a PVC drain?

_____

9   If using chemicals to assist in removing blockages, what are two safety precautions?

_____

_____

10   Why would you ask a home owner whether chemicals have been used in an attempt to clear a
     blockage?

_____

_____

11   Why is a plunger the most common device used to clear blockages in a drainage system?

_____

_____

_____

WORKSHEET 2

Student name: _____

Enrolment year: _____

Class code: _____

Competency name/Number: _____

Task: Review the sections 'Planning for work' and 'Locating and clearing blockages', then complete the following.

1   After completing work to clear a blockage, how would you plan to test the drainage system to ensure its correct operation and that the blockage has been cleared?

_____

_____

2   If it is suspected that there is a blockage in a water pipe, what should first be eliminated before investigating the pipes?

_____

_____

3   If it is suspected that there is a blockage in a gas pipe, what are three things that must first be eliminated before investigating the pipes?

_____

_____

_____

4   What should be fitted to gullies to prevent blockages?

_____

5   What information do you need to determine the location of the point of connection to the authority's sewer main?

_____

6   Name four access points that may be used to gain access to clear a blockage in a private sanitary drainage system.

_____

_____

_____

_____

7 Research your local authority. Name the options that are available to you to obtain a property sewerage plan so as to locate the access points.

_____

_____

8 If you were not able to clear the blockage, and could not find the property sewerage drain, what action would you take?

_____

_____

9 In the diagram below, label the four access points.

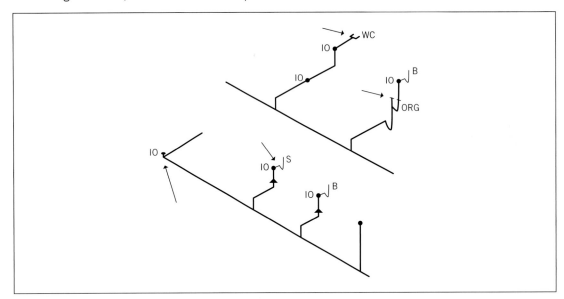

10 Explain three problems that could occur when locating access for clearing a blockage.

_____

_____

_____

11 State two parties who could be responsible for a blockage in a sanitary drain.

_____

_____

12 What do the following abbreviations mean?

BT _____

DG _____

IS _____

IO _____

 **WORKSHEET 3**

Student name: _____

Enrolment year: _____

Class code: _____

Competency name/Number: _____

Task: Review all sections in Unit 7, then complete the following.

1   State two parties who could be responsible for stormwater drain blockage.

_____

_____

2   With the aid of a neat, fully labelled diagram, show how a sewer plunger is constructed.

3   Describe how a plunger is used when clearing a blockage.

_____

_____

_____

_____

4   Why should a plunger be used before using machine-driven equipment?

_____

_____

5   Who must be contacted if a blockage is found to be in a property other than the property you are working in?

_____

_____

_____

6   Name three types of drain-cleaning attachment tools that may be used, and identify in what situation you think they would be used.

_____

_____

_____

7   Briefly describe the use of mechanically operated clearing equipment.

_____

_____

_____

WORKSHEET 4

Student name: _____

Enrolment year: _____

Class code: _____

Competency name/Number: _____

Task: Practical exercise

All questions in the previous three worksheets must be completed and checked by your trainer/teacher before you attempt the following practical exercise.

Your trainer/teacher should provide preliminary guidance, conduct an induction and ask underpinning knowledge questions.

Using both manual tools and mechanical equipment, you are required to locate and clear a blockage from a section of drainage pipework.

Your trainer/teacher may choose to combine the above exercise with exercises in other units.

PART 2
INSTALLING SANITARY SYSTEMS

- **Unit 8** Installing sanitary plumbing
- **Unit 9** Installing and fitting off sanitary fixtures

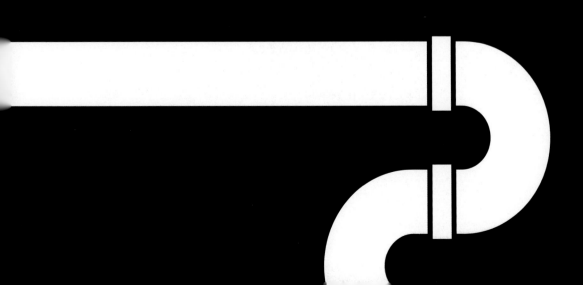

8 INSTALLING SANITARY PLUMBING

Learning objectives

This unit provides practical guidance for installing above-ground sanitary plumbing systems.
Areas addressed in this unit include:
- preparation and sequence of work
- determining positions for pipework
- allowances for expansion and movement of pipework
- assembly of pipework
- installation of discharge pipes
- fabrication and installation of sanitary stacks
- pipe support and clipping
- overcoming problems when installing pipework
- working with different materials
- testing of pipework.

Introduction

Plumbers are required to install sanitary plumbing systems to collect and convey sewage and waste from sanitary fixtures to the sanitary plumbing and drainage system. This type of work is carried out at all properties where there are sewage and waste discharges. It could be new work, or it could involve alterations or additions to existing systems.

Sanitary plumbing is generally regarded as an assembly of discharge pipes, stacks, vents, fixtures and appliances that are located above ground.

In Unit 5, we studied the requirements for planning the layout of a residential sanitary plumbing system. This unit is a progression from planning, and we will now study the procedures and requirements for fabricating and installing discharge pipes and sanitary stacks.

While studying this subject, you will be required to respond to questions and complete exercises referring to AS/NZS 3500.2 Sanitary plumbing and drainage.

AS/NZS 3500.2 SANITARY PLUMBING AND DRAINAGE

Before studying this subject, you should have completed training in work health and safety (WHS) as outlined in Chapter 3 of *Basic Plumbing Services Skills*.

Preparing for work

Plans and specifications

Before work can commence on-site, you must have obtained all plans and specifications, including finding out the requirements of the regulatory authorities. To review the requirements discussed in Unit 5, refer to the following checklist:

- plans and specifications that identify the location, number, type and colour of fixtures
- plans and specifications that identify or reveal options for the position of discharge pipes, stacks, and vents such as pipe ducts, wall and ceiling cavities
- plans and specifications that identify the type and size of materials to be used, location of pipes and type of fixing devices to be used
- plans and specifications that identify other services and obstructions
- applications and notifications to the network utility operator and/or plumbing regulatory authority (e.g. permits and/or work notices)
- plumbing regulatory and Australian Standard requirements (e.g. pipe sizing and clipping)
- WHS regulations (e.g. confined space entry and working at heights)

- site inspection (e.g. work conditions, slope of land, obstacles, access)
- manufacturers' specifications (e.g. for installation of fixtures and fittings).

Plans and specifications are necessary for planning the position of fixtures and fixture discharge pipes and stacks to connect to the drain. It may be necessary to lodge applications and work notices with state, territory or local authorities, and these may need to be submitted at certain stages of the job.

Safety and environmental requirements

As discussed in Unit 5, you must have planned for the job and obtained all of the necessary equipment to comply with WHS and environmental regulations. The examples given in Unit 5 included:

- working at high levels
- using electrical equipment in wet conditions
- preventing entry of stormwater, sand, silt and rubbish to the sanitary drainage system.

Each state and territory has WHS regulations that require employers and self-employed people to identify hazards and assess and control risks at the workplace in consultation with their workers.

If working at high levels, then, as discussed in Unit 5, this must be identified as a hazard, and an elevating work platform or scaffolding must be available on-site before work commences.

If you can foresee wet conditions, it is best practice to maintain flexibility to rearrange the work schedule. Electric power leads and tools must only be used in dry conditions, due to the possibility of electrocution.

You will also need to consider any environmental requirements. When connecting to the sanitary drain, you will need to be prepared to provide protection to the connection points near ground level to prevent the entry of any stormwater, sand, silt and rubbish. Stormwater must not be allowed to enter the sanitary drain, and *must only be discharged to an approved point of discharge for stormwater.*

Quality assurance

As discussed in Unit 5, there may be requirements for quality assurance. You must be aware of the project requirements for quality assurance, such as the necessary inspections and tests required by the regulatory authority. You should also be aware of your company's particular quality assurance requirements.

When preparing for your work, you must have planned when the sanitary plumbing, or section of the plumbing, will be completed, and have the necessary reference numbers and phone numbers to book inspections and tests with the regulatory authority.

In some areas, inspection of sanitary plumbing systems is not required by regulatory authorities. However, if you are unfamiliar with local requirements, or are moving to a different area, you must research the local regulations and standards.

In some areas, the regulatory authority will require that you book the sanitary plumbing for inspection with at least 24 hours' notice. Inspections are carried out by regulatory authorities to ensure that the sanitary plumbing has been installed correctly in accordance with local regulations and AS/NZS 3500.2, and local authority requirements.

AS/NZS 3500.2 SANITARY PLUMBING AND DRAINAGE

On larger building sites throughout Australia, additional requirements for quality assurance of plumbing work are common, such as sign-off on portions of work for progress payments, and additional tests and inspections. In these cases, quality assurance requirements may involve 'hold points' where your work must stop, allowing time for the necessary tests and inspections (see Table 8.1).

TABLE 8.1 Examples of tests and inspections

Examples of tests	Examples of inspections
Hydrostatic or air testing	Pipe clipping and support
Testing of sanitary fixtures by subjecting them to normal use. Fixtures traps must retain a water seal of at least 25 mm.	Provision for expansion
	Pipe gradient
	Pipe sizing

All pipes and fittings used in a sanitary plumbing system must comply with the *National Construction Code (NCC) Volume 3, Plumbing Code of Australia* and bear WaterMark and Australian Standards certification.

Later in this unit, we will discuss the specific procedures for inspection and testing of sanitary plumbing.

In review, some of the issues mentioned in Unit 5 that should be considered when preparing to install a sanitary plumbing system are:

- conformance with company operating procedures
- ensuring the necessary applications are lodged with the authorities
- arranging for all necessary inspections and tests
- ordering materials and products in accordance with job plans and specifications
- receiving and inspecting materials and products for conformance with orders
- storage of materials and products to prevent deterioration
- operation and maintenance of tools and equipment
- compliance with the contract work schedule
- maintaining records of job variations.

LEARNING TASK 8.1

Undertake research on the internet to find the Australian Building Codes Board (ABCB):
- WaterMark Certification Scheme Schedule of Products
- WaterMark Certification Scheme Schedule of Excluded Products.
1 What is the Australian Standard for polyethylene (PE) fittings?
2 Are self-sealing traps required to have WaterMark?
3 Are sinks required to have WaterMark?

Sequencing of tasks

In Unit 5, we discussed the coordination of your work with other trades. Before commencing work on-site, it is best practice to arrange for a site meeting with the construction supervisor and trade supervisors. In preparing to install a sanitary plumbing system, it is important that tasks are sequenced so that the building project will run smoothly. Each trade on a building site has a job to do, and understanding the project schedule and communicating with others is fundamental to the success of a plumbing business.

A good site meeting should provide, among other things, the proposed construction schedule and contact details of trade supervisors.

Many building projects have a fairly loose construction schedule. This is when communication with other trades is especially important, as problems can arise when there is a clash of trades or work is not carried out in the correct sequence. This cannot be emphasised too strongly.

For example, it was previously mentioned that when preparing for work, you might note that scaffolding is required to install high-level pipework. In that case, you will need to know the date of commencement of plasterwork that might cover the piping, so that you will have time to arrange for the scaffolding and to install and test the pipework before that date.

Another example is that tiling might be required for shower bases, and so you will need to know the date when the tiling is to commence so that you have time to complete the waste pipe connection.

Aside from the sequencing of work in coordination with other trades, you will need to prepare to install the sanitary plumbing in the correct sequence, and this will be discussed in greater detail later in this unit.

By familiarising yourself with construction schedules, and planning the sequence of work, you will be able to meet the schedules on one or more projects, utilising your time in the most efficient way for the successful operation of your plumbing business.

COMPLETE WORKSHEET 1

Tools and equipment

Safety equipment

An important safety consideration when installing sanitary plumbing is the correct selection and use of equipment when working at heights, such as elevating work platforms, scissor lifts, access towers and scaffolding. Therefore, it may be necessary to undertake training in the assembly or operation of this equipment. Check with your instructor or trainer on the requirements in your area.

Types of personal protective equipment (PPE) that may be required when installing sanitary plumbing include:

- overalls
- boots
- safety glasses or goggles
- ear plugs or ear muffs
- dust masks or respirators
- gloves
- hard hat.

Tools

Tools and equipment typically used for installation of sanitary plumbing may include:

- ruler
- measuring tape
- hacksaw with a range of blades
- wood saw
- mitre box
- cordless drill with a range of bit tools for self-drilling screws
- power drill with a range of hole saws
- silicone sealant cartridge gun
- range of round and flat files
- string line
- plumb-bob
- spirit level
- plugging chisel
- cold chisel
- mash hammer
- claw hammer.

When fabricating and installing sanitary plumbing, you might be required to use lifting or load-shifting equipment such as:

- hand trolleys
- rollers
- forklifts
- chain blocks
- hoists
- jacks.

Special equipment – selection and operation

Special equipment is required for some pipe systems such as high-density polyethylene (HDPE). The manufacturers of these systems provide special equipment, which must be selected and operated in accordance with their strict guidelines to achieve watertight, sound joints.

An example of an HDPE product is the Geberit HDPE system, which uses butt welding and electro-fusion jointing equipment. To research Geberit HDPE products, go to the Geberit website at https://www.geberit.com.au. Refer also to Unit 6.

Installation requirements

Materials and fittings

The PVC (polyvinyl chloride) fittings chart in Unit 5 (see Figure 5.5) provides illustrations of some of the fittings that are most commonly used in the installation of sanitary plumbing systems. For more information on PVC materials and fittings, go to the Vinidex website at http://www.vinidex.com.au.

When installing pipes and fittings, you must consider the characteristics of the materials and material installation requirements as outlined in Unit 5 – for example, you must provide for the expansion and correct support of PVC pipe. You should be aware that the installation, expansion and clipping requirements differ for different pipe materials.

Determining discharge pipe positions

As outlined in Unit 5, the manufacturers of fixtures normally provide specifications on the exact fixture dimensions to determine the location of fixture outlets – for example, for water closet pans, the vertical measurement above floor for a 'P' trap pan or the horizontal distance from a wall for an 'S' trap pan. Remember also to allow for the additional thickness of finished surfaces such as tiles.

Fixture discharge pipes should be located where they do not cause obstruction to doors and windows, or the operation of any equipment. Discharge pipes may be installed either concealed or exposed to view, but this will depend largely on the circumstances of the job. For example, within a factory it may be quite acceptable for a pipe to be installed on the outside of a wall in full view. But if the discharge pipe is to be installed in a bathroom within a house, then it is best practice to conceal the pipe by locating it inside a cupboard, wall or under the floor. It is important to make the right decision on the positioning of fixture discharge pipes for each job. It is best practice to discuss the location of pipework with the property owner/builder before commencing any work.

Whenever a discharge pipe is to be installed in a wall, below the floor or in a ceiling, it is important to remember to install the pipe in such a way as not to cause any unnecessary damage by weakening the building structure. Wherever possible, pipes should be located in the centre of noggings, top and bottom plates, and not to the side, because this can weaken the wall structure. If you are unsure about cutting holes and notches in studs and plates, ask the project manager, or refer to AS 1684.2 Residential timber-framed construction.

AS 1684.2 RESIDENTIAL TIMBER-FRAMED CONSTRUCTION

Determining stack and branch positions

When you have planned the layout for the sanitary plumbing system as outlined in Unit 5, the final installation requirements for determining the stack and branch positions will primarily relate to:

■ the correct orientation of the branches

■ clearance from restricted zones.

The normal branch orientation should be to a vertical stack, and any branches to steep offsets should be avoided. Double 'Y' junctions must only be installed in vertical stacks.

AS/NZS 3500.2 specifies the types of junctions that may be used, as mentioned in Unit 5. Read AS/NZS 3500.2 and study:

■ zone restrictions for stack connections

■ connections in or near restricted zones.

AS/NZS 3500.2 SANITARY PLUMBING AND DRAINAGE

Ventilation requirements

After planning the layout for the sanitary plumbing system, as described in Unit 5, you will have determined what sections of the system must be vented. The following is an overview of venting requirements:

■ Every property with a direct-type connection should have a sanitary drain with at least one open upstream drainage vent of minimum size DN 50.

■ Every property with a boundary trap connection should have a sanitary drain with at least one open downstream and one open upstream drainage vent, each of minimum size DN 50.

■ Stack vents must be correctly sized, extending to outside the building and terminating in an approved location clear of openings to the building, air duct intakes, etc. Alternatively, for stacks up to 10 floors, an air admittance valve may be used to terminate a stack vent, providing that the property drain has at least one upstream vent (see **Figure 8.1**).

■ A relief vent must be correctly sized, connected to the stack vent or extended separately to outside the building and terminating in an approved location clear of openings to the building, air duct intakes, etc. Alternatively, it may be replaced by the installation of a pressure attenuator. Where an air admittance valve has been installed as a stack vent, the associated relief vent must be either extended separately to open air or installed as a pressure attenuator (see **Figure 8.1**).

■ Trap vents and group vents may also be extended to outside the building, or may be replaced by air admittance valves (see **Figure 8.1**).

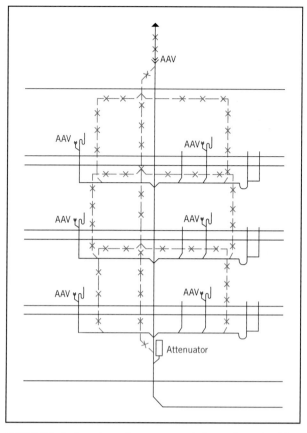

FIGURE 8.1 Options to replace open vents in sanitary plumbing

Source: Grant Weymouth, Studor

■ All vents and air admittance valves must be correctly sized in accordance with AS/NZS 3500.2.

AS/NZS 3500.2 SANITARY PLUMBING AND DRAINAGE

Allowances and ordering materials

Material quantities

At the end of Unit 5, material quantities were obtained from an example project and recorded in a materials list (see 'Determining material quantities' in Unit 5). While two plumbers could arrive at the same solution for a layout for the sanitary plumbing, estimates of material quantities can vary depending on the combination of fittings used. There is no right or wrong way of selecting fittings, providing that the installation complies with regulations and standards.

When purchasing pipe, it is good practice to order full lengths of pipe, and to order more than is required. PVC pipe is normally sold by the 6 m length.

GREEN TIP

There will always be some off-cuts and material waste, although this should be avoided as much as possible. Plan to set aside full lengths of pipe for main runs and exposed (visible) installations.

GET IT RIGHT

VENTILATION

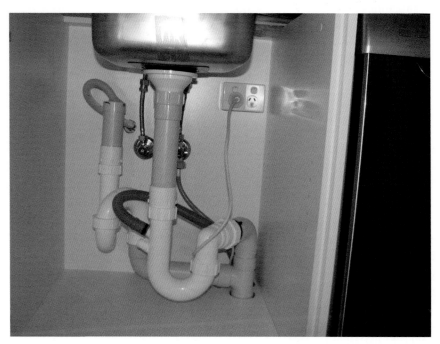

FIGURE 8.2 Waste pipe connections without ventilation

Failure to provide for correct ventilation can result in loss of trap seal. In **Figure 8.2**, two connections have been made to a DN 50 waste pipe, and there is no ventilation. Remember that DN 40 and DN 50 waste pipes are for connection to a single outlet unless the pipe is ventilated. Installation of an air admittance valve (AAV) is one way to solve this problem.

Research AS/NZS 3500.2.

l   Where should an AAV be installed in this situation?

_____

_____

_____

2   What size AAV should be installed?

_____

_____

_____

3   At what level should the AAV be located in relation to the fixtures/traps?

_____

_____

_____

It is also good practice to order a range of different angle bends. Pipe junctions must be made using the correct angle bends, and pipes must not be forced into fittings as this will place them under stress.

Depending on the building structure, other materials will also be required and added to the materials list, such as self-drilling screws, masonry anchors, wall sleeves and roof penetration flashings. Also, you will need to add consumables such as drills and cutting blades.

Ordering materials

Ordering materials is a simple matter of completing the materials list as per the example at the end of Unit 5 (see Table 5.3), and obtaining quotes from plumbing suppliers, or ordering them from your preferred supplier.

Identifying materials

Sanitary plumbing pipes and fittings can be identified by standards markings. For example, the standard for PVC-U (commonly referred to as 'PVC') is AS/NZS 1260 PVC-U Pipes and fittings for drain, waste and vent application.

AS/NZS 1260 PVC-U PIPES AND FITTINGS FOR DRAIN, WASTE AND VENT APPLICATION

If you are asked to work with different materials, you can refer to the list of 'Acceptable pipes and fittings' in AS/NZS 3500.2.

AS/NZS 3500.2 SANITARY PLUMBING AND DRAINAGE

Collecting and checking the delivery

When receiving the delivery of pipes, fittings, clips and brackets, and other materials, you should count the number and type of items and compare them with your materials list order. Ensure that you have identified all of the items as correct before accepting the order. There is nothing worse than setting up to begin work, only to discover that there are fittings missing or the wrong fittings have been supplied.

FROM EXPERIENCE

The ability to manage your time to ensure that you have all of the necessary pipes, fittings, brackets, clips and consumables to complete the job can influence the efficiency and costs of the job.

COMPLETE WORKSHEET 2

Fabrication, assembly and testing

Fabrication and assembly of discharge pipes

Setting out

After preparation, work can commence on installing discharge pipes. Fixture discharge pipes can be installed before or after the installation of the fixture. Depending on the circumstances, it may be necessary to install the discharge pipe before the fixture has even been installed. This may be due to different reasons, the most common being that the pipe needs to be hidden from view and therefore installed within the wall or under a floor.

If the new pipework to be installed needs to be joined to existing live pipework, it may be necessary to isolate the new installation. This will ensure that you are able to complete your task without interference from unwanted discharge from existing fixtures connected to the same line. In the case of a small job, this may only require you to instruct the occupier of the building to stop using some fixtures. In the case of larger installations, it may be necessary for you to shut off water to fixtures, disconnect fixtures or place 'DO NOT USE' signs in appropriate positions.

Remember, the pipe system must be installed without damaging or distorting the pipework, and without damaging the surrounding environment or services.

Before installing any pipework, it is important to set out where you are to run the pipe. It is best practice to set out the path of the pipework using the shortest possible route and the minimum number of bends in accordance with AS/NZS 3500.2. This would normally ensure that a minimum of material is used at least cost.

AS/NZS 3500.2 SANITARY PLUMBING AND DRAINAGE

Once the path of the pipework has been determined, brackets, hangers and pipe clips may be installed. They must be installed allowing for the required gradient of the pipe, as discussed in detail in Unit 5.

The gradient must be calculated (see Example 8.1).

EXAMPLE 8.1

A DN 50 discharge pipe that is 2 m (2000 mm) long must be installed at a minimum gradient of 2.5%.

$$\frac{25}{100} \times \frac{2000}{1} = 50 \text{ mm}$$

Therefore, the required fall over 2 m is 50 mm.

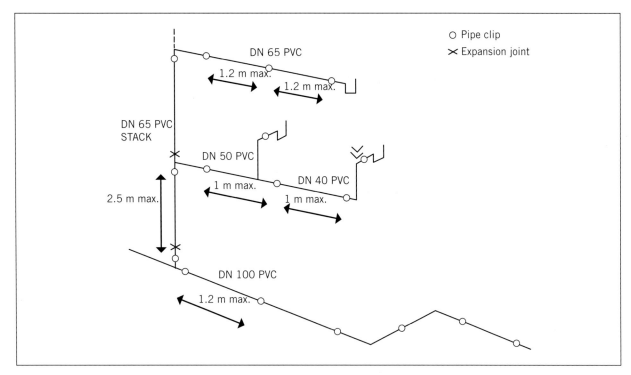

FIGURE 8.3 Example application of maximum support spacings for PVC piping

Fixings and supports

All sanitary plumbing systems must be securely supported and clipped to the building structure in accordance with the pipe manufacturer's specifications and AS/NZS 3500.2.

AS/NZS 3500.2 SANITARY PLUMBING AND DRAINAGE

Sanitary plumbing must be supported by means of brackets, saddles, hooks, straps, hangers or other approved methods. The support system must:
- be capable of supporting the pipes when they are filled with sewage
- firmly hold the piping in the intended position to prevent movement of the piping, while allowing for expansion
- have pipe support material that is compatible with the pipe material. Pipe clips for PVC are usually coated with compatible plastic material.

When installing the support system, the weight of a full pipe of sewage should be considered, as this will be a reality if the pipe is blocked. One metre of DN 100 PVC pipe filled with sewage will weigh approximately 10 kg. Therefore, *one 6 m length of DN 100 PVC pipe with five pipe brackets and clips could weigh almost as much as a man – that is, at least 60 kg.*

The maximum spacing of supports for PVC pipe are:
- DN 40 to DN 50: 1 m for graded pipes, and 2 m for vertical pipes

- DN 65 to DN 100: 1.2 m for graded pipes, and 2.5 m for vertical pipes.

Figure 8.3 illustrates the application of maximum support spacings.

Figure 8.4 provides examples of support bracket, clip and hanger combinations for different pipe materials. There are many other bracket and clip types and combinations.

Some pipe systems, such as polypropylene push-fit sanitary plumbing systems, must be installed with both fixed and guide point brackets and clips. This ensures that pipes are able to expand with changes in temperature, while pipes and fittings are restrained to prevent joint separation.

For further information on the support and fixing of pipework, refer to AS/NZS 3500.2, and the relevant pipe system manufacturer's installation instructions.

AS/NZS 3500.2 SANITARY PLUMBING AND DRAINAGE

For product research, visit the Abey website at http://www.abeytrade.com.au.

Installation of, and joining, pipes and fittings

To ensure that the pipework will be straight, use a string line to align the clips in the path you have chosen, and then fix the clips in position so that the pipe will be straight when inserted into the clips. This is particularly important if the clips are to sit away from the building, such as where using hanging brackets (see **Figure 8.5**).

a) Hanging bracket with DWV pipe clip head

b) Hanging bracket with powder coated split clip head

c) Stand-off bracket with DWV pipe clip head

d) Stand-off bracket with powder coated split clip head

e) Double bolted clip and welded shank

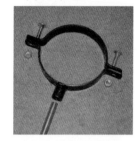

f) 10 mm all thread rod with welded nut double bolted clip head

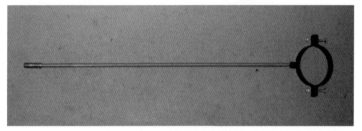

g) 10 mm all thread rod with welded nut double bolted clip head. End fitted with expanding metal masonry anchor.

h) Expanding metal masonry anchor

i) Hanging bracket with DWV pipe clip head

j) Powder coated saddle pipe clip

FIGURE 8.4 Examples of support brackets, clips, hangers and fixings

There are various methods for measuring the length of pipe required. A simple technique for measuring pipe is described as follows and is shown in **Figure 8.6**. Measure the length of pipe required from the centre of the vertical pipe at the outside of the building. The centre of drain entry point is found by using a plumb line that is suspended over the centre of the entry point.

When you have found the centre of the drain entry point, measure the pipe section required from the centre of the drain entry point to the depth of socket in the 'P' trap. Fit a bend to the end of the pipe, then mark and cut the pipe to suit (see **Figure 8.7**). One method of cutting PVC pipe square is by using a panel saw and mitre box.

When joining the pipes and fittings, it is important to prepare the pipe to ensure it can be pushed all the way into the pipe sockets. For this to happen, the end of the pipe must be cut square and any burrs removed. This will also remove projections that might cause blockages.

When solvent cement jointing PVC pipes and fittings, both spigot and socket must be prepared by applying priming fluid (red) to the areas to be within

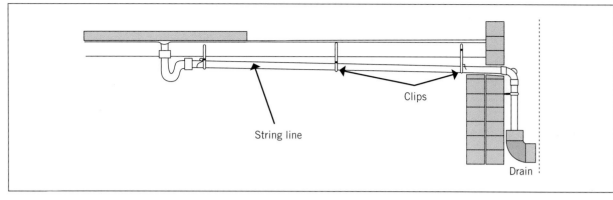

FIGURE 8.5 Clips fitted with fixture discharge pipe

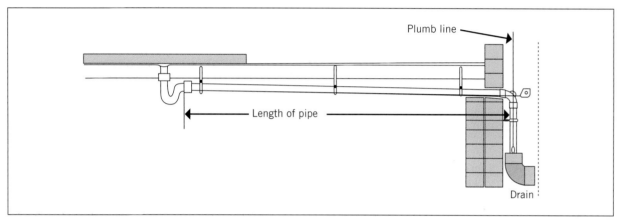

FIGURE 8.6 Finding the centre of the drain entry point using a vertical plumb line

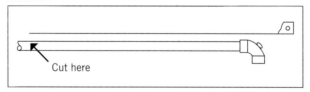

FIGURE 8.7 Measure and cut the pipe at the position marked

the joint. This should be done carefully so as not to spill and stain other pipe surfaces, particularly if the pipes will be in an exposed, unpainted location. Figure 8.8 shows a basin with a PVC fixture discharge pipe.

Once the priming fluid has been applied, the solvent cement is applied in the same manner.

Priming fluid and solvent cement for joining PVC pipe and fittings contain methyl ethyl ketone and acetone. They are flammable chemicals. Avoid skin contact and inhalation. Read the worksite material safety and data sheet (MSDS), and use appropriate safe work practices.

FIGURE 8.8 Basin at the rear of a shop, with DN 40 PVC fixture discharge pipe to DN 65 PVC discharge pipe

Overcoming problems

Occasionally, problems arise with obstacles within the building construction.

Figure 8.9 illustrates an example problem encountered while setting out to install a discharge pipe to a first-floor 'S' trap closet pan. A support beam is located through the centre of the floor. A solution is to install a 'P' trap closet pan, but in this situation there is another obstacle – a timber stud. It will be necessary

FIGURE 8.9 Overcoming problems. An 'S' trap closet pan cannot be installed due to a support beam beneath the centre of the first floor; therefore, a 'P' trap closet pan will be installed after the timber stud is relocated.

to arrange for the builder to relocate the timber stud to allow the discharge pipe to connect to the 'P' trap closet pan.

Fabrication and assembly of stacks

Setting out

Stacks and discharge pipes can be installed in exposed positions when fitting additional fixtures. However, stacks and discharge pipes in new buildings are usually installed during the rough-in stage within internal pipe ducts or wall cavities that are part of the building design.

After you have planned the layout for the sanitary plumbing system, and determined the position of the discharge pipes, stack and stack branches, you can then fabricate, assemble and install the stack (see Figure 8.10).

The stack should be positioned clear of windows, doors or other openings, and should be located so that the fixture discharge pipes do not exceed the maximum length specified in AS/NZS 3500.2 for the type of plumbing system that you are installing. If the maximum length of fixture discharge pipe is exceeded, then it must be fitted with a vent.

AS/NZS 3500.2 SANITARY PLUMBING AND DRAINAGE

When installing stacks and elevated sanitary plumbing, it is often necessary to locate the piping within the Posi-Strut system (see Figures 8.11 and 8.12). *Never drill, cut or modify the Posi-Strut system in any way, as this might weaken the system.* As in the example mentioned above, problems do occasionally arise, and if you have difficulty achieving the necessary pipe gradient, or in complying with other regulations

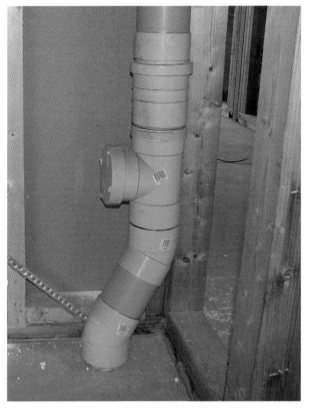

FIGURE 8.10 Base of a DN 100 PVC stack with testing opening and expansion joint. *Note:* Expansion joint not yet clipped.

FIGURE 8.11 Sanitary piping installed within the Posi-Strut truss system

and standards, then you must discuss this with the project supervisor or local authority to arrive at a solution that does comply.

Other set-out requirements for stacks are similar to those mentioned earlier for discharge pipes.

Fixings and supports

The requirements for fixing and support are similar to those mentioned earlier for discharge pipes. Brackets and clips must be set at the gradients and maximum

FIGURE 8.12 Sanitary piping being installed within the Posi-Strut truss system – allowing space for gradient and connection of fixture discharge pipes

FIGURE 8.13 DN 50 PVC vent pipe installed through nogging, top plate and tile roof

support spacings specified in AS/NZS 3500.2. They should be fixed with screws into floor joists or studs.

AS/NZS 3500.2 SANITARY PLUMBING AND DRAINAGE

When fixing to brickwork, bracket shanks can be inserted into brickwork mortar joints and fixed by plugging. Alternatively, stand-off brackets or 10 mm all-thread rod can be fixed to brickwork and masonry using expanding metal anchors.

Fabrication, installation, and jointing pipes and fittings

The fabrication, installation and jointing techniques are similar to those mentioned earlier for discharge pipes. Figures 8.13 and 8.14 depict PVC pipes and fittings assembled and installed prior to concealment.

If the project specification is for copper or copper alloy (brass), then pipework is partially prefabricated in sections off-site from plans and specifications, and then installed by the plumber on-site (see Figure 8.15). PVC and cast-iron pipework is usually fabricated and installed on-site.

Working with different materials and pipe systems

Occasionally, project specifications require materials that are different from PVC, and in these cases you

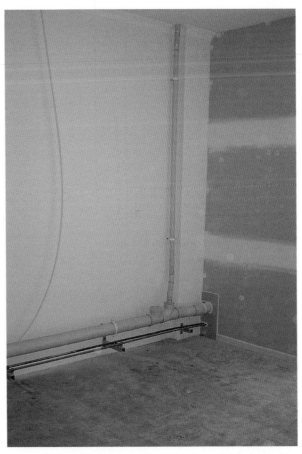

FIGURE 8.14 DN 100 PVC discharge pipe and DN 50 PVC vent installed in the rear of a shop. The piping will be concealed.

FIGURE 8.15 Installing copper alloy (brass) piping

FIGURE 8.16 Working with different pipe systems: the Blucher 'stainless-steel push fit' drainage system

FIGURE 8.17 Combining different pipe systems and materials: the Blucher 'stainless-steel push fit' drainage system floor-waste trap combined with copper alloy (brass) riser and fixture discharge pipes

will need to research the installation requirements (see Figures 8.16 and 8.17). Some aspects that you may need to be aware of are:

- different material characteristics, as mentioned in Unit 5
- different requirements in providing for expansion and contraction
- different pipe support and clipping distances
- methods of adapting and joining to other pipe systems
- compatibility with other materials.

Working with acoustic pipe systems

Some projects require sanitary plumbing materials other than PVC so that the sound of waste travelling through the pipes is minimised – for example, Rehau Raupiano Plus®, which is a push fit polypropylene (PP) pipe and fitting system. The pipe support and clipping requirements must comply with the manufacturer's recommendations, to ensure the system complies with the acoustic performance of the product.

Sound-dampening brackets with vibration-decoupling system for vertical stacks, and fixed and expansion brackets for horizontal lines, must be rubber-lined to reduce noise transmitted from the sanitary plumbing (see Figure 8.18).

FIGURE 8.18 Rehau Raupiano Plus® sanitary plumbing

Source: Rehau/Shane Ross

While PVC can be lagged to reduce sound transmission, it could then take longer to install. Therefore, there may be time-saving benefits in installing a push-fit acoustic sanitary plumbing system versus installing a lagged PVC sanitary plumbing system. Also, the push-fit system will require less space. Watch the time-lapse video, which compares the installation of a Rehau Raupiano Plus® system, to a conventional lagged PVC system: https://www.youtube.com/watch?v=E5s2Cz3ZSWE.

Requirements for fire collars

Passive fire protection (PFP) is an integral component of structural fire protection and fire safety in a building. The purpose of PFP is to attempt to contain fires or slow the spread of fire, through use of fire-resistant walls, floors and doors.

When installing plastic pipes (e.g. PVC) through walls and floors in buildings that require spaces to have PFP between different compartments, such as multi-level apartment buildings, it is necessary to install fire collars to prevent fire from passing from one area to another.

Fire protection of service penetrations through fire-rated elements is a requirement of the Building Code of Australia (BCA). The protection method at the penetration must comply with AS 4072.1 and AS 1530.4 and have achieved the required fire resistance level (FRL).

Compliance with the BCA is not just about fire collars. All the components listed must be part of a tested system.

The parts of the system include the:
- size and material of plastic pipe used
- type, size, construction and FRL of building substrate that the pipe passes through and to which the collar is affixed
- fire collar that has been installed
- orientation of the pipe and collar
- fixing method used to apply the collar to the building substrate.

For example, a test for a DN 100 PVC pipe through a two-hour fire-rated plasterboard wall (2 x 16 mm each side of 70 mm stud, total wall thickness of 134 mm) can only be applied to an installation that uses the same wall, pipe size and collar. If the wall itself is thinner, or there is a different pipe material and size, then the protection requirements may differ.

In summary, compliance with fire-rating requirements is not a simple matter of installing a fire collar on a pipe. If you are unsure, consult with a fire protection engineer and/or supplier of passive fire protection systems.

Testing

After the sanitary plumbing system has been installed, it should be tested to ensure it complies with AS/NZS 3500.2. This testing may need to be completed in the presence of a plumbing inspector. It is best practice to test the system before it is enclosed in an inaccessible position, such as a duct or wall cavity.

The traditional method of testing is the hydrostatic (water) test. However, water conservation is driving a change towards air testing, and we will discuss both methods for testing PVC sanitary plumbing systems. If you are unsure, ask the local regulatory authority which method is required.

Hydrostatic testing

Hydrostatic testing involves filling the entire pipe system, or sections of the system, with water to the flood level.

Procedure

- An expandable testing plug is fitted into the downstream end of the system, or the section of the system to be tested through an inspection opening, and tightened to expand the plug and seal it against the walls of the pipe. This prevents the passage of water.
- All other open ends of the system to be tested should also be plugged.
- A large piece of timber should be placed behind the test plug and tied to ensure that if the plug does come loose, the timber will prevent the plug slipping away down the pipe.
- The system is filled with water to flood level.
- The head pressure of water on the lowest pipe must not exceed 3 m.
- The test will pass if there is no loss of water over a period of 15 minutes. (Read AS/NZS 3500.2.)

Air test

One limitation on the hydrostatic test is that the maximum allowable head pressure on any section of pipe is 3 m. This limitation does not apply to air testing. Therefore, aside from water conservation, a benefit of air testing is that long, vertical sections of sanitary plumbing can be tested.

An air test may be applied to the completed work either in its entirety or in sections.

Procedure

- An expandable testing plug is fitted into an inspection opening and tightened to expand the plug and seal it against the walls of the pipe. This prevents the passage of air.
- All other open ends of the system to be air tested should also be plugged and made airtight. For safety, tie all the plugs.
- An air pump and connection incorporating a pressure gauge is attached to the testing plug. Air is then pumped into the pipe system, or section of the system, to achieve a test pressure of 15 kPa. (*Note:*

Some regulatory authorities may specify a different test pressure.)

■ Normally, at least three minutes is allowed for the temperature in the system to stabilise.

■ For most installations where DN 100 piping is less than 120 m long, the test period will be two minutes after the temperature has stabilised, and during the test, the pressure should not drop more than 3 kPa. The test acceptance criteria is specified in AS/NZS 3500.2.

■ To find any leaks, use soapy water.

AS/NZS 3500.2 SANITARY PLUMBING AND DRAINAGE

See the important safety note in the 'Caution' box.

Important safety note: Compressed air has the energy potential to cause serious injury. There have been cases of injury due to the sudden, unexpected release of testing plugs under pressure. Also, when releasing air, there could be particles in the air that can cause eye injury. Therefore, when air testing, always securely fix and tie testing plugs and wear safety glasses.

Background

Air (unlike water) is compressible and, as a result, much more energy is required to raise its pressure. In fact, at the pressure ranges normally used for testing pipe systems, 200 times more energy is stored in compressed air as compared to water at the same pressure and volume.

Should a joint, pipe or any other component fail under test pressure when using compressed air, the energy can be released with explosive and deadly force. This method of pressure testing is dangerous. Any persons in the vicinity of such pressure testing are placed at risk.

Preventative measures

Due to the hazards and risks associated with high-pressure air testing, wherever practicable the pressure testing of pipelines, whether they are for water or sewerage, should be undertaken using low pressure.

Employers, including principal contractors and sub-contractors, must ensure that, *before* commencing pressure testing of water or sewer pipes, those persons undertaking the testing:

■ receive appropriate training in pressure testing

■ are instructed in the appropriate test procedures issued by the regulating authority

■ fully understand the specifications and procedures of the relevant regulating authority and the need to comply with them at all times.

Never pressure test any sanitary plumbing or drainage system above 50 kPa.

If there are exceptional circumstances that necessitate the use of compressed air to a pressure of greater than 50 kPa, protective measures are required to be implemented and must be inspected and certified by a competent person, such as a qualified engineer. These protective measures may include using a designated test zone, an exclusion zone, appropriate procedures, equipment, materials and pipe end supports.

Operation of sanitary fixtures

After the sanitary plumbing and fixtures have been installed, all fixtures should be tested by operating them normally. There should be no leaks, and the residual water seal in any fixture trap must be no less than 25 mm. If any water seal is less than 25 mm, the situation must be rectified. Unit 5 provides information on the six main causes of trap seal loss and solutions to this problem.

Clean-up

Work area

After installing the sanitary plumbing system and reinstating surfaces, it will be necessary to clean up the work area. Clean-up includes:

■ removal and storage of usable off-cuts

■ removal, disposal and recycling of waste, and unusable off-cuts

■ restoration of any soiled building surfaces to a clean condition

■ restoration of the worksite, where possible, to its original condition

■ disposal of rubbish and used containers into recycling bins or trash.

Tools and equipment

The procedure for cleaning tools and equipment could include the following:

■ removing power tools and equipment and carrying out service work to maintain it in good condition

■ inspection, cleaning, maintenance and storage of tools

■ disposal of unserviceable tools

■ disassembling and storing of scaffolding

■ maintaining and returning hire equipment in accordance with the hire agreement.

Documentation

This section is relevant to documentation required after the work to install sanitary plumbing is complete.

Generally, sewerage and regulatory authorities do not require 'as-constructed' plans of sanitary plumbing; however, requirements vary around Australia. Therefore, you will need to research your local authority's requirements.

For large projects, it is a common requirement for plumbers to submit 'as-constructed' plans to the builder or client on completion. This may be required in computer-drafted format.

Sewerage or regulatory authorities in many areas often require notifications when the project is complete to carry out an inspection or audit.

In some areas, plumbers are required to lodge a Certificate of Compliance within a certain period of time after the job is complete.

COMPLETE WORKSHEETS 3 AND 4

SUMMARY

- When installing sanitary plumbing, it is necessary to follow the plans and specifications that specify the location of fixtures, dimensions of fixtures, and floor and wall finishes. Also, familiarise yourself with local regulations and inspection requirements, and ensure that the pipes, fittings and fixtures have WaterMark as required.

- Installation also includes the use of serviceable tools and understanding the operation of special equipment that might be required for specific materials or pipe systems. Most fixtures, pipes and fittings must be approved for installation by the relevant regulatory authorities. The pipework must be installed through the building in such a way that it will not compromise the strength of the building structure.

- The pipes and fittings will need to be installed with correct gradient, orientation and ventilation to achieve a sanitary plumbing system that is free draining and not prone to blockages or siphonage of fixture traps.

- When fabricating, assembling and installing pipes, make provisions for expansion and adequate clipping and support. PVC pipes should be joined with solvent and cement, considering WHS and MSDS requirements in handling toxic and flammable substances. Either during or after installation, the sanitary plumbing should be air tested or hydrostatically tested. Plumbing fixtures should be tested by operating them normally.

- After installing the sanitary plumbing system and reinstating surfaces, it will be necessary to clean up the work area and tools, and to complete any relevant documentation.

WORKSHEET 1

Student name: _____

Enrolment year: _____

Class code: _____

Competency name/Number: _____

Task: Review the section 'Preparing for work', then complete the following.

1   Identify three possible locations for stacks and discharge pipes.

    a   _____

    b   _____

    c   _____

2   Complete the following sentence.
Fixtures traps must retain a water seal of at least _____.

3   Complete the following sentence.
All pipes and fittings used in a sanitary plumbing system must comply with
_____ and bear _____ marking.

4   Research your local regulations and state what inspections are required for sanitary plumbing.

    _____

    _____

    _____

    _____

5   What information should be obtained from a site meeting?

    _____

    _____

WORKSHEET 2

Student name: _____

Enrolment year: _____

Class code: _____

Competency name/Number: _____

Task: Review the sections 'Tools and equipment' and 'Installation requirements', then complete the following.

1   Why is it necessary to have training in the assembly or use of access towers and scaffolding?

_____

_____

_____

2   Imagine that you are required to install a 6 m length of DN 50 PVC pipe on grade. Research Unit 5 and AS/NZS 3500.2, and then complete the following:

a   The maximum distance between supports is:

_____

b   The minimum gradient is:

_____

c   If one end is fixed, what is the minimum length of unclipped offset leg required for expansion?

_____

Draw an isometric sketch and label your answers to the above.

3   When installing a closet pan, what must you allow for when positioning the discharge pipe?

_____

_____

4 Research AS/NZS 3500.2. What is the distance of a restricted zone above the base of the stack that extends only two floor levels?

_____

_____

5 Research AS/NZS 3500.2 and draw an elevation of stack branches for the following:

a Restricted entry zones for a DN 65 discharge pipe connecting to a DN 100 stack using a sweep junction.

b Restricted entry zones for a DN 80 discharge pipe connecting to a DN 100 stack using a square junction.

6 Research AS/NZS 3500.2. What is the minimum horizontal distance between a vent terminal and an openable window?

_____

7 Research AS/NZS 3500.2, and then draw an isometric sketch of a basin discharge pipe, including a trap connected to the plumbing system, 'elevated pipework using drainage principles'. Label the maximum length of discharge pipe, and state the maximum number of bends in both the vertical and horizontal planes.

 **WORKSHEET 3**

Student name: _____

Enrolment year: _____

Class code: _____

Competency name/Number: _____

Task: Review the section 'Fabrication, assembly and testing', then complete the following.

1   What is the maximum spacing of pipe supports for a DN 50 PVC waste pipe?

_____

_____

2   Research and calculate the minimum fall that is required for the following:

   a   A DN 100 discharge pipe on grade that is 5.0 m long.

_____

_____

   b   A DN 40 discharge pipe on grade that is 4 m long.

_____

_____

3   What is one method of making a square cut through PVC pipe?

_____

_____

_____

4   Why is it necessary to remove burrs from PVC pipe ends before jointing? Give two reasons.

   a   _____

   b   _____

5   If the maximum length of a fixture discharge pipe is exceeded, what must be fitted?

_____

_____

6   Why must you never drill, cut or modify a structural support?

_____

_____

7   How would you attach a pipe bracket to a timber structural support?

_____

_____

8   If you have problems achieving the minimum regulation pipe gradient, how would you solve the problem?

_____

_____

_____

9   What are two methods of attaching pipe brackets to brickwork?

a   _____

b   _____

10 When installing a push-fit acoustic system of sanitary plumbing, what are two benefits it has compared with PVC?

a   _____

b   _____

11 Research your local requirements and find out what testing is required of sanitary plumbing.

a   Is testing done by plumbers?

_____

b   Is hydrostatic or air testing done?

_____

c   Does your authority witness testing?

_____

 **WORKSHEET 4**

Student name: _____

Enrolment year: _____

Class code: _____

Competency name/Number: _____

Task: Practical exercises

All questions and exercises in the previous three worksheets must be completed and checked by your trainer/teacher before you attempt the following practical exercises.

Your trainer/teacher should conduct an induction and ask underpinning knowledge questions before you attempt the practical exercises.

1   Connect one closet pan and one basin on a ground floor direct to a below-ground unventilated branch drain using PVC pipe and fittings. The basin discharge pipe must include a section of copper pipe.

2   Connect one closet pan and one basin on a first floor to elevated pipework using drainage principles and PVC pipe and fittings. The basin discharge pipe must include a section of copper pipe. An expansion joint must be fitted within the elevated pipework.

3   Fabricate and install fixture discharge pipes from a bath and closet pan to connect to a below-ground drain incorporating an overflow relief gully (ORG).

4   Fabricate and install a sanitary stack in accordance with an approved system of plumbing to connect fixtures from two floors of a building. On each floor, discharge pipes must be provided to connect to a closet pan, bath, basin, shower and floor waste gully. In this exercise, you must fabricate at least two DN 50 branches in copper tube. As part of the fabrication, the stack may incorporate copper, polymer, cast iron or other approved materials.

Your trainer/teacher may choose to combine the above exercises; however, all these exercises are critical for assessment. It is recommended that, upon completion of one of the above exercises, the installation should be either hydrostatic or air tested.

9

INSTALLING AND FITTING OFF SANITARY FIXTURES

Learning objectives

This unit provides information needed to install and fit off sanitary fixtures. Areas addressed in this unit include:

- an explanation of Australian product compliance requirements
- ordering materials
- detailed notes on installing:
  - basins
  - baths
  - bidets and bidettes
  - clothes-washing machines
  - dishwashing machines
  - shower bases
  - sinks
  - spa baths
  - troughs
  - wall-hung urinals
  - water closets.

Introduction

When sanitary fixtures and appliances are used or operated, there is a discharge to the sanitary plumbing system. Plumbers are required to install sanitary fixtures and appliances correctly for the protection of public health and safety.

The work for installation of sanitary fixtures and appliances can involve one or more of the following:

■ secure fixing and support
■ connection of fixture discharge pipes
■ installation of water supply tapware
■ connection of water from flushing valves or cisterns (except waterless fixtures)
■ connection of small bore macerators (for more information on macerators, see Unit 12).

This type of work is carried out at all properties where there are sewage and waste discharges. It could be new work, or may involve alterations or additions to existing systems.

In this book, Units 5 and 8 describe planning the layout for, and installation of, sanitary plumbing systems, which is incidental to the installation and fit off of sanitary fixtures.

While studying this subject, you will be required to respond to questions and complete exercises referring to AS/NZS 3500.2 Sanitary plumbing and drainage.

AS/NZS 3500.2 SANITARY PLUMBING AND DRAINAGE

Before studying this subject, you should have completed training in work health and safety (WHS) as outlined in Chapter 3 of *Basic Plumbing Services Skills*.

Preparing for work

When preparing for work, it is important to understand two terms commonly used in the plumbing trade: *rough in* and *fit off*. Both terms are used in this unit to describe plumbing work.

■ Rough-in work may involve the installation of above-ground sanitary plumbing, drainage and water pipes within wall, ceiling and floor cavities, and must be carried out before the wall sheeting and finishes are installed. At this stage of work, plumbers must be aware of the fixture and appliance connection requirements, allowing for the thickness of wall sheeting, and wall and floor finishes.

■ Fit-off work involves installation of fixtures, connection of discharge pipes, and installation and connection of tapware to complete the fixture installation. This work is done after wall and floor finishes have been installed.

In this unit, the focus is on the installation of fixtures and connection of discharge pipes. The installation of tapware is covered in *Basic Plumbing Services Skills – Water Supply*.

Plans and specifications

Generally, all plans and specifications must be obtained when planning the layout of sanitary drainage and plumbing to connect sanitary fixtures and appliances, as discussed in Units 4 and 5. If the specifications for fixtures and/or appliances change after the rough in has been completed, it may be necessary to alter the position of discharge pipes, water supply pipes and other connections. This can be costly and time-consuming, especially if walls and finishes have been completed.

In Unit 8, the section titled 'Overcoming problems' outlines an example of an 'S' trap closet pan being changed to a 'P' trap closet pan due to building constraints. The same problems can occur when the owner or builder changes the fixture specification after walls and finishes have been completed. In these situations, as a professional plumber, you will need to consider all options to minimise the need to disrupt walls, floors and finishes.

For example, imagine that the property owner asks for two basins in a bathroom where you have installed a discharge pipe for only one basin. After reading Unit 5 and AS/NZS 3500.2, you will be aware that you have the option of installing two fixtures to a single trap as 'fixture pairs', thus providing part of the solution where there is no need to install another discharge pipe.

AS/NZS 3500.2 SANITARY PLUMBING AND DRAINAGE

Plans and specifications are necessary for determining the position of fixtures and fixture discharge pipes. The following is a checklist of information required for the installation and fit off of sanitary fixtures:

■ plans and specifications that identify the brand, type and colour of fixtures
■ plans and specifications that identify or reveal options for the position of discharge pipes, stacks and vents
■ manufacturers' installation instructions and specifications that identify fixture and appliance support and fixing methods
■ plans and specifications that identify other services and obstructions
■ applications and notifications to the network utility operator and/or plumbing regulatory authority (e.g. permits and/or work notices)

- plumbing regulatory and Australian Standards requirements that specify how fixtures and appliances are identified as approved for installation
- WHS regulations
- plumbing regulatory requirements for inspection and testing.

Safety and environmental requirements

Generally, when installing and fitting off sanitary fixtures and appliances, the safety and environmental considerations are related to working inside with power tools and in awkward positions.

Each state and territory has WHS regulations that require employers and self-employed people to identify hazards and assess and control risks at the workplace in consultation with their workers.

Some fixtures and appliances present unique problems, and while these will be related to the different experiences of installers, the following are some examples:

- Do not over-tighten screws and nuts on vitreous china fixtures, as these fixtures may crack into very sharp fragments.
- Be aware of sharp edges on stainless-steel fixtures.
- Some fixtures have spring connections.

 When installing and fitting off sanitary fixtures, measures must be taken to avoid injury, such as wearing gloves and safety glasses, particularly when using power tools.

Quality assurance

Products

The regulations for plumbing and drainage products vary from state/territory to state/territory.

All pipes and fittings used in a sanitary plumbing system must comply with the *National Construction Code (NCC) Volume 3, Plumbing Code of Australia* and bear WaterMark and Australian Standards certification.

As described in Unit 5, WaterMark is a product certification mark that confirms a plumbing product complies with the requirements of the *Plumbing Code of Australia* and the specifications listed in relevant Australian Standards. WaterMark relates to product quality, aspects of health and safety, and warrants that the product *can be legally installed*. The mark, resembling a 'W', is used on sanitary apparatus, and plumbing and drainage fixtures. The WaterMark logo, the relevant product standard and the licence number must be marked on the product. WaterMark is *not* required for a product to be *legally sold*.

While most sanitary fixtures must bear WaterMark, some are excluded. You should familiarise yourself with two documents that are available on the Australian Building Codes Board (ABCB) website (https://www.abcb.gov.au):

- WaterMark Certification Scheme Schedule of Products
- WaterMark Certification Scheme Schedule of Excluded Products.

Quality system mark, which is a mark resembling five ticks, is sometimes used on sanitary apparatus, plumbing and drainage fixtures.

Another mark is WELS (Water Efficiency Labelling Scheme), which relates only to the water efficiency of products and appliances. WELS products *must* carry a WELS label in order to be *legally sold*, but some of these products may not have a WaterMark.

Before purchasing a WELS-labelled product that does not carry a WaterMark, you should ask your regulatory authority or supervisor if it can be legally installed.

Figure 9.1 shows examples of WELS marks. They display the number of stars and the water consumption or flow. WELS marks are used on showers, taps, toilet and urinal equipment, clothes-washing machines and dishwashers.

GREEN TIP

According to AS/NZS 3500, Parts 1 and 4, the maximum permitted flow rate of cold or heated water from an outlet at a shower, basin, kitchen sink or laundry trough is 9 litres per minute. This does not apply to baths.

Installation work

It is important to be aware of the project requirements for quality assurance. Most regulatory authorities require inspections and tests.

When preparing for your work, you must have planned for the complete installation of sanitary fixtures, and have the necessary reference numbers and phone numbers to book inspections and tests with the regulatory authority (where required).

In some areas, final inspection of the sanitary plumbing is not required by regulatory authorities. However, if you are unfamiliar with local requirements, or are moving to a different area, you must research the local regulations and standards.

In some areas, the regulatory authority will require that you book the sanitary plumbing for final inspection with at least 24 hours' notice, and that you arrange for access to the building. Inspections are carried out by regulatory authorities to ensure that the sanitary plumbing has been installed correctly in accordance with local regulations and AS/NZS 3500.2. In some areas, the regulatory authority will require an inspection of all, or a percentage, of completed plumbing work, and this is a form of quality assurance of your work for the sewerage authority and property owner.

<na...

Lavatory 4-star label – pan

Lavatory 4-star label – cistern

FIGURE 9.1 Caroma and Fowler Smartflush® toilet suites have a minimum 4-star WELS rating

Source: © Department of Agriculture, Water and the Environment 2020. Licensed under a Creative Commons (CC) Attribution 4.0 International licence

 **AS/NZS 3500.2 SANITARY PLUMBING AND DRAINAGE**

On larger building sites throughout Australia, additional requirements for quality assurance of plumbing work are common, such as sign-off on portions of final plumbing work for progress payment, and additional tests and inspections (see Table 9.1).

TABLE 9.1 Examples of tests and inspections

Examples of tests	Examples of inspections
Hydrostatic or air testing	Correct fixture (e.g. for people with disability)
Testing of sanitary fixtures by subjecting them to normal use. Fixtures traps must retain a water seal of at least 25 mm	Fixture support and fixings
	Approved fittings (e.g. removable grates where traps are not accessible)
	Fixture gradient (shower bases and drainage channels)

Later in this unit, we will discuss the specific procedures for final inspection and testing of sanitary plumbing.

Some of the issues that should be considered when preparing to install sanitary plumbing fixtures are as follows:

■ conformance with company operating procedures
■ ensuring necessary applications are lodged with authorities
■ arranging for all necessary final inspections and tests
■ ordering fixtures, appliances, materials and products in accordance with job plans and specifications
■ receiving and inspecting fixtures, appliances, materials and products for conformance with orders
■ storage of fixtures, appliances, materials and products to prevent damage or deterioration
■ selection, operation and maintenance of tools and equipment
■ compliance with contract work schedules
■ maintaining records of job variations.

LEARNING TASK 9.1

Undertake research on the internet to find the ABCB:
• WaterMark Certification Scheme Schedule of Products
• WaterMark Certification Scheme Schedule of Excluded Products.
1 What is the Australian Standard for cisterns?
2 Are urinals required to have WaterMark?
3 Are baths required to have WaterMark?

Sequencing of tasks

New buildings

When installing sanitary plumbing fixtures, you must consider the order in which the work will be completed. In new installations, the plumber generally does the work in two stages – the rough in and the fit off.

In renovations and additions, greater flexibility is sometimes required because there may be a need to minimise disruption to existing walls, floors and finishes.

Renovations and additions

It is often more difficult to install a sanitary fixture, such as a basin or bath, in an existing building than in a new building under construction. Problems in existing buildings can include:

- installing noggings for new fixture supports
- gaining access to lugged (back plate) elbows and tees in existing walls
- gaining access to piping in existing walls and floors
- existing piping in poor condition
- restoration of tiled surfaces and other wall cladding and surfaces to their original condition.

Water connection

Generally, sanitary fixtures need a water supply. Some fixtures, such as water closets, urinals or dishwashers (with a built-in water heater), require only cold water. Other sanitary fixtures, such as showers, baths or hand basins, need to have both heated and cold-water connections. Some fixtures may need both drinking water and non-drinking water supply connections.

The type of water connection required for each sanitary fixture will be described later in this unit in the relevant section for each fixture.

Discharge pipe connection

Discharge pipes must be connected to all sanitary fixtures mentioned in this unit. The connections must meet the requirements of AS/NZS 3500.2 and local regulations.

AS/NZS 3500.2 SANITARY PLUMBING AND DRAINAGE

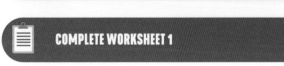

COMPLETE WORKSHEET 1

Tools and equipment

Safety equipment

An important safety consideration when installing sanitary fixtures is the correct selection and use of safety equipment. Types of personal protective equipment (PPE) that are used when installing sanitary fixtures and appliances include:

- fire-fighting equipment
- overalls
- boots
- safety glasses or goggles
- ear plugs or ear muffs
- gloves
- dust masks or respirators.

Tools

Tools and equipment typically used for installation of sanitary fixtures and appliances include:

- ruler
- measuring tape
- hacksaw with a range of blades
- wood saw
- mitre box
- cordless drill with a range of bit tools for self-drilling screws
- cordless multi-function tool with blades for different cutting applications
- power drill with a range of drill bits and hole saws
- range of screwdrivers
- shifting spanners
- basin wrench
- tap sockets
- copper tube bending tools
- silicone sealant cartridge gun
- range of round and flat files
- string line
- plumb-bob
- spirit level
- plugging chisel
- cold chisel
- claw hammer.

When moving and installing sanitary fixtures and appliances, you might be required to use lifting or load-shifting equipment such as:

- hand trolleys
- rollers
- forklifts
- chain blocks
- hoists
- jacks.

Special tools and equipment

Special equipment is required to install some sanitary fixtures and appliances. Special equipment must be used in accordance with manufacturers' installation instructions. Examples include:

- special tools for servicing and disassembly of cistern button modules
- special tools for installing and commissioning appliances.

Some manufacturers supply special tools and equipment, including fixings, and if these are not used for the installation, the product warranty may be void.

Installation requirements

Fixtures and materials

Sanitary fixtures and appliances that you may be required to install or fit off include:

- basin
- bath or spa bath
- bidet or bidette
- clothes-washing machine
- dishwashing machine
- shower
- sink
- trough
- urinal
- water closet pan.

The abbreviations commonly used for these fixtures are listed in AS/NZS 3500.2.

AS/NZS 3500.2 SANITARY PLUMBING AND DRAINAGE

Discharge pipe materials used to connect sanitary fixtures include:

- cast iron
- copper
- copper alloy (brass)
- high-density polyethylene (HDPE)
- polyvinyl chloride (PVC)
- polypropylene (PP)
- stainless steel.

Other approved materials are listed in AS/NZS 3500.2.

AS/NZS 3500.2 SANITARY PLUMBING AND DRAINAGE

Note: All materials and sealants must be compatible. For example, linseed and other oil-based putty must not be used with plastics.

Determining the position of sanitary fixtures

The position of sanitary fixtures is determined in accordance with the project plans, specifications and site requirements, as discussed in detail in Units 5 and 8.

Allowances and ordering materials

Material quantities

At the end of Unit 5, material quantities needed to construct the sanitary plumbing system were obtained from an example project and recorded in a materials list (see 'Determining material quantities'). The list of materials for the sanitary plumbing system should include all necessary materials to complete the installation and connect the fixtures and appliances. The materials are usually purchased by the plumbing contractor.

Depending upon the contract, the plumbing fixtures and appliances might be purchased by the plumber, builder or owner. Plumbing fixtures and appliances are usually supplied with the necessary fixings to complete the installation, but, ultimately, the plumber must be aware of any additional materials that may be required.

Plumbing fixtures and appliances must be approved for installation, bearing the relevant standards marks mentioned earlier in this unit. The quantity, make, model and colour of fixtures and appliances can usually be taken from the project plans and specifications. The *complete* list of materials required is sometimes referred to as the material take off list (MTOL).

Note regarding materials: Water supply pipes, fittings and tapware are mentioned in this unit, but the materials, standards and procedures for installation are outside the scope of this book.

Ordering materials

When ordering and purchasing fixtures and appliances, it is good practice to plan ahead for secure storage to prevent them from being damaged or stolen. After you have confirmed that the fixtures and appliances are approved and comply with standards and local regulations, they can be ordered from a plumbing supplier by make, model, configuration, colour and so on, as detailed in your MTOL.

Identifying materials

Sanitary fixtures and appliances can be identified by product packaging and standards markings. Products with a WaterMark are usually also marked with manufacturer warranty details. Materials and products with a certification mark, but without a warranty, are not authorised products.

Collecting and checking the delivery

When receiving the delivery of fixtures and appliances, you should count the number and type of items and compare them to your MTOL. Ensure that you have identified all of the items as correct and undamaged before accepting the order. If you have the wrong model or colour fixture/appliance, it may take some time to exchange and order the correct item.

 COMPLETE WORKSHEET 2

Installing and fitting off sanitary fixtures

Basins

Basin types

There are four types of basins available:

1 wall-hung basins
2 bench top or vanity basins
3 semi-recessed basins
4 pedestal basins.

Basins are manufactured from:

- vitreous china
- stone composite
- pressed steel with vitreous enamel finish
- stainless steel
- plastic.

Securing and flashing the basin

All basins must be rigidly mounted so there is no strain placed on the water and discharge pipe connections. They must also be flashed in accordance with the regulations to prevent water damage to the building. Vanity-type basins must be sealed to the bench tops in which they are fitted.

Wall-hung basins

There are a number of ways for fitting wall-hung basins.

Fixing brackets

Wall-hung basins are designed to be supported by brackets. These brackets must be solidly connected to the building structure if they are to support the basin properly. Because there are so many different basin and bracket designs, it is critical that you obtain specific information about the basin you are going to install from the manufacturer before doing the rough-in work. Most manufacturers supply pamphlets and installation details for their products.

Figures 9.2, 9.3, 9.4, 9.5 and 9.6 show some examples of common brackets for wall-hung basins.

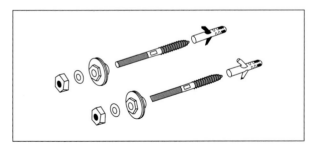

FIGURE 9.2 Caroma D.200 wall basin fixing kit

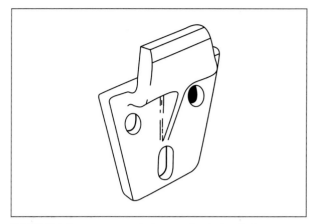

FIGURE 9.3 Caroma concealed corner basin bracket

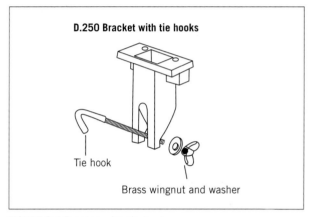

D.250 Bracket with tie hooks

Tie hook

Brass wingnut and washer

FIGURE 9.4 Bracket with tie hooks

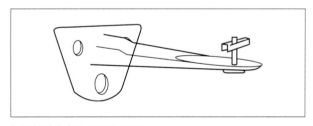

FIGURE 9.5 Side arm bracket

Height

A wall-hung basin is usually installed about 865 mm (depending on the type and style of basin) above the finished floor. This measurement is taken from the finished floor level to the front edge of the basin.

Fixing devices

When installing a wall-hung basin on a timber or metal stud wall, the brackets will need to be attached to a support nogging that is securely attached to the building structure. To find the height and width of support noggings, refer to the manufacturer's

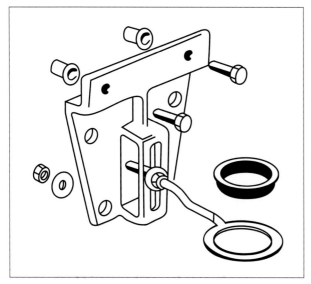

FIGURE 9.6 Concealed hanging bracket

Source: Department of Education and Training (http://www.training.gov.au) © 2013 Commonwealth of Australia.

information (see the example for Caroma Liano in Figure 9.7).

When fixing brackets to timber noggings, use either heavy-gauge wood screws or coach screws to prevent brackets from coming loose if someone leans heavily on the basin. *Do not use nails!*

When fixing brackets to either brick or concrete masonry walls, manufacturers recommend the use of good-quality expansion fasteners of a size to match the holes in the brackets (see Figure 9.8). The size and number of fasteners is listed for each type of bracket in the manufacturer's literature. If the heads of the fasteners will be visible after the installation of the basin, chrome plated heads should be used.

Flashing

The term 'flashing' in relation to sanitary fixtures refers to protection of the building and finishes from water spillage or splashing, which may occur when the fixture is used. If flashing is not provided around sanitary fixtures, it may result in permanent damage to plaster sheeting or the timber used to make the wall frame. Some wall-hung basins have an integral flashing up-stand that can be fitted to any type of wall. If the basin is designed without an integral flashing, the wall must be clad with an impervious (water-resistant) surface. The cladding must extend to at least 150 mm above and for the full width of the basin.

Bench top or vanity basin

Vanity basins can be moulded in one piece and fitted into an impervious bench top. Basins of this type with a 50 mm up-stand require no further flashing.

Vanity basins that are designed to be fixed in or under an impervious bench require flashing in accordance with AS 3740 Waterproofing of domestic areas and the building regulations. Manufacturers provide various methods of securing the basin in position. These usually require the installation of special metal hooks or brackets that attach to the basin and are screwed to the underside of the bench top. The manufacturer supplies these fixings with the basin. Refer to the manufacturer's installation information to ensure that the basin is correctly fixed (see the example of Caroma vitreous china counter basins in Figure 9.9).

AS 3740 WATERPROOFING OF DOMESTIC AREAS

Vanity basin rims must be sealed to the counter top with an appropriate silicone sealant. *Epoxy-type glues must not be used, as this may lead to cracking of the basin.*

Basins are often supplied with a template that must be used for the counter top cut-out.

Pedestal basins

Fixing

Pedestal basins are made in two sections: the basin itself and a concealment column that sits on the floor and is positioned beneath the basin (see Figure 9.10). The pedestal must *not* be used to support the basin.

Flashing

Pedestal basins are usually provided with a 50 mm integral flashing so that they may be placed against a wall. Some free-standing 'tabletop'-type basins must stand 75 mm clear of the wall to prevent water damage, and to provide access for cleaning.

Basin installation

When installing a basin, the support nogging in stud wall installations must be correctly located and firmly fixed to the building structure. Noggings can be made of timber or metal, and are usually 90 mm × 38 mm in size. They can be nailed or screwed into position. In most installations, the face of the nogging should be flush with the face of the stud. A separate recessed nogging is fixed to attach water supply back plate (lugged) elbows. They must be firmly screwed to the nogging and installed level and square to the wall surface.

Basins are available with none, one, two or three tap holes, depending on the type and style of taps required.

The basin outlet may be connected via:

- a direct connection to a sanitary drain
- a direct connection to a stack
- connection to an overflow/disconnector gully
- a floor waste gully.

Liano wall basin

Ceramic wall basin/Shroud
Basin size: 420 mm x 470 mm (nominal).
Basin capacity: 4.8 litres to overflow.

The **Liano** wall basin has been designed to meet the needs of the designer market for prestige projects. An architectural minimalist style was adopted for the square **Liano** incorporating a generous bowl area and distinctive tapware platform. The basin is suitable for domestic and commercial applications.

Tap holes: Basin available in no taphole and one taphole (below) options.

Shroud: (Optional) The shroud completely conceals P-trap waste pipe fittings. Use **Concealed Fixing Bracket** suitable and 40 mm Caroma plastic P-trap.

Note: It is essential for the installation of the shroud to the basin that clearance is provided for the shroud in the plumbing setout from water inlet pipes and connections.

Overflow: Overflow available only. Maximum tap flow rate of 15 litres/minute.

Fixing: D.200 Basin Fixing Kit (supplied) – bolts directly to finished wall.

Installation: Refer to Important Information for Plumbers section.

Colours: White only.

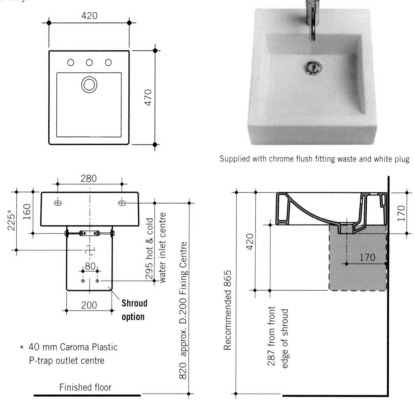

FIGURE 9.7 Caroma Liano basin with shroud option

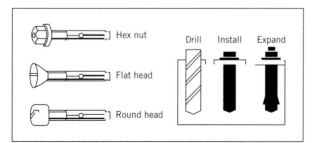

FIGURE 9.8 Suitable types of expansion fasteners

The basin connection method must meet the requirements of AS/NZS 3500.2 and relevant local regulations. For further detail, read Unit 8.

AS/NZS 3500.2 SANITARY PLUMBING AND DRAINAGE

When considering the options for waste pipe connection, the location of the fixture will be the main determining factor. Generally, the objective is to select the most economical method.

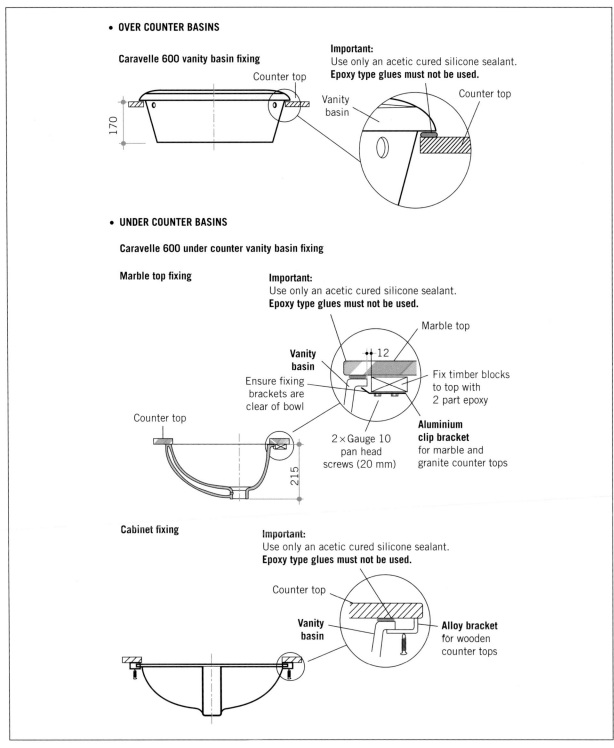

- **OVER COUNTER BASINS**

Caravelle 600 vanity basin fixing

Counter top

Important:
Use only an acetic cured silicone sealant.
Epoxy type glues must not be used.

Counter top

Vanity
basin

170

- **UNDER COUNTER BASINS**

Caravelle 600 under counter vanity basin fixing

Marble top fixing

Important:
Use only an acetic cured silicone sealant.
Epoxy type glues must not be used.

Marble top

**Vanity
basin**

12

Ensure fixing
brackets are
clear of bowl

Fix timber blocks
to top with
2 part epoxy

Counter top

2 × Gauge 10
pan head
screws (20 mm)

**Aluminium
clip bracket**
for marble and
granite counter tops

215

Cabinet fixing

Important:
Use only an acetic cured silicone sealant.
Epoxy type glues must not be used.

Counter top

**Vanity
basin**

Alloy bracket
for wooden
counter tops

FIGURE 9.9 Methods of fixing vitreous china counter basins

Source: Reproduced with permission from Caroma.

Because some basins are manufactured with rounded sides and bottoms, forming an almost funnel shape, the discharge of water can be very fast with a small trailing waste. This means that the basin trap is more likely than most other fixtures to lose its water seal by syphonage. For more information on trap seal syphonage solutions, read Unit 5.

Baths

Modern baths are produced in a wide range of materials, sizes, shapes and colours to suit their particular function (see **Figures 9.11** to **9.13**). They are usually designed to be built in against walls and fitted with a surrounding apron. The support system may be timber, metal or brick; they may also be suitable for island installation and can be free-standing.

FIGURE 9.10 Caravelle wall basin and pedestal

Source: Reproduced with permission from Caroma.

FIGURE 9.11 Caroma Shark bath (enamel steel)

Source: Reproduced with permission from Caroma.

FIGURE 9.12 Caroma Blanc bath

Source: Reproduced with permission from Caroma.

FIGURE 9.13 Caroma Mayfair shower bath (enamel steel)

Source: Reproduced with permission from Caroma.

Baths are made from stone composite, cast iron, pressed steel with a porcelain or enamel coating, or acrylic.

The shape and type of bath will normally have been determined, but the plug, tap fittings and connecting pipes must be appropriate to the needs of the client and the nature of the installation. Baths must be firmly secured into position using adequate support, and must be installed according to the manufacturer's installation instructions.

Installing a bath

To install a bath, knowledge of AS/NZS 3500.2 and the manufacturer's approved fixing methods is required.

AS/NZS 3500.2 SANITARY PLUMBING AND DRAINAGE

The bath may be placed into position at the 'rough in' stage (when the building is at the frame stage), or it may be left out until the floor, ceiling and walls have been clad. If installed at the 'rough in' stage, other trades must take care not to damage the bath during construction. It is the plumber's responsibility to ensure that the bath is adequately protected from damage during construction. One way of achieving this is to cover it with cardboard.

Usually the builder is responsible for installing (levelling, flashing and securing) the bath, but it is best to determine who will be responsible for the bath installation before work commences.

Bath fixing

The following installation instructions are taken from the *Caroma Technical Handbook* for the installation of a pressed steel bath and can be used as a guide for the installation of this type of bath (see also **Figure 9.14**). Before installing any bath, you must read and follow the bath manufacturer's installation instructions.

1 Protect the bath from accidental damage in handling and installation. Check the porcelain enamel surface carefully on removal of packaging materials, both prior to and again after installation.

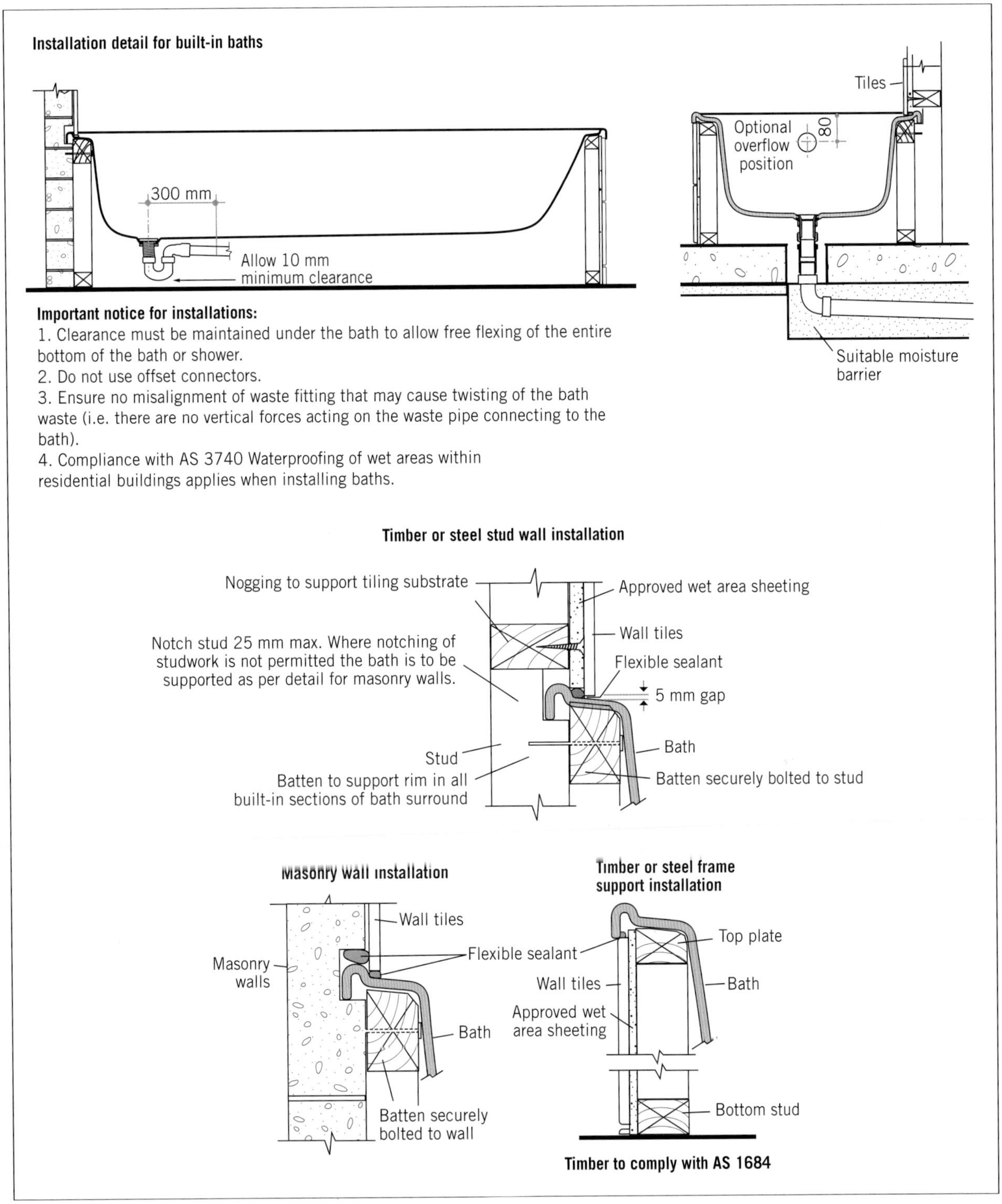

Installation detail for built-in baths

300 mm

Allow 10 mm minimum clearance

Tiles

Optional overflow position

80

Suitable moisture barrier

Important notice for installations:
1. Clearance must be maintained under the bath to allow free flexing of the entire bottom of the bath or shower.
2. Do not use offset connectors.
3. Ensure no misalignment of waste fitting that may cause twisting of the bath waste (i.e. there are no vertical forces acting on the waste pipe connecting to the bath).
4. Compliance with AS 3740 Waterproofing of wet areas within residential buildings applies when installing baths.

Timber or steel stud wall installation

Nogging to support tiling substrate

Notch stud 25 mm max. Where notching of studwork is not permitted the bath is to be supported as per detail for masonry walls.

Stud

Batten to support rim in all built-in sections of bath surround

Approved wet area sheeting

Wall tiles

Flexible sealant

5 mm gap

Bath

Batten securely bolted to stud

Masonry wall installation

Masonry walls

Wall tiles

Flexible sealant

Bath

Batten securely bolted to wall

Timber or steel frame support installation

Top plate

Wall tiles

Approved wet area sheeting

Bath

Bottom stud

Timber to comply with AS 1684

FIGURE 9.14 Caroma built-in bath installation

Source: Reproduced with permission from Caroma.

2 *Accurate set-out for the bath waste is essential.* Set out the pipework for the bath and check the alignment with the waste before installation.

3 Build in the bath support structure as per the manufacturer's instructions for timber, steel or masonry installations, and grade the bath sub-floor to local authority requirements.

4 The bath *must* be supported continuously and evenly under the entire rim, and the bath rim support must positively locate the bath. Plumbing connections must not be used to restrain the bath against movement (see **Figure 9.15**).

5 *Metal pipework installation:* to avoid damage to the porcelain enamel surface, force must not be applied

to the bath by excessively rigid or misaligned pipework. *Do not support the bath with the pipework.* Metal waste pipework must incorporate a suitable flexible coupling, equivalent to that shown in **Figure 9.15**, which will accommodate a slight misalignment. For trap installations above the floor level, ensure that the bottom of the trap or pipework has a minimum 10 mm clearance from the floor level below the bath (see **Figure 9.15**).

6 *Plastic waste pipework installation:* to avoid damage to the porcelain enamel surface, forces must not be applied to the bath by excessively rigid or misaligned pipework. *Do not support the bath with the pipework.* Plastic waste pipework *must* be

unrestrained for a length of 300 mm from the waste outlet fitting or a flexible coupling should be used (see **Figure 9.15**). For trap installations above the floor level, ensure the bottom of the pipework has a 10 mm clearance from the floor.

7 Before enclosing and tiling the bath, ensure that all the connections and pipework are watertight. Ensure that the bath drains fully prior to completion of installation. The air space beneath the bath should be ventilated to the requirements of the local authority.

8 Apply an approved flexible sealant to all exposed edges.

9 Enclose and tile in the bath.

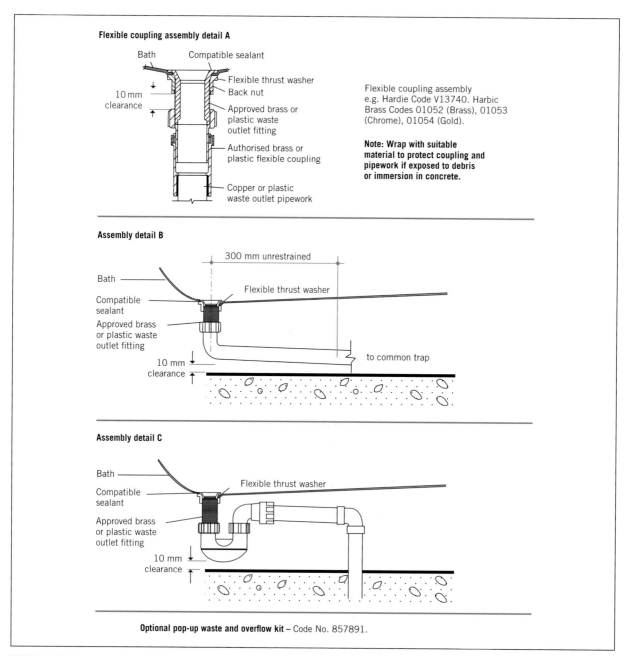

FIGURE 9.15 Caroma bath waste installation

AS 3740 WATERPROOFING OF DOMESTIC AREAS

Cleaning and maintenance

Use only neat detergent or non-abrasive cream cleanser as recommended by the manufacturer for porcelain enamel. Apply on a soft cloth and hand-rinse clean. If the bath has an optional 'Sure Step' surface, clean it with a stiff polyester or nylon brush, as well as with liquid cleaning detergents.

After cleaning the bath or adjacent tiles, or using bath salts, always rinse the bath clean with water to remove any chemical residues.

Do *not* use an abrasive cleaner to remove surface grime.

Important note: Avoid contact with sharp objects, and do not drop heavy or hard objects onto the bath surface. Always fill the bath before adding acidic or alkaline bath salts.

AS 1684 RESIDENTIAL TIMBER FRAMED CONSTRUCTION

Levelling the bath

Baths, when manufactured, have a graded base to the outlet, and the top surrounding apron flange must be kept level. Therefore, it is essential that a spirit level be placed on the surrounding apron flange when installing the bath (see **Figure 9.16**).

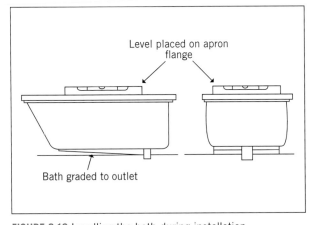

FIGURE 9.16 Levelling the bath during installation

Source: Reproduced with permission from Caroma.

Water connection to a bath

Water is supplied to a bath through taps, or a combination of taps and an outlet known as a bath set. The bath set can be fitted in varying positions:

■ in a hob at the plug end (see **Figure 9.17**)
■ on an end wall (see **Figure 9.18**)
■ at the centre of the bath on the back or rear wall (see **Figure 9.19**)
■ in combination with a free-standing bath filler.

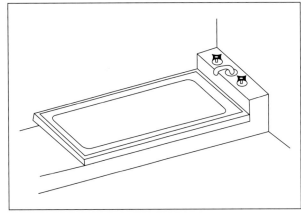

FIGURE 9.17 Bath with hob tap set

Source: Department of Education and Training (http://www.training.gov.au) © 2013 Commonwealth of Australia.

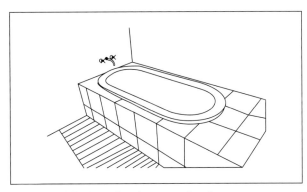

FIGURE 9.18 Bath set fitted on an end wall

Source: Department of Education and Training (http://www.training.gov.au) © 2013 Commonwealth of Australia.

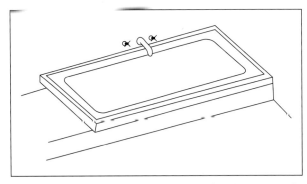

FIGURE 9.19 Bath set fitted at the centre of bath on rear wall

Source: Department of Education and Training (http://www.training.gov.au) © 2013 Commonwealth of Australia.

When the tap set is secured over the end of a bath, the taps are easier to operate and easier to adjust from outside the bath, but this is usually decided by the builder or owner. The height of the taps is approximately 150 mm above the flashing rim of the bath (see **Figure 9.20**).

Connecting the bath discharge pipe

The bath discharge pipe may be connected during either the rough-in or the fit-off stage, depending on

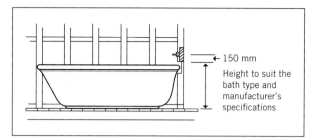

FIGURE 9.20 Position of the bath tap set

access to the plug and washer. If there is no access – for example, in the case of a concrete slab floor – the branch drain must be installed and tested before the slab is poured. If the floor is of timber construction and there is access under the floor, or if there is access to the plug and washer in a ceiling, then the plumber can install the discharge pipe when it is most convenient or essential to do so.

Pipe sizes for a bath discharge pipe

Traditionally, bath outlets and discharge pipes were usually DN 50, but conversion of these to DN 40 is necessary to comply with AS/NZS 3500.2. The reason for the reduction of size is to reduce the rate of waste discharge. If the plug and washer is installed and its diameter is DN 50, it must be fitted with a DN 50 × DN 40 reducer before installing the trap (see Figure 9.21).

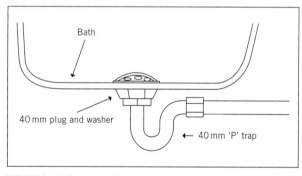

FIGURE 9.21 Size of bath trap and waste

AS/NZS 3500.2 SANITARY PLUMBING AND DRAINAGE

The trap must be installed so that its water seal is as close as possible to the bath outlet. It should never be more than 600 mm from the outlet. This will assist in protecting the trap water seal from syphonage. Under normal operating conditions, the trap must retain a water seal of not less than 25 mm.

Note: It is essential when using a trap extension that the vertical distance between the bath outlet and the trap water seal be kept to a minimum and not exceed 600 mm.

In many cases, when installing the discharge pipe from the bath (especially in a ceiling or when a wooden floor is present), it may be necessary to install the trap below the bearer for ease of connecting the discharge pipe (see Figure 9.22).

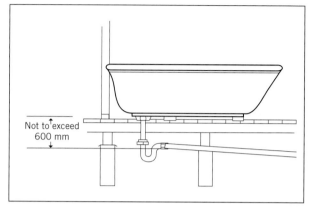

FIGURE 9.22 Installing a bath waste trap under a floor

The bath waste pipe may be connected via a:
- direct trapped connection to a sanitary drain
- direct trapped connection to a stack
- trapped connection to an overflow/disconnector gully
- trapped or untrapped connection to a floor waste gully.

Normally, bath traps will not be accessible, so therefore, as with any fixture where the trap will not be accessible, the outlet grate must be removable.

Bidets and bidettes

A bidette has an over-the-rim supply of warm and cold water, and is available with one and three tap hole configurations. The Caroma Cube bidette depicted in Figure 9.23 has a mixer tap with an adjustable nozzle and a plated pop-up waste. Tapware is not supplied with the bidette.

FIGURE 9.23 Caroma Cube bidette

Bidettes are usually supplied with a fixing kit. They can be either wall or floor mounted, although floor mounting is more common.

It is essential when installing a bidette that the waste pipe position be located accurately from the finished wall to allow correct placement. It is recommended when fixing the bidette to the floor with a bidette fixing kit that it is bedded in an acetic cured silicone sealant to allow for easy removal and servicing.

Bidets are similar to bidettes, except the water supply outlet of the bidet is below the rim of the fixture. For this reason, the water supply connections to bidets must be fitted with high-hazard backflow prevention. Both cold and warm water supply to a bidet must be through either a registered break tank or a reduced-pressure zone device.

Bidets and bidettes may have an integral trap; if not, they must be fitted with a DN 40 trap.

For further information, see manufacturers' specifications and installation instructions.

Clothes-washing machines

Most homes, units and flats have a clothes-washing machine (CWM). Semi- and fully automatic machines have specific requirements for the water supply and the drainage connections. As a plumber, you must be able to install these machines correctly.

Installation of a clothes-washing machine

A CWM must be installed according to the manufacturer's specifications. Normally, this will require connection to the water supply (either cold, for machines with a water heater, or hot and cold), and connection to a waste discharge pipe.

Type of water supply

Most CWMs require a supply of both heated and cold water. Some special machines use only cold water. Refer to the manufacturer's installation instructions.

Maximum working pressure

The manufacturer's instructions may state a maximum working pressure. The minimum pressure is normally 150 kPa and the maximum is normally 500 kPa. Low pressure can result in a long fill period, whereas high pressure can result in excessive water hammer and burst hoses.

To find whether the mains pressure is too high, enquire at the local water authority or measure the pressure using a pressure gauge. One method of measuring the pressure is shown in **Figure 9.24**.

The apparatus shown in **Figure 9.24** includes both a pressure gauge and a Hedland EZ-View® flow-measuring device. (*Note:* The flow-measuring device is optional, but may be useful in also determining whether flow rates are acceptable.)

If the mains pressure is too high, a pressure-reducing or pressure-limiting valve should be fitted,

FIGURE 9.24 Measuring the water supply pressure

ensuring that pressure to the appliance is controlled, as necessary, for both hot and cold water.

Hose connections

Many manufacturers supply restrictors fitted in the inlet of the water supply hoses. However, these restrictors only limit the flow. The water supply hoses are still exposed to the static mains pressure. Mesh filters are also fitted to strain out any grit to prevent it entering the CWM's valves (see **Figure 9.25**).

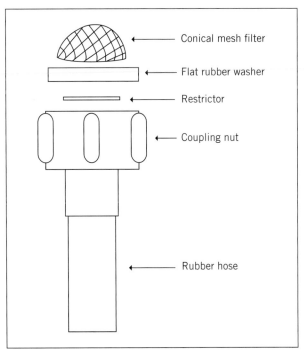

FIGURE 9.25 Typical components of a CWM water supply hose

Source: Department of Education and Training (http://www.training.gov.au) © 2013 Commonwealth of Australia.

Installation of the water supply

Most CWMs are installed close to the laundry trough. A separate water supply, controlled by CWM control valves, is usually provided for the CWM. If the taps are not available, it may be possible to cut branches into the existing trough water supply piping for new CWM control valves (see Figure 9.26).

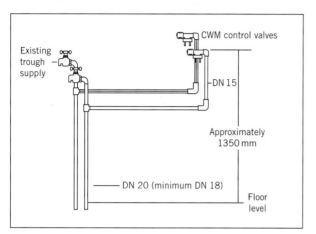

FIGURE 9.26 Cutting branches into trough water supply for a CWM

Source: Department of Education and Training (http://www.training.gov.au) © 2013 Commonwealth of Australia.

Control valves

CWM valves are designed specifically for the connection of the heated and cold inlet water hoses of the CWM. They enable the machine to be positioned closer to the rear wall than if hose bib taps were used. They are right-angled stop taps with large hose threads to which the nut and tail of each hose is screwed.

The control valves should be installed in an accessible location above the floor level so that they are clear of the CWM.

Connection of the clothes-washing machine waste

The waste hose from a CWM can be connected in several ways. Refer to AS/NZS 3500.2.

AS/NZS 3500.2 SANITARY PLUMBING AND DRAINAGE

Method 1: Waste discharge hose over the rim of a laundry trough

In this method, the waste hose is simply looped over the rim of the laundry trough and the waste water is discharged directly into the trough (see Figure 9.27).

This arrangement has two disadvantages:

1 The trough cannot be used for other washing while the CWM is being used.
2 If the plug is left in the trough, or the plughole is blocked, waste water from the CWM can fill the trough and overflow onto the laundry floor.

Method 2: Waste discharge through a trough bypass

The waste is discharged through a trough bypass (see Figure 9.28). This waste arrangement can be built into the trough when it is installed.

A standard trough without a bypass can be modified to include a bypass, but the trough waste

piping has to be lowered to include the special fitting (see Figure 9.29).

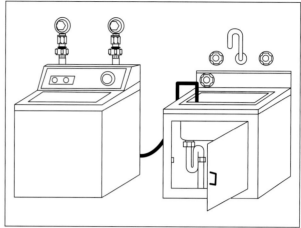

FIGURE 9.27 Waste discharge hose over rim of trough

Source: Department of Education and Training (http://www.training.gov.au) © 2013 Commonwealth of Australia

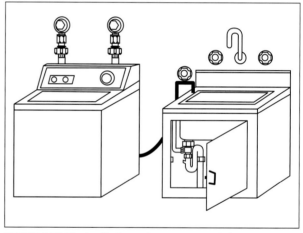

FIGURE 9.28 Waste discharge hose to a trough bypass

Source: Department of Education and Training (http://www.training.gov.au) © 2013 Commonwealth of Australia.

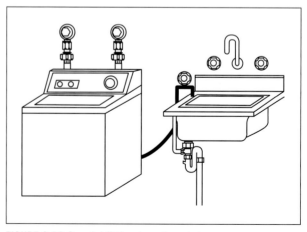

FIGURE 9.29 Special fitting for a trough bypass

Source: Department of Education and Training (http://www.training.gov.au) © 2013 Commonwealth of Australia.

Method 3: Separate trapped standing waste pipe

The trapped standing waste (minimum DN 40) can be connected to a disconnector gully, as shown in **Figure 9.30**. For the maximum length, refer to AS/NZS 3500.2.

The height of the standing waste pipe above the floor will be stated in the manufacturer's instructions, but must be a maximum of 600 mm above the water seal of the trap. (Refer to AS/NZS 3500.2.)

Alternatively, a CWM may also be connected to a floor waste gully (see **Figure 9.31**), provided that foaming will not cause a problem. (Refer to AS/NZS 3500.2.)

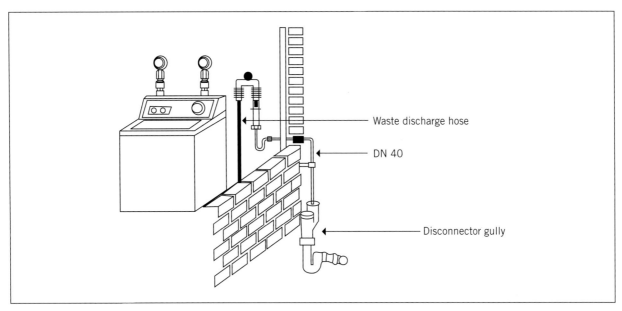

FIGURE 9.30 CWM standing waste pipe to disconnector gully

Source: Department of Education and Training (http://www.training.gov.au) © 2013 Commonwealth of Australia.

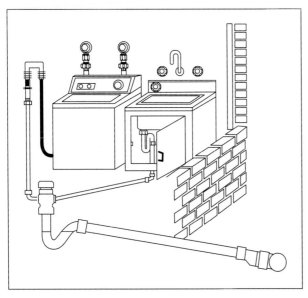

FIGURE 9.31 CWM standing waste to floor waste gully

Source: Department of Education and Training (http://www.training.gov.au) © 2013 Commonwealth of Australia.

Suds discharging from the machine can foam up when they reach the water seal in the floor waste (FW). If the foam rises up and out of the floor waste gully grate and onto the laundry floor, it will cause a slip hazard. Therefore, the height of the FW riser must be sufficient to prevent the foam from reaching floor level.

For a CWM discharge entering a DN 80 FW riser, the minimum height from water seal in the FW to floor level is 200 mm. (Refer to AS/NZS 3500.2.) For the maximum length of CWM discharge pipe to a FW, refer to AS/NZS 3500.2.

AS/NZS 3500.2 SANITARY PLUMBING AND DRAINAGE

Installation and commissioning of the machine

After new taps and/or waste for a CWM have been installed in an existing building, it is usually the plumber's responsibility to carry out the manufacturer's installation and commissioning instructions.

Electrical connection

First, check that there is a three-pin general purpose outlet within 1.5 m on either side of the machine. No other appliance should share the power outlet when the washing machine is running. It is dangerous to use an extension lead to connect a CWM. Any electrical connection required should be done only by a qualified electrician.

Levelling the machine

A spirit level is required. The CWM must be positioned and the foot bolts adjusted until the level shows that the machine is level from front to back and side to side. Locking nuts on the bolts must then be tightened against the base so that vibration does not work the bolts loose.

Discharge hose connection to plumbing

The discharge hose needs to be looped into the standing waste or into the trough, free of kinks. Most CWMs have a plastic or aluminium guide through which the hose is looped.

Water supply connections to stop taps

Some water should be flushed through the wall stop taps to clear out any debris or sediment. The hoses can then be connected to the taps, making sure that the heated water supply hose is connected to the heated water tap. If the hoses supplied with the CWM are too short, they should be replaced with longer lengths, rather than joined and extended.

Commissioning

Refer to the manufacturer's instructions regarding the removal of any transit bolts, strapping or clips that may have been attached to parts of the CWM for protection during transport and delivery.

The CWM should be operated through a complete cycle to check its functions and inspect for leaks. The customer should then be instructed in its correct operation and advised to turn off the water supply control valves after each use of the machine. This is to stop the build-up of static pressure, which may cause the hoses to burst.

Dishwashing machines

Installation of domestic dishwashing machines

A dishwashing machine must be installed according to the manufacturer's specifications. Normally, this will require connection to both the water supply (either cold for machines incorporating a water heater, or both heated and cold) and to the sanitary plumbing. The manufacturer's or supplier's instructions specify the connection requirements.

If the machine will allow the use of heated or cold water, the choice of which water to connect may depend upon the type of water heater installed in the house. If the heated water is from a gas hot water unit, it is probably better to use the heated water supply, since it is cheaper to use gas to heat the water, rather than the electric heating element in a dishwashing machine. If the domestic heated water supply is from an electric storage water heater, it may be better to use a cold water supply.

While the electric storage water heater may use electricity at the cheaper night tariff (where available), using the limited supply of heated water available from the storage cylinder may limit the amount of heated water available for other uses.

Connection of heated water to the dishwashing machine may result in lower running costs and faster cycles.

The heated water supply temperature must not exceed the manufacturer's recommendation. The connection options should be discussed with the property owner before commencing installation.

Connection of water supply to the dishwashing machine

Most dishwashing machines are installed next to the sink, so it is usual to take a branch from the sink heated and/or cold water supply. The branch/es must have a *minimum diameter* of DN 15, and each must have a separate control valve.

The type of water supply control valve – either a stop tap or gate valve – may be specified in the manufacturer's instructions. Using the wrong type of valve may void the manufacturer warranty.

Most authorities do not permit the water supply control valve/s to be located behind the dishwashing machine, due to the lack of access and visibility to operate the valve/s in an emergency. The valves are best located in the cupboard next to the dishwashing machine (see Figure 9.32).

Connection of the water supply

Maximum working pressure

Read the manufacturer's instructions to find out the maximum working pressure of the dishwashing machine. To check whether the mains pressure is too high, ask the

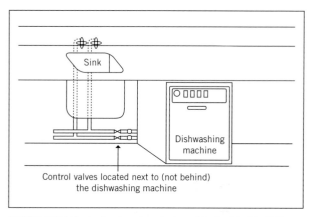

FIGURE 9.32 Control valves located next to (not behind) the dishwashing machine

Source: Department of Education and Training (http://www.training.gov.au) © 2013 Commonwealth of Australia.

local water utility or use a pressure gauge. This must be done for both the heated and cold water supplies.

A simple arrangement for measuring the static water pressure is described and illustrated earlier in this unit for clothes-washing machines.

If the mains pressure is too high, a pressure-reducing (or limiting) valve must be fitted in the supply line/s. This pressure-limiting valve is best fitted in the water service pipe near the water meter. The valve in this position gives a safe working pressure for other applications in the house, and can assist in reducing water hammer.

Minimum working pressure

Some dishwashing machines use a time cycle for filling. If the water pressure is too low, not enough water will enter the machine for it to reach the correct working level.

Manufacturers of this type of dishwasher may specify a minimum water pressure in their installation instructions.

Connection from water supply control valve/s to the dishwashing machine

If flexible hoses are supplied with the machine for the connection, the water supply should ideally be turned off after each use of the machine. This will avoid the possibility of water pressure bursting the hose and flooding the home. Some manufacturers recommend fitting a special valve between the control valve and the hose to prevent flooding should the hose burst.

A more durable connection is one made from copper pipe (minimum diameter DN 15), but it should be connected in a manner that facilitates disconnection of the pipe and removal of the dishwasher.

Most dishwashing machines have filters in the water supply hose connections to prevent dirt entering valves in the machine. The water supply connecting hoses and piping should be flushed out to remove any dirt before connecting them to the dishwashing machine.

Some machines have a restrictor nipple in the water supply connection to reduce the static water pressure. However, if the water supply to the machine is from a low-pressure storage hot water cylinder, the restrictor nipple must be removed. Refer to the manufacturer's instructions.

Electrical connection

The electrical installation required for the washing machine must be done by a qualified electrician and is usually via a three-pin general purpose outlet.

Connection of discharge pipes

Domestic dishwashing machines must discharge by one of the following methods:

- connection to a DN 40, or larger, trapped waste pipe (see **Figure 9.33**)
- connection above the water seal of a DN 50 sink trap (see **Figures 9.34** and **9.35**)
- connection through a food waste disposal unit (see **Figure 9.36**).

For further information, read AS/NZS 3500.2.

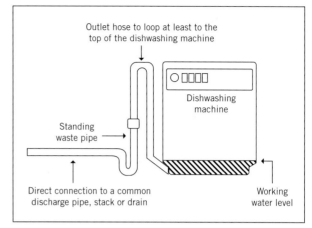

FIGURE 9.33 Connection of dishwasher through a trapped standing waste pipe

Source: Department o f Education and Training (http://www.training.gov.au) © 2013 Commonwealth of Australia.

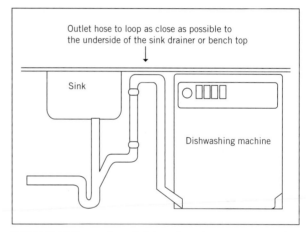

FIGURE 9.34 Connection of dishwasher above water seal of DN 50 sink trap

Source: Department o f Education and Training (http://www.training.gov.au) © 2013 Commonwealth of Australia.

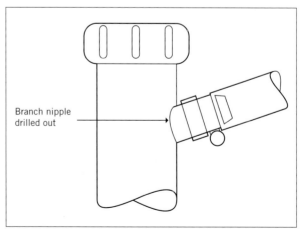

FIGURE 9.35 Connection of a waste hose to a fixture trap nipple

Source: Department o f Education and Training (http://www.training.gov.au) © 2013 Commonwealth of Australia.

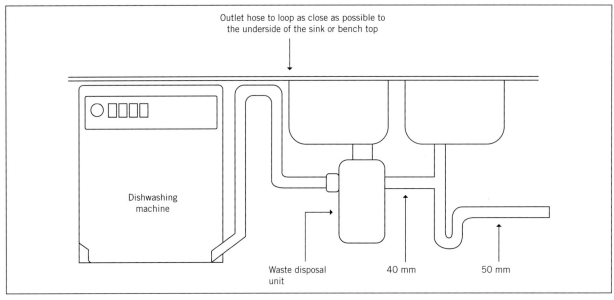

Outlet hose to loop as close as possible to the underside of the sink or bench top

Dishwashing machine

Waste disposal unit

40 mm

50 mm

FIGURE 9.36 Connection through a food waste disposal unit

Source: Department o f Education and Training (http://www.training.gov.au) © 2013 Commonwealth of Australia.

 **AS/NZS 3500.2 SANITARY PLUMBING AND DRAINAGE**

Commissioning the dishwasher

After the dishwasher has been installed, check the following:

- The power lead is plugged in and the power switch is on.
- Mains water pressure is within the range specified by the manufacturer.
- Recommended valves have been fitted.
- Hoses are connected to the correct water supply.
- Water supply valves are open.
- Hoses and fittings do not have leaks.
- The customer has been instructed in the correct use of the machine, or leave the operating instructions with the customer.
- Ensure installation items – screws, fittings, etc. – are not left in the dishwasher.
- Be sure the filter is properly fitted.
- Ensure connection to the fixture trap nipple is drilled out cleanly and free of obstruction.
- Ensure the waste hose is properly connected and clamped to the fixture trap nipple.
- Select 'rinse' cycle and let the machine run to ensure it fills and empties correctly. (If the unit does not pump out, look for a kinked drain hose.)
- Check for any water leaks from the door seals, etc.
- Make sure the warranty requirements have been completed.

Shower bases

Showers may comprise any of these types of installations:

- shower over baths, whether fixed or hand-held
- individual shower compartments
- group showers.

These installations may be either pre-formed or built in-situ. They are generally constructed from acrylic (polymer) and may also be fibreglass or pressed metal.

A shower base may be classed as a slab floor installation or above-floor installation. Before installing any shower base, you must check with the relevant manufacturer's installation instructions.

Slab floor installation

The following installation instructions have been taken from the *Caroma Technical Handbook* and are only a guide (Reproduced with permission from Caroma).

1 Determine access requirements for the particular shower base arrangement and ensure the concrete formwork conforms (see **Figure 9.37**). Allow adequate depth for pipework (normally 300 mm).
2 Remove protective film from tiling bead and waste before installation, and remove the remaining film on completion.
3 Assemble the shower outlet and trap to base.
4 For an optimum installation option, it is recommended to bed the shower base with a 3:1 sand/cement mix.
5 Connect the discharge pipe to the waste outlet.
6 Fill access space with sand and backfill with concrete.
7 Fasten 20 mm packing battens to the wall and ensure the wallboards sit 10 mm minimum over the tiling bead (see **Figure 9.37**).
8 Leave the 'Cleaning and Maintenance Instructions' label in a position for the householder to read.

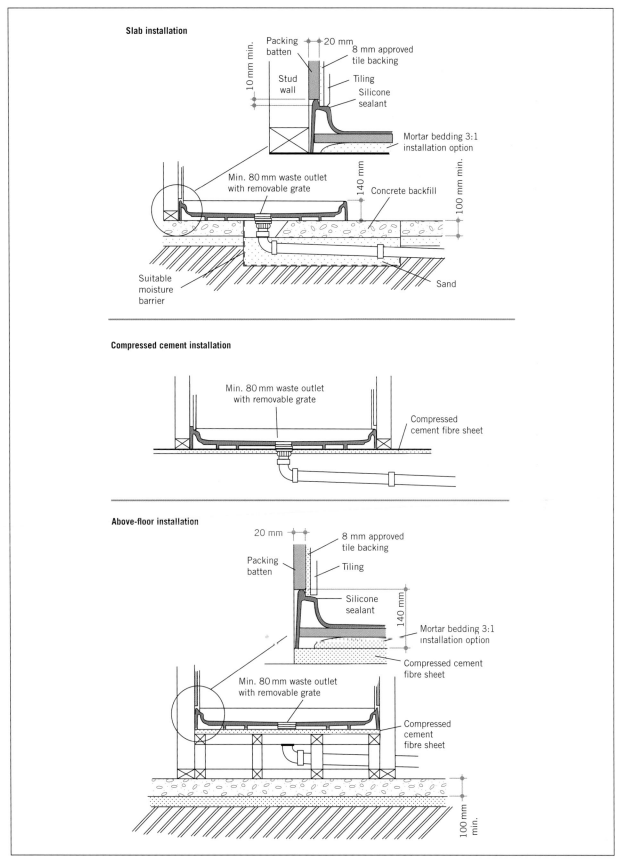

FIGURE 9.37 Shower base installation methods

Source: Reproduced with permission from Caroma.

Above-floor installation

The following installation instructions have been taken from the *Caroma Technical Handbook* and are only a guide (Reproduced with permission from Caroma).

1 Ensure adequate space has been provided under the shower base for pipework.
2 Remove protective film from tiling bead and waste before installing, and remove the remaining film on completion.
3 For an optimum installation option, it is recommended to bed the shower base with a 3:1 sand/cement mix. *Note:* Due to the rigidity of the shower base, provide and install support pads on the underside of the shower base bonded to the floor with a suitable approved adhesive.
4 Connect the discharge pipe to the waste outlet.
5 Fasten 20 mm packing battens to the wall and ensure the wallboards sit 10 mm minimum over the tiling bead (see **Figure 9.37**).
6 Leave the 'Cleaning and Maintenance Instructions' label in a position for the householder to read.

Installing a pre-formed shower base

When installing the pre-formed shower base, it must first be positioned to ensure adequate flashing. The base must be levelled to ensure the correct gradient to the outlet. Place a spirit level on the flashing lip of the edges of the base, as illustrated in **Figure 9.38**.

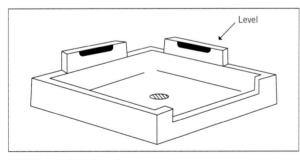

FIGURE 9.38 Levelling the shower base

Source: Department of Education and Training (http://www.training.gov.au) © 2013 Commonwealth of Australia.

Note: Always ensure that the base will drain by discharging water on to the base.

Note: When installing a pre-formed shower base (see **Figure 9.39**), you must take care when handling and lifting the base to prevent damage.

In-situ shower base

In-situ showers made on a waterproof base with tiles must have gradient and comply with AS 3740 Waterproofing of domestic wet areas.

Falls in shower floors must be sufficient to prevent:

- surface water being retained on the shower floor (except for minor water)
- water discharging outside the shower area.

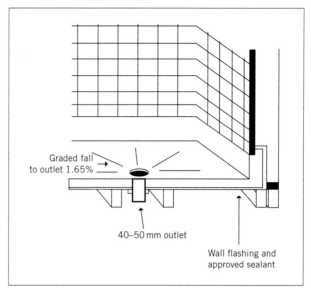

FIGURE 9.39 Pre-formed shower base

Source: Department of Education and Training (http://www.training.gov.au) © 2013 Commonwealth of Australia.

For shower areas with a separation between the shower area and general wet area, such as a shower screen, hob, step-down or water stop, the fall to the waste must be at least 1:100. For other shower areas (such as unenclosed showers), the fall to the waste must be at least 1:80.

AS 3740 WATERPROOFING OF DOMESTIC WET AREAS

Discharge pipe

The discharge pipe may be connected during either the rough-in or fit-off stage. If the floor is a concrete slab, the discharge pipe must be installed and tested at the rough-in stage, before the slab is poured. If the floor is timber, then the discharge pipe may be installed at either stage – whichever is most convenient for the plumber – provided access can be gained to under the floor.

The built in-situ base (see **Figure 9.40**) is usually poured during the 'rough in' stage. These bases are constructed from concrete (four parts aggregate, two parts sand and one part cement) and poured where the shower base is required.

A puddle flange is required on the PVC riser to achieve a seal between the PVC and the concrete. This assists in preventing leaks through the shower base.

It is normally the plumber's responsibility to ensure that there is adequate provision for flashing. The work of checking out the wall to accommodate the shower base may be done by the carpenter.

Water connections to a shower

The entry point to the shower compartment must be considered when locating the shower tap set.

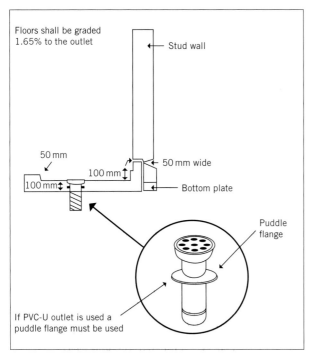

FIGURE 9.40 Built in-situ shower base

Source: Department of Education and Training (http://www.training.gov.au) © 2013 Commonwealth of Australia.

Within the figure: Floors shall be graded 1.65% to the outlet / Stud wall / 50 mm / 100 mm / 50 mm wide / 100 mm / Bottom plate / Puddle flange / If PVC-U outlet is used a puddle flange must be used

FIGURE 9.41 Dorf Stayfast shower

Source: Reproduced with permission from Caroma.

The tap set is positioned near the entry to the shower compartment to permit easy access to operate the taps from outside the compartment. If the tap set were positioned far into the compartment or directly under the shower rose, the user could be exposed to heated or cold water when operating the taps.

The height of the taps above the shower base is determined by the ease of operating the taps. Taps are normally positioned so they align with the horizontal tile joints – for example, taps at 1100 mm above the shower base at 200 mm centres.

In showers, when there is a heated water tap and a cold tap, the heated water tap must be the left-hand or upper tap.

When selecting or recommending shower roses and tapware, water efficiency should be considered for water and energy conservation. For further information, read *Basic Plumbing Services Skills – Water Supply*.

Positioning the shower rose

There are three basic types of shower roses:

1 universal shower rose
2 fixed shower rose
3 telephone shower.

A universal shower rose is a multi-directional rose, which allows flexibility in terms of height and direction (see **Figure 9.41**).

A fixed shower rose does not allow the same flexibility, so it must be positioned between 1.65 and 1.8 m above the shower base, depending on

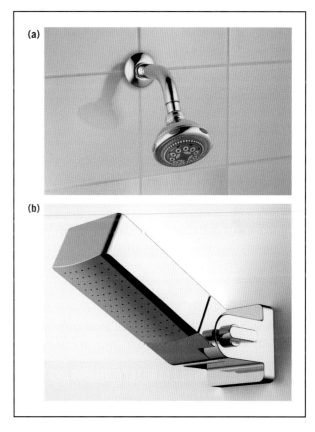

FIGURE 9.42 Fixed shower roses: (a) Dorf Nutra fixed wall shower; (b) Dorf Myriad fixed wall shower

Source: Reproduced with permission from Caroma.

the customer's requirements. Generally, 1.8 m is the suggested height (see **Figures 9.42** and **9.43**).

A telephone shower allows for greater flexibility than the universal shower, as it may be handheld or attached to a fixed point. The fixed point may be adjusted to a preferred height (see **Figure 9.44**).

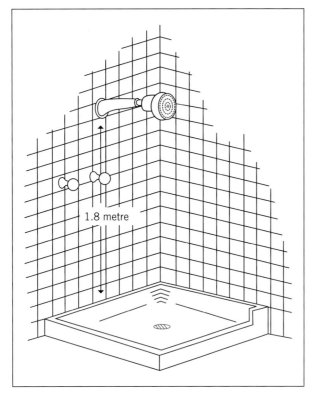

FIGURE 9.43 Position of a fixed shower rose

Source: Department of Education and Training (http://www.training.gov.au) © 2013 Commonwealth of Australia.

Connecting the discharge pipe to a shower

The shower outlet may be connected via a:

- direct trapped connection to a sanitary drain
- direct trapped connection to a stack
- trapped connection to an overflow/disconnector gully
- trapped or untrapped connection to a floor waste gully.

The discharge pipe connection must be in accordance with AS/NZS 3500.2 and local authority requirements.

AS/NZS 3500.2 SANITARY PLUMBING AND DRAINAGE

Sinks

When selecting a sink for installation, there are a number of considerations, including:

- available space to accommodate sink dimensions
- single, double or triple bowl
- raised or flat rim
- available support in a bench top cupboard for brackets and clips.

Sinks come in a variety of shapes and sizes, and there are different options for the position and type of

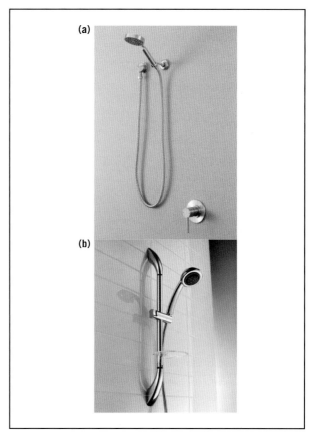

FIGURE 9.44 Telephone-type showers: (a) Caroma Titan stainless steel hand shower; (b) Dorf Nutra rail shower

taps. Before ordering a sink, it is best to clarify all of these issues with the builder or client.

Sinks are manufactured from plastic (hard resins) or stainless steel, with one or three tap holes, depending on the type and style of taps required.

Flat-rim (see **Figure 9.46**) and under-mounted sinks are the two most common types of sink.

Some authorities may have specific requirements for the location of sink taps. If you are unsure, ask your supervisor or trainer.

Sinks are usually set at a height of approximately 900 mm plus 25 mm flashing rim. The height of the taps is usually 75 mm above the flashing rim of a sink.

The vertical distance (air gap) between the tap outlet and the spill level of the sink must be at least 40 mm. This is to prevent the backflow of water from the fixture into the water supply.

When selecting or recommending sink tapware, water efficiency should be considered for water and energy conservation. For further information, read *Basic Plumbing Services Skills – Water Supply*.

GET IT RIGHT

SHOWER OUTLETS

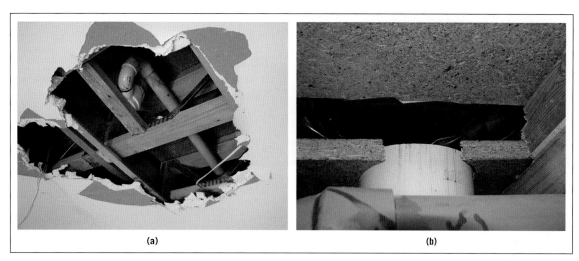

FIGURE 9.45 Fixture discharge pipes

Fixture discharge pipes must be properly connected to waste outlets. The two images in **Figure 9.45** show a DN 100 × DN 50 mm trap installed, and a DN 50 shower waste outlet that has simply been located over the trap without being properly connected. When a blockage occurred, waste overflowed into the ceiling space.

1   Would this sanitary plumbing installation have passed a hydrostatic test?

2   Is this trap the correct fitting to use in this situation?

3   How would you have installed the shower waste?

FIGURE 9.46 Bench-mounted flat-rim sink

Before fitting a sink with taps, you must be aware of the specifications for the tap set. Mixer taps are commonly specified, in which case the sink may require only one hole. Raised-rim sinks are usually installed with wall-mounted taps.

Ensure that water supply to sinks complies with WELS requirements.

The sink tap set is a breech that connects both heated and cold water. Sets are available with side or bottom connections. If a double-bowl sink is installed, the outlet point of the tap set should be positioned to service both bowls.

When fitting off a tap set, the flanges, handles and outlet are installed. They will have been selected by the owner or builder prior to fixing, and are sometimes supplied by the owner or builder.

Taps for flat-rimmed fixtures may be installed either through the fixture or bench top. It is also acceptable to have a tap set installed in the wall. The connection of a hob-mounted tap or tap set is done at the fitting-off stage. The connections are made from lugged (back plate) elbows, which are set in the wall below the sink.

Waste pipe connection

The DN 50 sink waste discharge pipe may be connected to a:
- direct trapped connection to a sanitary drain
- direct trapped connection to a stack
- trapped connection to an overflow/disconnector gully.

The sink waste discharge pipe must be installed in accordance with AS/NZS 3500.2 and local regulatory requirements.

AS/NZS 3500.2 SANITARY PLUMBING AND DRAINAGE

Installation and fixing a sink

The fixing of a sink into a bench top is usually carried out by the builder; however, on some occasions, the plumber may carry out the work.

The instructions below have been supplied by Clark Sinks for an over-mount sink and are to be taken only as a guide. Prior to installing any sink you must first read the manufacturer's installation instructions and check with the bench top manufacturer for the recommended method of sealing between the top and the underside of the sink.

Over-mount sink installation instructions

1 Use method a or b to draw a cut-out line on the bench top.
   a Using a knife or scissors, cut out the relevant carton template. Position the template on the bench top and trace around it with a pencil.
   b Place the product on the bench top and draw a line around the sink. Then remove the product and draw 12 mm *inside* the line.
2 Check the size against the overall cut-out dimensions before commencing cutting. Cut out the bench top carefully, following the inside edge of the pencil line. An exact cut is required to obtain the best fit.
3 For normal thickness bench tops, the supplied fixing clips should be suitable. Lightly hammer the fixing clips onto the clip brackets at equal points along the product rim. (If the thickness of the bench top exceeds the range of the fixing clips, then extension clips can be purchased and mounted on to the clip brackets). (See Figure 9.47.)
4 Position the sink in the cut-out hole and align properly. The fixing clips can be bent out of the way to assist installation (see 'Figure C' in Figure 9.48). Check that the sink sealing tape will seal evenly around the edge of the hole.
5 Remove the sink and, if required, install the waste assemblies and the tap. It is suggested that silicone sealant be applied between the plug and washer and the sink outlet hole.
6 Remove the protective backing from the sealing tape, apply a continuous bead of silicone sealant between the sealing tape and the edge of the sink, and quickly place into position.
7 Bend the fixing clips straight so that the legs clear the hinge area. Tighten the clips evenly, using a low torque setting (see 'Figure D' in Figure 9.48). *Note:* Over-tightening the clip may strip the thread.
8 Fit off the tap set and waste pipe.
   Note: For thin bench tops, special Clark clips (A0167) are available.

Spa baths

The installation and fit-off requirements for a spa bath are very much the same as for a normal bath (see Figure 9.49). The difference is in the installation of the pump and blower, which must be installed in accordance with the manufacturer's installation instructions.

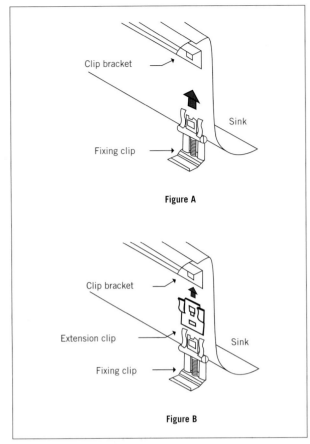

FIGURE 9.49 Spa bath

Typical considerations for the location of the pump and blower are:

■ the elevation (height of the pump and blower in relation to the bath water level)
■ access to service and maintaining the pump and blower
■ suitable electrical connection
■ locating the pump and blower to minimise undesirable noise.

Troughs (laundry)

When selecting a trough for installation, there are a number of considerations, including:

■ available space to accommodate the trough's dimensions
■ single or double bowl
■ raised or flat rim
■ available support in a bench top cupboard for brackets and clips
■ whether a washing machine bypass connection is required.

Troughs are available in a variety of shapes and sizes, and there are numerous options for the position of taps. Troughs are manufactured from plastic (hard resins) and stainless steel.

When ordering materials and arranging for fit off, it is important to confirm the tapware specifications and other requirements, as mentioned above.

Some trough types are:

■ flat-rim trough
■ raised-rim trough
■ double-bowl trough.

Figures 9.50 and 9.51 show two common types of trough installation.

When fixing a trough cabinet, screw the inside edges of the cabinet to the wall studs, or fit a piece of timber or plate across the back of the cabinet and screw it to the wall (see Figure 9.52). *Note:* Some troughs are supplied with wall brackets.

FIGURE 9.47 Fixing method for an over-mount sink

Source: Reproduced with permission from Caroma.

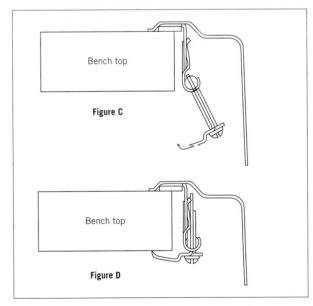

FIGURE 9.48 Fixing an over-mount sink to a bench top

Source: Reproduced with permission from Caroma.

FIGURE 9.50 Flat-rim trough

FIGURE 9.51 Trough cabinet

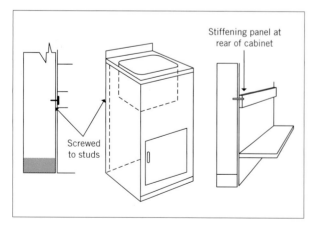

FIGURE 9.52 Fixing of a trough cabinet

The trough waste discharge pipe may be connected to a:

- direct trapped connection to a sanitary drain
- direct trapped connection to a stack
- trapped connection to an overflow/disconnector gully
- trapped or untrapped connection to a floor waste gully, provided that foaming is not likely to cause a problem.

Urinals

Wall-hung urinals

The following information is taken from the *Caroma Technical Handbook for Torres Urinals*. Prior to installing any wall-hung urinal, you must follow the manufacturer's installation instructions and read the requirements for installing urinals in AS/NZS 3500.2 (see the example in Figure 9.53).

AS/NZS 3500.2 SANITARY PLUMBING AND DRAINAGE

Figure 9.54 shows a variety of wall-hung urinals. The Torres is a general-application wall-hung urinal with integral trap. It has a spreader to wash down the bowl on each flush, and has a stainless-steel grate.

Inlet

Standard options include a top inlet for exposed sparge pipe with compression fitting, or a back inlet for concealed pipes with DN 20 inlet for connection with capillary or compression fittings (see Figure 9.55).

Outlet

Two rubber connection seals are supplied for a push-on fit with DN 50 brass tube or DN 50 DWV plastic pipe.

Flush action

An important feature is the chrome-plated adjustable spreader, which allows cisterns to be installed at increased head height.

Fixing

The Torres urinal is fixed to the wall with a concealed support bracket. It has a cast aluminium natural finish, and is supplied with cadmium-plated locating bolts and fasteners.

Installation height

The front lip is approximately 600 mm above the floor. In public buildings that may be used by children, one urinal in a group should be fixed lower than normal height; this height should be 400 mm from floor to the lip.

The floor of any room containing one or more wall-hung urinals must be graded to a floor waste gully. The floor waste gully must be charged in accordance with the requirements of AS/NZS 3500.2. Untrapped (dry) floor drains must not be used in a room containing a wall-hung urinal.

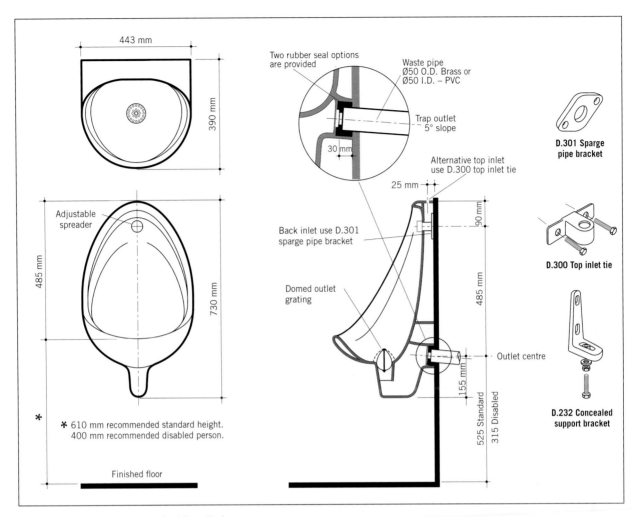

FIGURE 9.53 Caroma Torres urinal installation

Source: Reproduced with permission from Caroma.

AS/NZS 3500.2 SANITARY PLUMBING AND DRAINAGE

It is best practice to locate the floor waste adjacent to a wall and not in the middle of the floor.

Wall-hung urinals may be flushed using either a mains supply or a gravity flush valve.

Wall-hung urinals are usually fabricated with an integral trap (see **Figure 9.56**); if there is no such trap, a DN 40 minimum trap must be connected directly to the urinal outlet.

Urinal outlets must be fitted with a removable domed grating. The Caroma Torres grate is fixed, as shown in **Figure 9.57**.

The length of the urinal fixture discharge pipe from the crown of the trap to the adaptor fitting in the drain must not exceed 2.5 m.

Stall urinals

Stall (or slab) urinals are usually stainless steel up to 5 m long (see **Figure 9.58**). They must be securely supported and connected to a trap and discharge pipe at least DN 65. If the length of the stall is over 5 m, additional outlets are required.

Classification of urinal discharge

Because the urinal is used for the collection of body waste, it is classified as a soil fixture and must be connected directly to a drain or soil discharge stack.

The floor waste gully within a urinal compartment must not be connected to a disconnector gully.

Urine has a corrosive action on some materials, and for this reason, urinal discharge pipes must not be constructed using prohibited materials. Prohibited materials include copper, brass and galvanised steel pipes.

The discharge pipes from waterless (non-flushing) urinals should not be constructed of metallic materials. It has been found that where flushing urinals have been replaced with non-flushing urinals, copper alloy (brass) waste pipes have corroded through within two years.

Note: Where a waterless urinal discharge pipe is connected to a sanitary plumbing system, the common discharge pipe or drain should be flushed by fixture/s

FIGURE 9.54 Wall-hung urinals: (a) Caroma Leda urinal; (b) Caroma Integra urinal; (c) Caroma Torres urinal; (d) Caroma H2Zero™ Cube urinal; (e) Caroma Cube 0.8L Smartflush® urinal

Source: Reproduced with permission from Caroma.

of at least two upstream fixture units. This will assist in minimising the build-up of uric scale in the pipework.

Wall and floor area around a urinal

The wall and floor area around a wall-hung urinal must be easy to clean. The wall and floor area must be of an approved impervious (waterproof) material covering at least the:

- wall from the floor to a height of 50 mm above the top of the urinal and 225 mm to each side of the urinal
- floor at least 400 mm beyond the front edge of the urinal and 225 mm to each side of the urinal.

There may be local variations to the above waterproofing requirements. Always follow local regulations, AS/NZS 3500.2 and the manufacturer's installation instructions.

AS/NZS 3500.2 SANITARY PLUMBING AND DRAINAGE

Water closets

Water closets are soil fixtures that are manufactured in a wide variety of styles and designs to match job specifications. They are manufactured from vitreous china and stainless steel (vandal-resistant).

Most old existing toilet suites are single flush. All new suites are either 6/3 or 4.5/3 litre dual flush. Whenever either an old cistern or pan is replaced, both cistern and pan must be replaced together. The main reason is that new cistern flushing volumes will not clear an old pan.

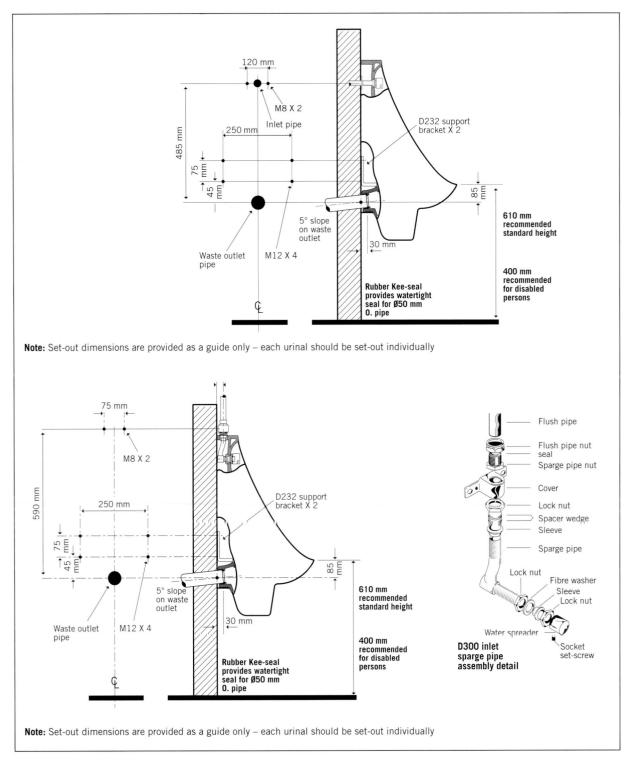

Note: Set-out dimensions are provided as a guide only – each urinal should be set-out individually

Note: Set-out dimensions are provided as a guide only – each urinal should be set-out individually

FIGURE 9.55 Caroma Torres urinal top and back inlet installation

Note: It is now mandatory that all new or replacement cisterns are of the dual flush type. This has greatly reduced water usage.

Caroma has developed a 4.5/3 litre dual flush toilet suite; however, the cistern can only be installed on the range of pans designed for them, and cannot be installed on a 6/3 litre pan.

Water closet suites are available in a variety of styles and designs. Some typical cistern designs include:

■ low level

■ close coupled and connector (see **Figures 9.59** and **9.60**)

■ concealed.

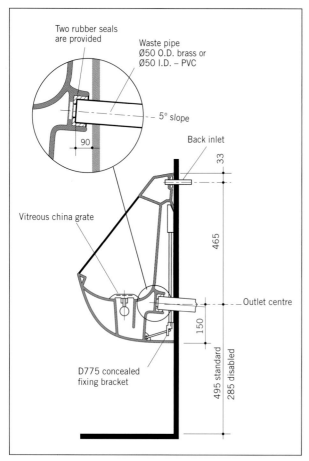

FIGURE 9.56 Integral trap within the Caroma Integra urinal

Source: Reproduced with permission from Caroma

Two rubber seals are provided

Waste pipe
Ø50 O.D. brass or
Ø50 I.D. – PVC

5° slope

90

Back inlet

33

Vitreous china grate

465

Outlet centre

150

D775 concealed fixing bracket

495 standard
285 disabled

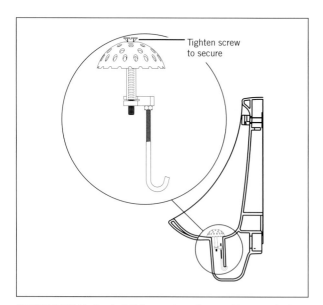

Tighten screw to secure

FIGURE 9.57 Method of fixing grate – Caroma Torres urinal

Source: Reproduced with permission from Caroma

FIGURE 9.58 Stall urinal during rough-in

FIGURE 9.59 Caroma Aire Concord close coupled toilet suite

Source: Reproduced with permission from Caroma.

Low-level cistern

The cistern is fitted approximately 1 m above the floor, connected with flush pipe.

Close-coupled and connector cistern

The cistern is usually fitted directly to the water closet (no flush pipe) (see **Figure 9.60**). **Figure 9.61** shows a close coupled toilet suite.

Concealed cistern

Concealed cisterns are designed for installation in a duct or built into a wall cavity. Access for maintenance must be provided by either a removable panel on the front of the cistern or from within the service duct (see **Figures 9.62** and **9.63**).

FIGURE 9.60 Caroma Leda wall faced connector toilet suite

Source: Reproduced with permission from Caroma.

FIGURE 9.61 Caroma Regal II close coupled toilet suite

Source: Reproduced with permission from Caroma.

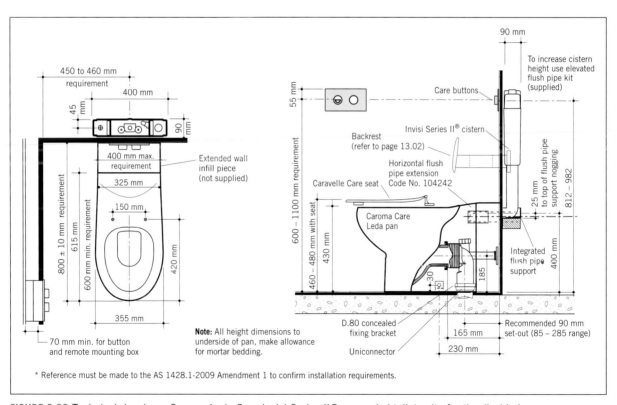

FIGURE 9.62 Technical drawing – Caroma Leda Care Invisi Series II® concealed toilet suite for the disabled

Source: Reproduced with permission from Caroma.

FIGURE 9.63 Installation of concealed closet cisterns at rough-in

FIGURE 9.64 Caroma Urbane wall hung pan with Invisi Series buttons

Disabled pan

Facilities for people with disabilities must comply with local regulations, AS 1428 Design for access and mobility and AS/NZS 3500.2.

AS 1428 DESIGN FOR ACCESS AND MOBILITY

AS/NZS 3500.2 SANITARY PLUMBING AND DRAINAGE

Position of the pan

For wheelchair access, the installed closet pan must be a minimum of 800 ± 10 mm from the front of the pan to the finished wall and provide a seat height of 460 to 480 mm (see Figure 9.62).

Water closet pan installation

Water closet pans are manufactured in various designs to suit different job specifications (e.g. see Figures 9.64 and 9.65). Common design characteristics include:

- integral trap (usually 'P', 'S' or skew trap)
- a flush pipe horn
- a flushing rim
- seat attachment holes
- fixing holes for mounting bracket.

Water closet pans are expensive fixtures, and are therefore installed near the completion of the job at the fit-off stage. Care must be taken when securing vitreous china pans because over-tightening or poor alignment of screws can stress the pan and cause it to break.

FROM EXPERIENCE

Learning to select correct fasteners and use appropriate torque on fasteners is critical to the correct installation of plumbing fixtures.

FIGURE 9.65 Caroma Care 800 Cleanflush Invisi Series II® toilet suite

When setting a pan on a tiled concrete floor, a section of the tiles and bedding should be removed to provide a keying surface, and the pan is then set on a mixture of 3:1 sand and cement. Fast-setting cement must not be used, as it can crack the foot of the pan. The maximum height of the closet pan above the finished floor level is 20 mm.

Closet pans must be securely fixed, using either brackets or corrosion-resistant fasteners.

Note: Silicone sealant alone is not an acceptable method of securely fixing closet pans.

Flush pipe installation

Flush pipes are used to deliver water from the cistern to the pan. Flush pipes are supplied with the cistern and are joined to it by means of a back nut tightened onto a rubber ring.

The connection to the pan is made by inserting the flush pipe into the pan connection point, and is made watertight by a synthetic rubber cone.

The manufacturer supplies each new cistern with the appropriate connection.

The close-coupled cistern is installed so that its outlet sits directly on the matching design pan with a rubber seal insert. If you are unsure of the connection requirements, refer to the manufacturer's installation instructions.

Note: In relation to buildings connected to an on-site treatment system (septic tank), some authorities may require that cisterns have external overflows.

Water closet seats

Water closet seats are manufactured in styles and designs to suit both domestic and heavy-duty commercial use.

Toilet seats are normally packaged in sealed plastic bags, which contain the fixing bolts, washers and nuts, along with installation instructions.

When installing seats, ensure that the seat fits evenly around the rim of the pan before tightening.

Closet pan outlet connection

Water closet pans must be connected directly to soil stacks or drains at least DN 80 in size.

Clean-up

Work area

After installing the sanitary fixtures and appliances, and reinstating surfaces, it will be necessary to clean up the fixtures and work area. Clean-up includes:

- disposal of rubbish that cannot be recycled
- recycling of cardboard packaging waste
- cleaning soiled internal walls and surfaces
- restoration of the area to its original condition.

After some fixtures have been installed, they may also need to be cleaned to an appropriate standard to complete the job. It is best practice to follow the manufacturer's instructions and recommendations for cleaning of fixtures and appliances. You must consider that the surface material of fixtures and appliances may be adversely affected by solvents, cleaners and abrasion.

Tools and equipment

Some appliances and fixtures are supplied with special tools for installation – for example, tapware servicing tools and tools to remove transit bolts. After completing installation work, leave these tools with the property owner. You must also:

- remove power tools and equipment, and carry out service work to maintain equipment in good condition
- inspect, clean, maintain and store tools
- dispose of unserviceable tools (they can be dangerous)
- maintain and return hire equipment in accordance with the hire agreement.

Documentation

This section is relevant to documentation required after the work of installing sanitary fixtures and appliances is complete.

The installation and maintenance instructions for fixtures and appliances should be left on-site with the property owner.

Generally, the sewerage and regulatory authorities do not require 'as-constructed' plans; however, requirements vary around Australia. Therefore, you will need to research your local authority requirements.

Sewerage or regulatory authorities in many areas may require notification of when the project is complete, in order to carry out an inspection or audit.

In some areas, plumbers are required to lodge a Certificate of Compliance within a certain period of time after the job is completed.

 COMPLETE WORKSHEETS 3 AND 4

SUMMARY

- When installing sanitary plumbing, it is necessary to follow the plans and specifications that specify the location and dimensions of fixtures, and floor and wall finishes. Also, familiarise yourself with local regulations and inspection requirements, and ensure that the fixtures have WaterMark approval as required.
- Installation also includes the use of serviceable tools and following the fixture manufacturer's instructions, using special tools if required.
- Generally, fixtures must be installed so that they can be operated without adverse effect on fixture fixings, and the surrounding surfaces and finishes.
- When installing and fitting off sanitary fixtures, they must be securely fixed so that the fixtures can be safely used. For all fixtures, you will need to ensure that you comply with AS/NZS 3500 and manufacturers' requirements for connecting discharge pipes.
- When work has been completed, clean up the fixtures and the surrounding work area. Ensure that the fixture installation and maintenance instructions are left on-site with the property owner. If required by the regulatory authority, lodge the necessary documentation, such as a Certificate of Compliance, and book a final inspection/audit.

WORKSHEET 1

Student name:_____

Enrolment year:_____

Class code:_____

Competency name/Number:_____

Task: Review the section 'Preparing for work', then complete the following.

1 At what stage of construction is the following work performed? Circle your answer.

Fit taps and install basin	Fit off	Rough in
Fit basin support nogging	Fit off	Rough in
Fit chrome copper pipe and wall plate to closet pan	Fit off	Rough in
Fit lugged (back plate) elbows under sink	Fit off	Rough in
Install sink discharge pipe	Fit off	Rough in

2 What are three pieces of information that can be obtained from plans and specifications to assist in installing sanitary fixtures?

a _____

b _____

c _____

3 Are all plumbing fixtures required to have WaterMark?

_____

_____

4 Explain how you would check that a closet pan and cistern can be legally installed.

_____

_____

_____

_____

5 Give four examples of inspections on completed plumbing work.

_____

_____

_____

_____

6 What are two problems commonly encountered when installing sanitary fixtures in existing buildings?

a _____

b _____

WORKSHEET 2

Student name: _____

Enrolment year: _____

Class code: _____

Competency name/Number: _____

To be completed by teachers

Student competent ☐

Student not yet competent ☐

Task: Review the sections 'Tools and equipment' and 'Installation requirements', then complete the following.

1 From research or discussion with your supervisor or trainer, what tools are commonly used when installing taps to a basin?

_____

_____

2 Research AS/NZS 3500.2 and list the abbreviations for the following fixtures.

   a Bidet

_____

   b Clothes-washing machine

_____

   c Pot sink

_____

3 What is the name of the materials list that is taken from plans and specifications?

_____

4 If special tools are required to install sanitary fixtures and they are not used, how might this affect the product warranty?

_____

5 What is a simple way of determining whether a sanitary fixture has a manufacturer's warranty?

_____

_____

WORKSHEET 3

Student name: _____

Enrolment year: _____

Class code: _____

Competency name/Number: _____

To be completed by teachers

Student competent ☐

Student not yet competent ☐

Task: Review the section 'Installing and fitting off sanitary fixtures', then complete the following.

1   What is the minimum height that an impervious wall surface must be extended above the back of a wall-hung basin?

_____

2   a   Why are some free-standing 'tabletop'-type basins installed clear of the walls?

_____

   b   On the diagram below, label the minimum distance between the wall and the basin.

3   Explain how a wall-hung basin must be fixed, and why.

_____

_____

_____

4   Study Unit 5 and research AS/NZS 3500.2, and then complete the following.

   a   State the clause that specifies the method of connection of basins to unvented branch drains.

_____

   b   State the clause that specifies the grade for basin discharge pipes.

_____

c State the clause that specifies the maximum distance between the trap seal and basin outlet.

_____

d State the clause that specifies the minimum size for an unvented branch drain.

_____

5 If a bath trap is not accessible, what type of outlet grate is required?

_____

6 What alterations must be made to the outlet of a bath that has been manufactured with a 50 mm outlet?

_____

7 What is the difference between a bidet and a bidette?

_____

_____

8 Explain how you would check that the water supply pressure is not too high or too low for a clothes-washing machine.

_____

_____

_____

9 What is one solution for reducing the water pressure?

_____

_____

10 What are two disadvantages of discharging a clothes-washing machine hose over the rim of a laundry trough?

a _____

b _____

11 Why must control valves for a dishwasher be in an accessible position?

_____

12 The maximum height of a clothes-washing machine standing waste is _____ mm measured from the _____ in the trap.

13 a Research AS/NZS 3500.2. Can a clothes-washing machine be connected to a floor waste gully if the discharge could cause a foaming problem?

_____

b Before connecting a clothes-washing machine waste to a floor waste gully, how would you know whether a foaming problem could occur?

_____

_____

14 What is the minimum vertical distance from the water seal of a floor waste gully to floor level when the discharge from a clothes-washing machine connects to the riser of the floor waste gully at 88½°?

_____

15 What is the minimum gradient for the tiled floor of an unenclosed shower?

_____

16 Explain how a shower base should be levelled and tested.

_____

_____

_____

_____

17 What type of flange must be fitted to seal the PVC waste to concrete in a built in-situ shower base?

_____

18 Name four considerations for the location of the pump and blower when installing a spa bath.

a _____

b _____

c _____

d _____

19 What must be provided adjacent to a wall-hung urinal?

_____

_____

_____

20 What is the minimum size of waste pipe for a stall urinal?

_____

_____

_____

21 What is the maximum thickness of sand and cement bedding under a closet pan?

_____

22 What are two ways that a water closet pan can be securely fixed?

_____

23 Can silicone sealant alone be used to fix down a closet pan? Why, or why not?

_____

 **WORKSHEET 4**

Student name:_____

Enrolment year:_____

Class code:_____

Competency name/Number:_____

Task: Practical exercises

All questions and exercises in the previous three worksheets must be completed and checked by your trainer/teacher before you attempt the following practical exercises.

Your trainer/teacher should conduct an induction, and ask underpinning knowledge questions, before you attempt the practical exercises.

1   Set out, install and fit-off the following sanitary fixtures:

   a   water closet

   b   shower base

   c   sink

   d   bath

   e   basin

   f   dishwashing machine

   g   wall-hung urinal.

2   Explain the installation and operation requirements of the dishwashing machine to your trainer/ teacher.

Your trainer/teacher may choose to combine the above exercises with the exercises in Unit 8; however, all exercises are necessary for your assessment.

PART 3

INSTALLING DRAINAGE SYSTEMS

- **Unit 10** Installing below-ground sanitary drainage systems
- **Unit 11** Installing stormwater and sub-soil drainage systems
- **Unit 12** Installing sewerage pump sets
- **Unit 13** Septic tanks and on-site treatment facilities

10

INSTALLING BELOW-GROUND SANITARY DRAINAGE SYSTEMS

Learning objectives

This unit provides practical guidance for installing below-ground sanitary drainage systems. Areas addressed in this unit include:
- preparing for the work, and coping with ground conditions and obstructions
- setting out and excavation
- pipework configuration
- construction and termination of inspection shafts
- cutting in a branch
- installing inspection openings and enclosures
- pipe bedding and backfilling
- drain laying in unstable ground conditions
- drain-laying procedures
- testing procedures.

Introduction

Plumbers and drainers are required to install below-ground sanitary drainage systems to convey sewage and waste from sanitary fixtures to an authority's approved point of discharge or an on-site disposal system.

In this unit, we focus on connection to an authority's approved point of discharge. On-site disposal systems are covered in Unit 13.

This type of work is carried out on all properties where there are sewage and waste discharges. It could be new work, or it could involve alterations or additions to existing systems.

Sanitary drains can be suspended below a floor above ground, or laid in the ground. In this unit, we will study below-ground sanitary drainage systems. The installation of above-ground drainage is covered in Units 5 and 8.

In Unit 4, we studied the requirements for planning the layout of a below-ground sanitary drainage system, and followed the development of an example project. This unit is a progression from planning, and we will now study the procedures and requirements for installing a drain.

While studying this subject, you will be required to respond to questions and complete exercises referring to AS/NZS 3500.2 Sanitary plumbing and drainage.

AS/NZS 3500.2 SANITARY PLUMBING AND DRAINAGE

Before studying this subject, you should have completed training in work health and safety (WHS) as outlined in Chapter 3 of *Basic Plumbing Services Skills*.

Preparing for work

Plans and specifications

Before work can commence on-site, you must have obtained all the plans and specifications. The following checklist reviews the requirements discussed in Unit 4:

- plans of proposed building/building alterations
- approval from local government authority
- approval from sewerage authority
- specifications and/or standards relating to drain materials and installation
- identified responsibility for payment of authority fees and charges
- plumbing permit or consent from sewerage authority
- sewerage/regulatory authority plumbing inspection requirements
- sewerage authority plans indicating location and depth of approved point of discharge, whether the

connection type is direct or boundary trap, and the location of easements
- information on the location of all existing underground services – for example, Dial Before You Dig (http://www.1100.com.au; telephone: 1100)
- depth of cover requirements
- soffit-level requirements
- other special local requirements that might determine the layout of the sanitary drain.

Safety and environmental requirements

As discussed in Unit 4, you must have planned for the job and obtained all of the necessary equipment to comply with WHS and environmental regulations in your state or territory. The examples given in Unit 4 included:

- trench support
- prevention of stormwater entry to the approved point of discharge.

If trench support is required, as discussed in Unit 2, then trench shields and/or timbers must be available and installed before drain laying commences. If, due to weather conditions or groundwater, it is possible for stormwater to enter the approved point of discharge, then you must be prepared to dewater the excavation in an approved manner. You must therefore have pumping equipment, and possibly ground spears, to pump trench water *in an approved manner to a point of discharge for stormwater approved by the local controlling authority*.

While you will have studied general safety considerations and equipment, there are some specific issues of safety when installing below-ground sanitary drains, and you should be prepared for them before commencing work on-site. Often, excavation work is carried out by at least two people – one person on the ground and another operating an excavation machine. It is very important that you have an established communication system for stopping machine excavation.

The person on the ground must be watchful for other services that may be in the ground. Sometimes during excavation, service pipes and conduits are exposed that are in a different position from that indicated on plans from Dial Before You Dig.

Regardless of whether trench support and shields are required, you must remain watchful for loose rocks and soil, even in shallow trenches, as these can cause serious injury.

Remember, each state and territory has WHS regulations that require employers and self-employed people to identify hazards and assess and control risks at the workplace in consultation with their workers. *It is everyone's job to identify hazards and control risks.*

GET IT RIGHT

DEWATERING TRENCHES

FIGURE 10.1 Trench being dewatered

Figure 10.1 shows a trench being dewatered by draining the water into an inspection opening at the sewer point of discharge.

1   Is this an approved way of dewatering a trench?

_____

_____

_____

2   What could happen if stormwater, soil and bedding material enters the sewer main?

_____

_____

_____

3   How would you dewater this trench?

_____

_____

_____

Quality assurance

As discussed in Unit 4, there may be requirements for quality assurance. All pipes and fittings used in a sanitary drainage system must comply with the *National Construction Code (NCC) Volume 3, Plumbing Code of Australia* and bear WaterMark and Australian Standards certification.

You must be aware of the project requirements for quality assurance, such as the necessary inspections and tests required by the regulatory authority. Inspection and testing will be discussed in greater detail later in this unit. When preparing for your work, you must have planned when the drain, or section of the drain, will be completed, and have the necessary reference numbers and phone numbers to book tests and inspections with the regulatory authority (see Table 10.1). Usually, the regulatory authority will require that you book the drain for inspection with at least 24 hours' notice.

TABLE 10.1 Examples of tests and inspections

Examples of tests	Examples of inspections
Hydrostatic, vacuum and air testing on sanitary drains	Drain gradient
Pressure testing rising mains from a wet well	Adequate bedding and concrete support
	Lengths of unventilated drain
	CCTV inspection of renovated drains

Inspections are carried out by regulatory authorities to ensure the sanitary drain has been installed correctly in accordance with AS/NZS 3500.2 and local authority requirements.

 **AS/NZS 3500.2 SANITARY PLUMBING AND DRAINAGE**

Sequencing of tasks

In Unit 4, we discussed the coordination of your work with other trades. Before commencing work on-site, it is best practice to arrange for a site meeting with the construction supervisor and trade supervisors. A good site meeting should provide, among other things, the proposed construction schedule and contact details of trade supervisors.

Many building projects have a fairly loose construction schedule, and this is when communication with other trades is more important, as significant problems can arise when there is a clash of trades or work is not carried out in the correct sequence. It is not unknown for a concrete slab to have been poured when the drain has not been laid!

Aside from the sequencing of work in coordination with other trades, you will need to plan and carry out the installation of the sanitary drain in the correct sequence. The best approach is first to establish and reveal the unknown factors, such as the precise location and depth of the approved point of discharge.

As discussed in Unit 1, before work commences the worksite must be prepared to ensure that the sanitary drainage system can be installed in an efficient and safe manner.

 COMPLETE WORKSHEET 1

Tools and equipment

Safety equipment

Safety equipment specific to trenching operations and installation of sanitary drains includes:

- hard hat
- ear muffs
- safety glasses (particularly when using jackhammers and drills)
- sunscreen
- gloves
- safety boots
- trench support (see Unit 2).

Tools

A range of hand and power tools and mechanical equipment is required. This may include:

- shovel, pick and crowbar, to expose the approved point of discharge
- string line, to ensure the drain is laid straight and to the correct gradient
- linear measuring equipment, including scale ruler, 8 m tape measure, 25–30 m tape measure
- level measuring equipment, including spirit level and laser level or similar, to establish site ground levels, adjust the vertical alignment of shafts and jump-ups, and ensure drains are laid at the correct gradient
- excavator (including rock breaker, if required)
- cutting tools, including hacksaw, mitre box and panel saw, files, pipe de-burring tool
- angle grinder or compression cutter, to cut into older pipe materials, if required
- jackhammer, to break rock and concrete.

Special equipment – selection and operation

Laser level

Laser levels are commonly used in drain-laying operations. All laser levels must have a compliance plate attached to the body of the instrument. Among other information, this plate will give the 'class' of the laser. You may only use a Class 1 laser. To use any other class of laser, you must have special training. Laser levels should be regularly calibrated.

Laser levels should be regularly calibrated to ensure accuracy. This instrument should always be treated with respect, and should be carefully stored when not in use. Laser levels are commonly available for hire. For more information, refer to Unit 3.

Installation requirements

Materials and fittings

As discussed in Unit 4, the materials and fittings are established when planning the layout of the sanitary drainage system. In Unit 4, a materials list was developed from the plan of the example project. A similar list can be made from any plan, using the procedures in Unit 4. The list is then used to obtain competitive prices from plumbing suppliers. Refer to AS/NZS 3500.2 for a list of approved materials.

AS/NZS 3500.2 SANITARY PLUMBING AND DRAINAGE

In situations where there are hot discharges or corrosive wastes, it will be necessary to lay trade waste drains that are constructed of materials suited to the application. For example, The Rehau Raupiano Plus acoustic wastewater system, which is made of mineral-filled Polypropylene (PP-MD), is suitable for gravity wastewater and kitchen trade waste with temperature resistance up to 90°C (95°C short-term), pH 2 to 12. It is approved for both above and below ground (see Figure 10.2).

Allowances and ordering materials

Joining pipes and fittings

For each pipe material, there are different methods for joining pipes to fittings. For examples, see Table 10.2.

The most common pipe and jointing method is PVC joined using priming fluid and solvent cement. This type of joint cannot easily be disassembled, so if you make a mistake in measurements, it is likely that the fitting will not be usable again, unless the solvent cement joint can be used elsewhere on the job.

When jointing pipes to fittings, it is important to be aware of the allowances that need to be made for fabrication and assembly. For example, when jointing PVC pipes, you must allow for all of the consumables that will be used, such as priming fluid, solvent cement and cutting blades.

Material quantities

The pipes and fittings may be estimated or calculated from the plan by starting at the authority's connection point and progressively following the drain upstream to

FIGURE 10.2 Polypropylene trade waste drain

Source: REHAU Pty Ltd/Shane Ross

TABLE 10.2 Pipe material and jointing methods

Material type	Joint types
Brass (copper alloy)	Silver brazing (minimum 1.8% silver)
Cast iron and ductile iron	Mechanical coupling joint Elastomeric seal spigot socket joint Spigot socket joint with stainless-steel locking segment
Copper	Silver brazed (minimum 1.8% silver)
HDPE (high-density polyethylene)	Electrofusion couplings Butt welds
PP (polypropylene)	Push fit using lubricant
PVC (polyvinyl chloride)	Joint prepared using red priming fluid, and joined using Type N blue solvent cement (glued joint) (Joining techniques and precautions are covered in Unit 11.)
Vitrified clay	Rubber ring joint (socket spigot with toroidal rubber seal)
Note: Refer to AS/NZS 3500.2 for other approved materials.	

each branch, adding the materials and fittings until you arrive at the upstream end.

In Unit 4, we developed a materials list from the plan of an example project. This requires practice and knowledge of the pipe materials, fittings and consumables.

At this point, your teacher or instructor should introduce you to various pipes and fittings.

Ordering, collecting and checking the delivery

Once the pipes, fittings and consumables have been identified, they are ordered. Suppliers of plumbing materials may also require an order number, as this is often used by plumbing businesses to allocate costs to each job. Before ordering pipe and fittings, always check the order list to ensure nothing has been left out.

When ordering, ensure only approved-type pipe and materials are being used. Pipe-fitting quantities should be ordered to meet the needs of the plan. Different pipe materials are manufactured in different lengths, and so you will need to allow for some waste.

Pipe materials are manufactured in the following lengths:

- cast iron: 3 m
- PVC: 6 m
- copper: 6 m
- vitrified clay pipe: 300 mm, 600 mm and 1200 mm.

On delivery, inspect the materials and quantities to ensure they are those that were ordered, are of the correct standard and have been delivered in an acceptable condition. Without this inspection, you might be installing products that are inferior or damaged, and this could cause breakdowns within the drainage system. Don't forget that you are required to provide a warranty on your work.

 **COMPLETE WORKSHEET 2**

Installing the below-ground sanitary drain

Setting out

In setting out the job, you will need to read and interpret plans and specifications and then mark out for the following:

- the proposed location of fixture discharge pipes
- the location of the downstream connection, or approved point of discharge, noting the depth from the plans
- the location of all services and obstructions
- the location of easements
- the type of connection, identified as either boundary trap or direct.

Remember that if any excavation is required on public property, it will be necessary to make applications, and to pay the appropriate fees, to the local council/controlling authority.

All sewer drains must be laid to the minimum gradients specified in AS/NZS 3500.2. The first objective in drain laying is to achieve an even trench base at the correct gradient. At the same time, it is best practice to keep the lines of the drain as short as possible, and to limit changes of direction to less than 90° wherever possible. This is because 90° bends create a greater resistance to flow than, for example, 45° bends.

 **AS/NZS 3500.2 SANITARY PLUMBING AND DRAINAGE**

When connecting drains on grade, only 45° junctions may be used.

After considering the above issues, you can then mark where the drain is to be laid. The line of drain is marked from the point of connection to the last fixture. This can be done by stretching a string along the surface of the ground and marking it with a pick, or by sprinkling some lime (white powder) along the proposed line of drain (see Figure 10.3).

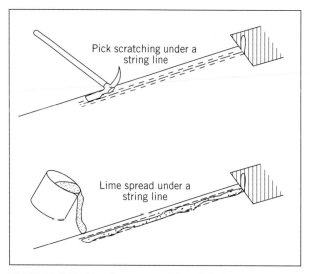

FIGURE 10.3 Marking the proposed line of drain

Source: Department of Education and Training (http://www.training.gov.au) © 2013 Commonwealth of Australia.

The final steps in setting out are:

- having ready the appropriate equipment to excavate a trench for the drain
- placing the pipes and fittings next to where they will be installed.

Excavating

You need to prepare the site by excavating to the existing drain (if any) or excavating to lay a new drainage system. Remember that, when excavating by machine, you cannot dig any closer than 600 mm to the authorised point of discharge. This is to prevent potential damage to the connection point. You must use hand tools to uncover the connection point.

After setting out the line of drain, you should first dig down and expose the authorised point of discharge. The purpose of this is to verify that the information on the plan is correct in relation to the depth and position of the point, and that drainage by gravity can be achieved.

In Unit 4, we also discussed the minimum distances required from excavation to the building footing. These minimum distances must be maintained, but, generally, it is good practice not to excavate within 600 mm of buildings. Where excavations are closer than 600 mm, there is an increased risk of foundation movement.

Excavations should be straight, as short as possible, and with the fewest possible changes of direction and gradient (see **Figure 10.4**). Excavated material should be placed at least 1 m away from the sides of the trench.

FIGURE 10.4 Excavations should be straight and as short as possible

Depth of cover

Before excavating, you must first determine where the drain is to be laid. These locations may include:

- a public thoroughfare
- a driveway
- elsewhere.

If the drain is laid in a public thoroughfare, it would have to be laid deeper for protection against damage than if it were laid in private property. Another important factor to consider is the drain material. For example, PVC requires more cover for its protection than 6.4 mm-thick cast iron.

The minimum cover may be reduced if the drain will have concrete cover, but approval for any reduced depth of cover may depend upon the local regulatory authority. (Depth of cover was discussed in Unit 4. Also read AS/NZS 3500.2.)

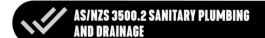

Excavation methods

Excavation of the sewer trench can be carried out in three ways:

1 by machine
2 by hand tools
3 by a combination of both.

Where access is not restricted, it is generally preferred to excavate by machine. Machine excavation is much faster and more economical than manual excavation. If machine access is not possible, then the excavation must be done manually. On some jobs, where access varies on the site, the main line for the drain is dug by machine and the branches are dug by hand.

When uncovering existing services and obstructions, it is important to be aware of the position, depth and type of material. It may be necessary to excavate by hand, particularly near other services, such as electrical, gas or communications, for example.

If the excavation is in unstable ground or exceeds 1.5 m in depth, then you must install shoring. If you are unsure of the requirements for trenches more than 1.5 m deep, it is your responsibility to contact your state/territory WHS authority and to comply with its requirements. (Read AS/NZS 3500.2.)

Types of property drain connection

As discussed in Unit 4, for new drainage installations you must install either a boundary trap or an inspection shaft near the connection point clear of easements and within the property. The sewerage authority will determine which method of connection is required.

The following text and related figures provide guidance for these installations.

Inspection shaft (direct) connection

In areas where no boundary trap is required, generally an inspection shaft is installed near the connection point. Inspection shafts must terminate at, or near, ground or surface level with a removable airtight inspection cap that must be the same diameter as the shaft or riser. Where the riser is likely to be damaged by vehicular traffic, the cap must be installed below surface level, with an approved protective cover.

The junction fitting or bend at the base of the shaft must be supported on concrete a minimum of 100 mm thick. The concrete supports the fitting from the load of the shaft and protects it from impacts during inspection and maintenance. *Note:* Concrete is not required under cast-iron fittings.

The shaft should not extend from any junction in a property drain laid in excess of a gradient of 1.65%. The reason for this is that the angle of the junction is manufactured only to accommodate this gradient. The shaft must rise vertically from the junction. If a steeper gradient were used, the shaft would not be vertical.

Termination of inspection shafts

Inspection shafts are extended from a junction to below surface level (see **Figure 10.5**). A PVC socket with screwed cap is joined by solvent cement to the top of the PVC inspection shaft, to seal the shaft just below surface level (see **Figure 10.6**). *Note:* Some regulatory authorities have differing requirements for the termination of inspection shafts.

Inspection shafts made from vitrified clay are terminated with a vitrified clay plug sealed with a rubber ring to the collar of the pipe (see **Figure 10.7**).

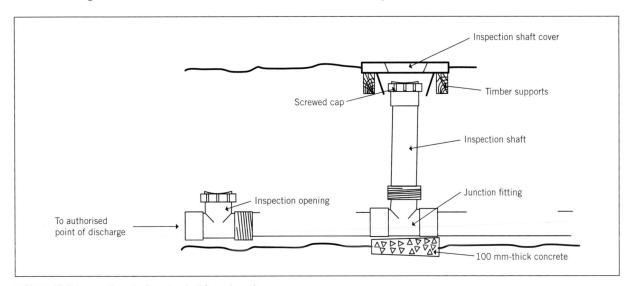

FIGURE 10.5 Inspection shaft extended from junction

Source: Department of Education and Training (http://www.training.gov.au) © 2013 Commonwealth of Australia.

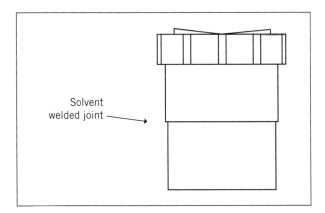

FIGURE 10.6 Screwed cap solvent welded

Source: Department of Education and Training (http://www.training.gov.au) © 2013 Commonwealth of Australia.

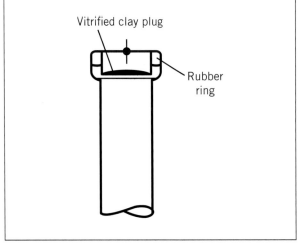

FIGURE 10.7 Vitrified clay plug sealed with rubber ring

Source: Department of Education and Training (http://www.training.gov.au) © 2013 Commonwealth of Australia.

Some fine-grained material (sand) should be packed between the inside of the collar and the plug to prevent the plug from rolling out. As an alternative to sand, a spring clip can be used to retain the plug.

Inspection shaft covers

Inspection shafts must be protected with either a concrete light cover (see Figure 10.8) or a cast-iron heavy cover (see Figure 10.9).

Heavy covers are used where the area is subject to vehicular traffic.

The minimum clearance between the top of the inspection shaft and the heavy cover is 75 mm. You must also ensure that the cover is adequately supported so that no loads at surface level are transmitted to the shaft and drain.

Provision for minor settlement

When laying drains of vitrified clay, the pipe or fitting supporting the inspection shaft must be bedded on concrete at least 100 mm thick. In these cases, a short pipe (600 mm) must be placed on grade (i.e. at its approved gradient) on either side of the concrete bedded fitting to allow for ground movement. Any movement may then be taken up in the rubber ring joints within 600 mm of the fixed point without fracturing pipes or fittings.

Note: Local authority requirements may differ. (Read AS/NZS 3500.2.)

Inspection shaft jump-ups

When the connection point is deeper than required for drainage by gravity, an inspection shaft may be extended from a jump-up. A jump-up is a vertical section in the drain used to connect two sections on a gradient (see Figure 10.10).

Boundary trap connection

As discussed in Unit 4, the purpose of a boundary trap (or BT) is to provide a water seal to prevent the flow of sewer gases from the sewer main entering the property drain. The water seal of the boundary trap is a barrier to the gases.

The boundary trap shaft can be constructed and terminated in the same manner as the inspection shaft; however, boundary traps must be provided with some form of downstream ventilation. The top of the shaft is usually constructed with a ground vent to provide this downstream ventilation (see Figure 10.11). Check with local authorities to meet their requirements.

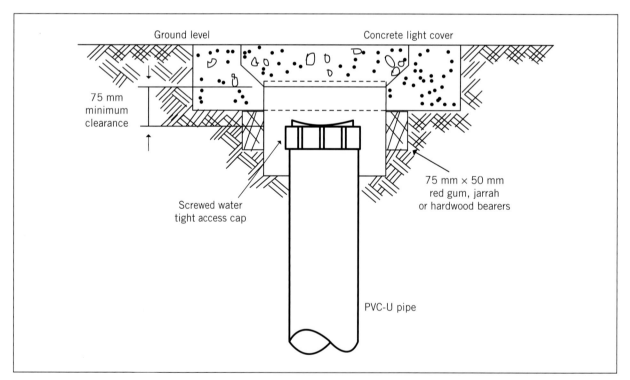

FIGURE 10.8 Inspection shaft light cover

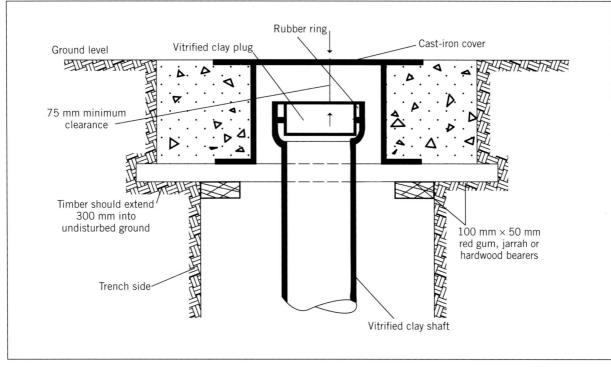

FIGURE 10.9 Inspection shaft heavy cover over vitrified clay inspection shaft

Source: Department of Education and Training (http://www.training.gov.au) © 2013 Commonwealth of Australia.

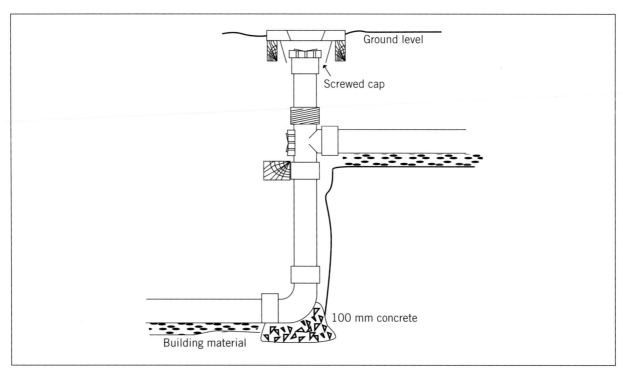

FIGURE 10.10 Inspection shaft jump-up

Source: Department of Education and Training (http://www.training.gov.au) © 2013 Commonwealth of Australia.

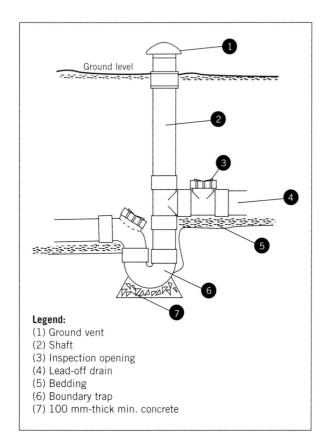

Legend:
(1) Ground vent
(2) Shaft
(3) Inspection opening
(4) Lead-off drain
(5) Bedding
(6) Boundary trap
(7) 100 mm-thick min. concrete

FIGURE 10.11 Boundary trap with ground vent

Source: Department of Education and Training (http://www.training.gov.au) © 2013
Commonwealth of Australia.

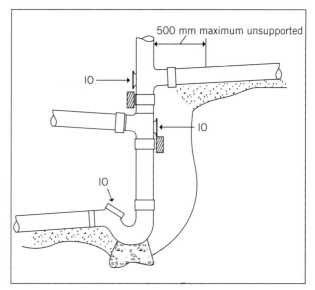

FIGURE 10.12 Boundary trap lead-off support

Source: Department of Education and Training (http://www.training.gov.au) © 2013
Commonwealth of Australia.

FIGURE 10.13 Boundary trap in PVC must have trench support within 500 mm of the boundary trap shaft

The boundary trap must be at least 100 mm in size and of a 'P' trap form with a minimum of 50 mm of water seal. Boundary traps may be constructed from PVC, cast iron or vitrified clay.

Location of boundary traps

Boundary traps are installed in the same location as inspection shafts, as close as possible to the sewer connection point, in an accessible position and clear of easements within the property.

The section of drain between the connection point and the boundary trap must be laid at a gradient to allow for the boundary shaft to be constructed vertical.

As discussed earlier for direct connections, the connection point is often deeper than required. In these cases, a lead-off from the boundary trap shaft may be used to connect the property drain.

PVC and cast-iron boundary trap installations

There are no restrictions on the depth of the boundary trap using cast iron or PVC. Two or more branch lead-offs can discharge into the shaft at different heights. If the shaft and lead-offs are constructed of PVC, the lead-off must be supported within 500 mm of the shaft (see **Figures 10.12** and **10.13**). If this cannot be achieved, the shaft and lead-off could be constructed in cast-iron pipe and fittings. Alternatively, the over excavation can be filled with crushed rock compacted to the original soil density, or filled with concrete providing 75 mm clearance from the drain.

Note: Regulatory authorities differ in their requirements for support of lead-offs from boundary trap shafts. It is best practice to accurately excavate and keep unsupported lead-offs less than 500 mm in length.

Vitrified clay boundary trap installations

The lead-off in vitrified clay must be a short pipe (600 mm), firmly supported for at least 150 mm on solid ground. When solid ground is not supporting this pipe for at least 150 mm, the shaft and lead-off must be constructed of 6.4 mm-thick cast iron or PVC.

Note: The requirements of local regulatory authorities differ in relation to support of lead-offs. Also, for materials other than PVC, cast iron and vitrified clay, it is best practice to seek the guidance of the local regulatory authority in relation to drain support.

A 75 mm × 50 mm or larger hardwood timber support should be laid across the excavation, extended at least 300 mm at each side into undisturbed ground and bearing against the back of the junction before the lead-off is installed. This is to prevent movement of the junction fitting while the lead-off is installed and settlement of the ground occurs. The timber is left in position and backfilled, where it will rot with time (see Figure 10.14).

Termination of boundary trap shafts

Boundary trap shafts can terminate at ground level in one of two ways:

1 The shaft is terminated with a ground (downstream) vent. *Note:* Some regulatory authorities place restrictions on this type of termination.

2 The shaft is sealed just below ground level with a cap or a plug and fitted with a light or heavy cover, in the same way as for inspection shafts (see Figures 10.8 and 10.9). In this case, downstream ventilation is provided in accordance with AS/NZS 3500.2.

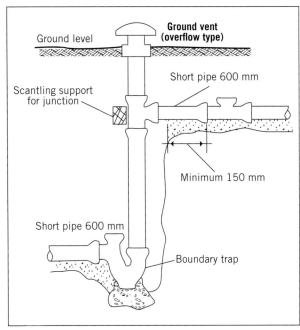

FIGURE 10.14 Vitrified clay boundary trap installation

Source: Department of Education and Training (http://www.training.gov.au) © 2013 Commonwealth of Australia.

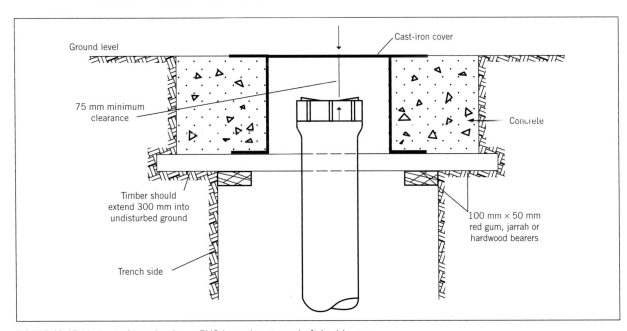

FIGURE 10.15 Method of terminating a PVC boundary trap shaft in driveway

Source: Department of Education and Training (http://www.training.gov.au) © 2013 Commonwealth of Australia.

Jump-ups and inclined drains

Jump-ups

Vertical jump-ups can be constructed at any point along the line of the drain. A jump-up has a bend at the base of the vertical section and is supported by a concrete footing (see Figures 10.16 and 10.17). At the top of the jump-up, an inspection opening (IO) is required. This may take the form of a bend or junction incorporating a full-size IO. Alternatively, a square junction IO may be installed immediately upstream.

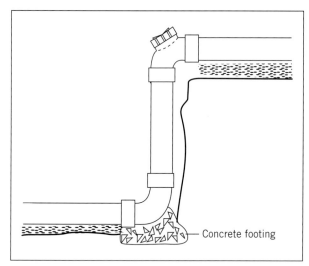

FIGURE 10.16 Jump-up constructed in PVC

Source: Department of Education and Training (http://www.training.gov.au) © 2013 Commonwealth of Australia.

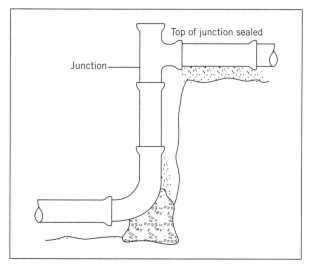

FIGURE 10.17 Jump-up constructed from vitrified clay

Source: Department of Education and Training (http://www.training.gov.au) © 2013 Commonwealth of Australia.

At the top of a jump-up in a vitrified clay drain, a junction fitting may be used, or an IO may be installed immediately upstream.

The purpose of the full-size IO or sealed junction is to provide access for testing and maintenance. (Read AS/NZS 3500.2.)

Steep grades and inclined drains

To minimise excavation, short inclined sections of drain can be laid to form a jump-up. Any drain that is laid on a grade of more than 20% (1:5) must have anchor blocks installed.

Bends of various angles can be assembled to form the change of gradient at both the upper and lower ends. (Read AS/NZS 3500.2.)

Anchor blocks are installed in inclined drains at the bend or junction fittings, at both the top and bottom, and between at intervals not exceeding 3 m.

Overflow relief

In Unit 4, the various options for overflow relief were discussed in detail. Generally, at least one overflow relief gully (ORG) must be installed in every property drain.

When constructing the ORG, it is important to ensure that the final location of the grate will be at least 150 mm below the outlet of the lowest fixture. Also, the grate must be either 75 mm above the unpaved natural surface level (see Figure 10.18), or above the paved surface and not subject to entry of stormwater.

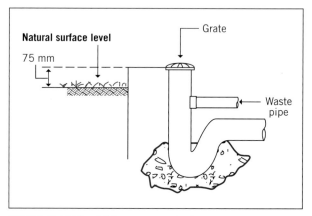

FIGURE 10.18 The ORG grate must be 75 mm above the unpaved natural surface level

Source: Department of Education and Training (http://www.training.gov.au) © 2013 Commonwealth of Australia.

The construction of the overflow gully must include at least 100 mm-thick concrete under the drainage trap.

Fixture discharge pipes may be connected to the ORG in the following ways:

- where above ground level, directly to the riser (see Figure 10.18)
- through an unventilated branch drain connected to a junction in the ORG riser
- where above ground level, to frogmouth or 100 mm surface fittings at either side of the riser (see Figure 10.19). The top finishing collar is fitted with a removable domed grate.

If the level requirements for an ORG cannot be achieved, it will be necessary to install a reflux valve (see Figures 10.20 and 10.21). For more information, read AS/NZS 3500.2. The installation requirements for reflux valves are discussed in detail in Unit 4.

FIGURE 10.19 ORG riser and surface fittings. The frogmouth is shown on the left, and the finishing collar is shown in the centre.

FIGURE 10.20 PVC 100 mm reflux valve, top view

AS/NZS 3500.2 SANITARY PLUMBING AND DRAINAGE

Ventilation of the drain

Alignment of drains

When using unequal junctions to join drains of different sizes on grade, you must ensure that the invert of the

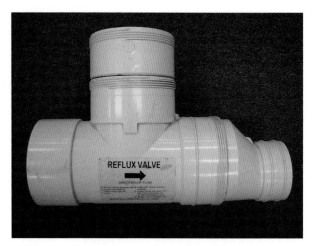

FIGURE 10.21 PVC 100 mm reflux valve, side view

branch drain is at least 10 mm higher than the soffit of the drain to which it connects. If an equal junction is used, and the branch drain is the same size as the main drain, the connection may be made on grade.

In Unit 4, we discussed ventilation of the drain in detail.

Remember, there are two types of drainage connections: direct (non-BT) and boundary trap (BT). The drainage ventilation requirements for each of these connections are different.

Note: The sewerage authority determines if the locality is a boundary trap area or a non-BT (direct) area. The plumber does not have a choice.

Ventilation of a drain in a non-BT area

In a non-BT area there is no water seal to disconnect the entry of sewer gases from the sewer main to the property drain. Air can move freely from one drainage vent through the property drain, sewer main and other property drains.

Ventilation of the sewer main is provided by the upstream drainage vents of numerous property drains.

The only vent that is required on a property drain in a non-BT area is an upstream vent, which must be located within 10 m of the upstream end of the drain. Also, any branch drain must not exceed 10 m without providing for ventilation. (Read AS/NZS 3500.2.)

AS/NZS 3500.2 SANITARY PLUMBING AND DRAINAGE

Ventilation of a drain in a boundary trap area

When a boundary trap is installed, there is a water seal in the boundary trap disconnecting the property drain from the sewer main. Therefore, it is necessary to provide downstream and upstream ventilation in the property drain. This is done by fitting a vent at both the lower (downstream vent) and upper (upstream vent) ends of the drain. This will provide airflow within the property drain to enable it to function properly.

We discussed the options for installation of downstream vents in Unit 4, and prior to installation you should have a plan for the installation of a downstream vent. At the installation stage, you should be aware of two common errors:

1 When installing a ground vent, you must ensure it will not be located within 3 m of the opening to a building, or 5 m from an air duct intake.
2 The downstream vent must be connected to the drain within 10 m of the boundary trap riser, and no other fixture is permitted to be connected to the drain between the boundary trap riser and the vent connection.

Inspection openings and enclosures

Inspection openings

Inspection openings in drains are installed to enable inspection, testing and maintenance, as follows:

■ inspection by the regulatory authority inspector, or CCTV camera
■ testing by hydrostatic (water) or air pressure
■ maintenance for clearance of blockages.

Prior to laying a property drain, you must make yourself familiar with the requirements for the position of IOs.

Inspection openings on drains may come in the form of separate fittings, or may be cast integrally into the bend or junction (see **Figure 10.22**). It is good practice (where possible) to install bends and junctions incorporating IOs, in preference to separate IOs, because this is more economical.

There are specific requirements for installation of inspection openings in sanitary drains. (Read AS/NZS 3500.2.)

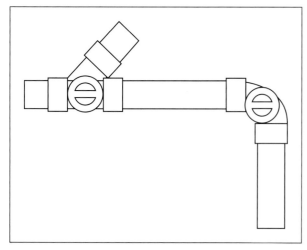

FIGURE 10.22 Bends and junctions incorporating IOs

Source: Department of Education and Training (http://www.training.gov.au) © 2013 Commonwealth of Australia.

AS/NZS 3500.2 SANITARY PLUMBING AND DRAINAGE

Prefabricated inspection openings and enclosures

Inspection openings may also take the form of an inspection chamber. Although more expensive to install, inspection chambers are often specified for projects where additional full access to drains is required.

Inspection chambers may also be used to provide access to reflux valves. Inspection chambers may be circular or rectangular, and can be prefabricated or constructed on-site from concrete.

The dimensions of inspection chambers are determined according to the depth required and their shape. (Read AS/NZS 3500.2.)

AS/NZS 3500.2 SANITARY PLUMBING AND DRAINAGE

Where an inspection chamber is more than 1.2 m deep, access such as ladders must be installed in accordance with AS 1657 Fixed platforms, walkways, stairways and ladders – Design, construction and installation and AS/NZS 4680 Hot-dip galvanized (zinc) coatings on fabricated ferrous articles.

AS 1657 FIXED PLATFORMS, WALKWAYS, STAIRWAYS AND LADDERS – DESIGN, CONSTRUCTION AND INSTALLATION

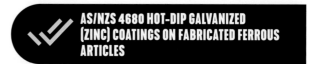
AS/NZS 4680 HOT-DIP GALVANIZED (ZINC) COATINGS ON FABRICATED FERROUS ARTICLES

When constructing an inspection chamber on-site, the construction must comply with AS/NZS 3500.2.

AS/NZS 3500.2 SANITARY PLUMBING AND DRAINAGE

When installing or assembling a prefabricated inspection chamber, the construction must comply with AS/NZS 3500.2 and the manufacturer's requirements. This is important, because there are restrictions on the installation of some proprietary inspection chambers, particularly in relation to the maximum depth and backfilling specifications.

✓ AS/NZS 3500.2 SANITARY PLUMBING AND DRAINAGE

Where holes are made in the walls of concrete inspection chambers for pipes or fittings, they must be made watertight by one of the following means:

1   preparing the concrete surface with BondCrete or a similar material, and filling the annular (ring-shaped) space between the concrete and pipe or fitting with a stiff 2:1 mix of sand-to-cement mortar

2   caulking and sealing the annular space with an epoxy-based, or other approved, sealant, such as Ferropre, or a similar product.

The drainage lines to any inspection chamber must be connected in accordance with AS/NZS 3500.2 for sanitary drainage.

✓ AS/NZS 3500.2 SANITARY PLUMBING AND DRAINAGE

Where different settlement may occur (i.e. mine subsidence areas) and the drain passes through the wall of an inspection chamber over 1 m deep, two flexible couplings should be installed external to and within 800 mm of the wall of the chamber. The pipe between the two flexible couplings must not exceed 600 mm in length. Figure 10.23 shows the base section of a pre-cast (rotationally moulded) polyethylene inspection chamber.

Polypropylene maintenance shafts are also available, and may be installed where permitted by the local sewerage authority (see Figure 10.24). Polypropylene maintenance shafts must be installed in accordance with the manufacturer's instructions.

Inspection chambers will need to be fitted with covers. The type of cover will be determined by the

FIGURE 10.24 Polypropylene maintenance shaft with PVC extension

loading and location. Where the inspection chamber is in an area subject to vehicular traffic, it should be fitted with a gas-tight cover made from concrete or cast iron (see Figure 10.25).

When arranging to install a pre-cast inspection chamber, it is best practice to have the excavation completed at the time of delivery, so the chamber can be installed directly into the excavation. This will save double handling.

Concrete forming for sewer inspection chambers

If you are using a pre-cast concrete chamber that does not have a base, then you will need to form up and pour a base. This will involve:

1   preparing a level sub-grade

2   erecting formwork

3   cutting, placing and tying reinforcement

4   placing and hand screeding concrete for a slab of a minimum depth of 150 mm to the required finished level and job specification.

Any concrete used in a channel of an inspection chamber must be properly formed in accordance with AS/NZS 3500 Plumbing and drainage, and the concrete should be made from sulphate-resistant cement. The wall of the chamber must also be keyed in and sealed to the concrete base.

FIGURE 10.23 Base section of pre-cast polyethylene inspection chamber

FIGURE 10.25 Access covers. Top left: Cast-iron and concrete heavy-duty access cover. Top right: Cast-iron heavy-duty lamp hole cover. Bottom left: Heavy-duty gatic cover. Bottom right: Heavy-duty concrete cover

Source: (top left, top right and bottom left) Department of Education and Training (http://www.training.gov.au) © 2013 Commonwealth of Australia; (bottom right) Gary Cook.

AS/NZS 3500 PLUMBING AND DRAINAGE

Bedding and backfilling

Trench dimensions

Trenches should be kept as narrow as possible, with a minimum clearance of 100 mm on each side of the drain.

Fill materials

Fill materials can be classified as either ordinary fill or selected fill. Ordinary fill is material that has been removed from the excavation and used to refill the trench. Selected fill is material chosen for its qualities to suit a particular application.

Bedding

Bedding is the material placed beneath the barrel of the pipe to support the drain. There are two types of bedding:

1   the undisturbed base of a trench if in stable ground, free from rocks or other hard-edged objects and tree roots
2   selected fill material if the trench is in loam, clay, rock, shale or gravel.

Types of bedding fill materials

Bedding fill materials should comply with AS/NZS 3500.2 and local authority requirements.

✔ AS/NZS 3500.2 SANITARY PLUMBING AND DRAINAGE

Installation of bedding material

Where selected bedding material is used, it should extend beneath the barrel of the pipe across the full width of the trench to a minimum depth of 75 mm. It should be compacted and provide even, continuous support to the drain (see **Figure 10.26**).

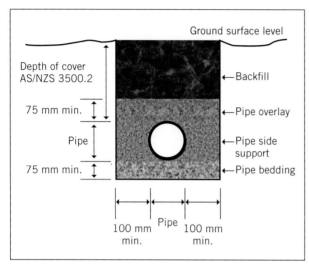

FIGURE 10.26 Trench dimensions and drain bedding

Where cement mortar is to be used, the drain should first be supported at intervals not exceeding 1.5 m. The mortar is then placed beneath the barrel of the drain to a minimum depth of 50 mm.

Note: Some authorities may not permit cement mortar bedding to be used to support drains in roadways and areas subject to vehicular traffic, or where the depth to the invert exceeds 1.5 m.

Pipe side support and overlay

Pipe side support and pipe overlay materials should be similar to the pipe bedding material.

Concrete support

Concrete at least 100 mm thick with a minimum characteristic compressive strength of 20 mPa (megapascals) must be used to support drains in the following locations:

- under drainage traps of material other than cast iron
- under junctions and bends at the base of an inspection shaft
- under all junctions and bends, and jump-ups larger than DN 65.

Note: Concrete must be kept at least 20 mm clear of flexible joints, and should not restrict the flexible joint movement.

Backfill material

Sanitary drains must be covered with at least 75 mm of compacted sand or fine-grained soil. The backfill material must not contain any bricks, rocks or other hard matter larger than 25 mm, or any pieces of soil larger than 75 mm.

In public roadways and other areas subject to vehicular traffic, the backfill should be compacted in layers. It is best practice to backfill areas subject to vehicular traffic with crushed rock.

Note: Local government authorities may have specifications for backfilling trenches in roadways and other public areas. Also, the manufacturers of pre-cast inspection chambers may have specifications for backfilling.

Drain support and provision for movement

Each new residential building site is classified in accordance with AS 2870 Residential slabs and footings. The building plans or geotechnical report will contain details of the site classification.

✔ AS 2870 RESIDENTIAL SLABS AND FOOTINGS

If there is no information available, a site investigation and report from a geotechnical engineer would inform you about the site classification. It is the responsibility of the installer to be aware of the site classification and soil conditions before commencing drainage work.

The following site classifications relate to movement due to moisture changes:

- Class A: Little ground movement (not reactive)
- Class S: Slightly reactive
- Class M: Moderately reactive
- Class H1: Highly reactive, possible high ground movement
- Class H2: Highly reactive, possible very high ground movement
- Class E: Extremely reactive, possible extreme ground movement.

Figure 10.27 illustrates the relationship between classifications, soil movement and costs.

Storm Plastics has developed a chart to give plumbers some drainage solutions to soil movement by installing expansion and swivel joints, and laying drains at specified gradients (see **Figure 10.28**). The referenced drawings may be found on the Storm Plastics website at http://www.stormplastics.com.au.

Drains through concrete edge beams and perimeter strip footings on highly or extremely reactive sites (i.e. Classes H1, H2 and E) must also have flexible joints within 1 m of the building to accommodate movement in any direction. The expansion and swivel joints must be installed in the correct configuration and

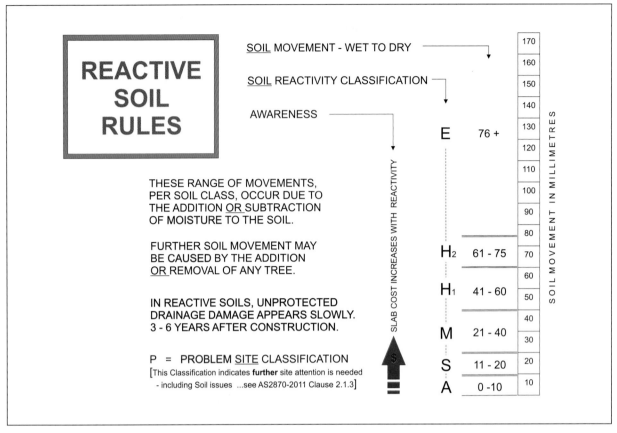

FIGURE 10.27 Reactive soil and soil movement

Source: Storm Plastics Pty Ltd., www.stormplastics.com.au

combination to prevent fracture of the drain when there is soil movement (see Figure 10.29).

Penetrations through concrete edge beams and perimeter strip footings must also have provision for movement. This can be in the form of flexible lagging, wrapped and waterproofed with polyethylene sheet around all sanitary plumbing drain pipe penetrations through footings. The lagging must be a minimum of 20 mm thick on Class M and Class H1 sites and 40 mm thick on Class H2 and Class E sites. Another option is to use sleeve pipes to provide for the same radial movement. The lagging must be flexible and waterproofed before concreting. The objective is to prevent concrete from permeating the lagging and annular space.

Note: Vertical penetrations through concrete slabs do not require lagging.

Problem sites: Class P or 'Problem' sites have potential for movement or collapse due to subsidence, landslip, erosion or uncontrolled fill. The method of protection and support for a drainage installation on a Class 'P' site must be determined by an appropriately qualified engineer.

LEARNING TASK 10.1

Research AS/NZS 3500.2 and answer the following questions:

1 Which clause refers to 'Drains in other than stable ground'?
2 What is the Australian Standard that specifies special design considerations for drains associated with residential slab or footing systems on moderately, highly or extremely reactive soils?
3 Based on the information in this unit, what are two features required of lagging around drains through concrete strip footings?

Installing the drainage system

After completing the excavation, the property drain can be installed. Figure 10.30 shows the process of drainage works being laid to football club shower rooms.

Before laying any pipes, you must remove the screwed cap or plug from the authorised point of discharge. In new estates, the connection point may be a screwed PVC cap. In some areas, the point may be sealed with a rubber ring jointed, vitrified clay plug. Figure 10.31 shows some examples of the way the vitrified clay plugs are fitted.

AS2870-2011 SOIL CLASSIFICATION	ON SITE SOIL CONDITIONS	DIFFERENTIAL MOVEMENT	SEWER GRADE	SWIVEL * (50mm Expansion)	SWIVEL/COMBO * (100mm Expansion)	EXPANDA JOINTS *	CREEP SLOPE SITES	DRAWING NUMBER
A	Most Sand & Rock sites	0 - 10mm		Not necessary	Not necessary	Not necessary		N/a
S	Slightly reactive Soils	11 - 20mm	1:60 Minimum	Not necessary	Not necessary	Not necessary		N/a
M	Moderately reactive soils	21 - 40mm	If any Trees 1:50 Minimum	Check yt on soil report ?	Check yt on soil report ?	Check yt on soil report ?	These are termed P sites and are referred to in Drawing SP 105	SP 100 & SP 101
H1	Highly reactive soils	41 - 60mm		a] Utilising 2 units separated by a 200mm pipe section outside the building footprint.[as shown] b] Expansion Joint at every horizontal junction c] A Swivel set on top of every riser under slab. Not applicable to suspended sub-floors	As necessary using either COMBO BEND or COMBO STRAIGHT UNITS	At Junctions within 1 mtr of internal building footprint and every 6 mtrs.		SP 102
H2	Very highly reactive soils	61 - 75mm	1:40 Minimum			As per Differential Movement		SP 102A
E	Extremely reactive soils	76 + mm		......unless suspended from slab.	......unless suspended from slab.	See AS2032-2006 Clause 6.4.2.2-4 for suspension requirements		SP 102A
P** SITE CLASSIFICATION	Soils affected by Abnormal moisture and conditions	From.. 20 + mm	As per SOIL Report					SP 105A

Sept. 2019, WPT.

NOTE: Engineer or local Authority details take precedence over this chart

To be read in conjunction with Storm Plastics drawings shown.

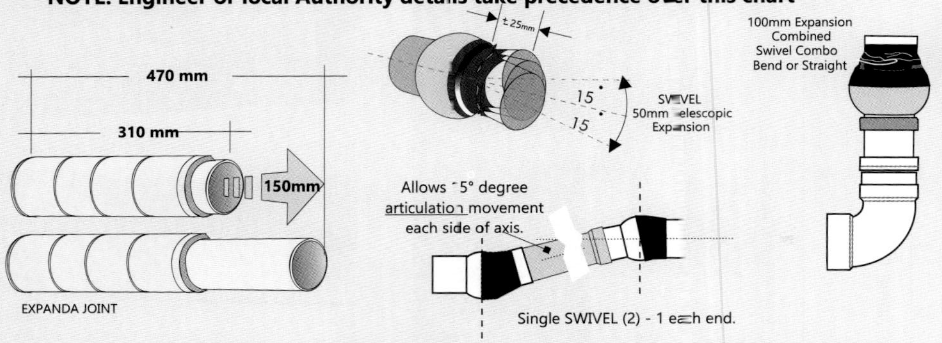

GRADE RATIO	FALL IN 10 mtrs	ANGLE	GRADE %
1:100	100 mm	.57	1.0
1:80	125 mm	.71	1.25
1:60	167 mm	.95	1.65
1:50	200 mm	1.14	2.0
1:40	250 mm	1.43	2.5

100mm Expansion Combined Swivel Combo Bend or Straight

SWIVEL 50mm telescopic Expansion

±25mm

.15 .15

Allows ˜5° degree articulation movement each side of axis.

Single SWIVEL (2) - 1 each end.

470 mm
310 mm
150mm

EXPANDA JOINT

* Unless specified otherwise, these joints are to be set at 50% of total telescopic movement.
** If the site has been called Class 'P' due to the provisions of Clause 1.3 of AS2870-2011 'Abnormal Moisture Conditions' i.e. trees, then the design 'ys' may actually increase generally in excess of 30–50% above the 'Normal Ys' provided within the Site Classification report. For example: A soil with a normal Characteristic Surface Movement 'ys' of 41–60mm (Class H1), which has been classified as Class P due to trees, may have an equivalent design ys of 60–90mm (Class H2 to E).

Source: Jason Bau, Managing Director, AW Geotechnics

FIGURE 10.28 Drainage solutions to soil movement by Storm Plastics

Source: Storm Plastics Pty Ltd., www.stormplastics.com.au

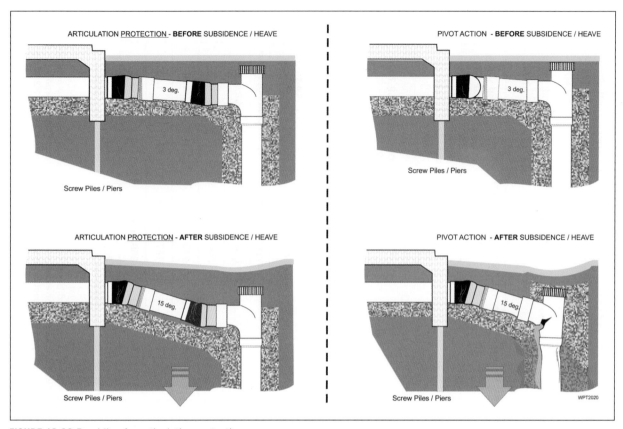

FIGURE 10.29 Providing for articulation protection

Source: Storm Plastics Pty Ltd., www.stormplastics.com.au

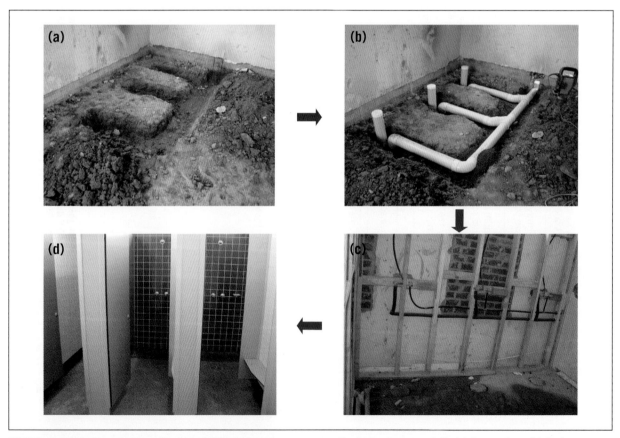

FIGURE 10.30 Sanitary drainage works to football club shower rooms. Clockwise from top left: (a) Excavation; (b) Drains assembled, bedding incomplete; (c) Rough in; (d) Complete.

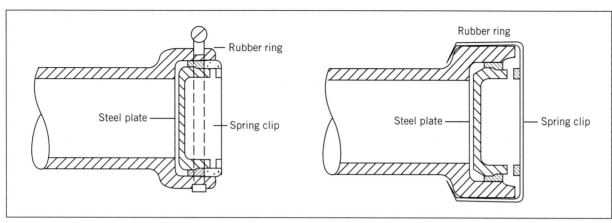

FIGURE 10.31 Sewer caps in vitrified clay drains

When removing the vitrified clay plug, a stainless-steel band can be fitted around the outside of the collar to prevent the earthenware collar from breaking. A plugging chisel can be used to carefully chip out the centre of the vitrified clay plug. When a hole has been opened, you may notice a steel disc under the plug. Continue chipping away from the centre outwards, until the plug has been removed. Remember that no water, soil, sand, rock or other substances are permitted to enter the connection point.

You can then remove the rubber ring and the steel disc. You need to take extra care when doing this, so as not to cause any damage to the surrounding material. If a spring clip is used, you must slide it off and then insert a hook into the holes in the side of the plug, pull it out and then remove the steel disc. The plug must not be removed until the drain is ready to be laid. The connection point should not be left open to entry of stormwater and soil. (Read AS/NZS 3500.2.)

AS/NZS 3500.2 SANITARY PLUMBING AND DRAINAGE

Having planned the layout for the drain, you will be aware of the location of any easements. All inspection shafts, boundary traps and jump-ups must be kept clear of easements, and the property drain must be laid to cross the easement at right angles (see Figure 10.32).

Laying procedures

Before commencing to lay the drain, ensure that the excavation is correct and true to line and grade. You may need to use boning rods to check for an even gradient.

Laying of the drainage pipes should commence from the authorised point of discharge and continue towards the head (upstream end) of the drain. It is best practice to install pipes so that the sockets of the pipe lengths face upstream. This will reduce the chances of matter building up in the joints. The joints will then be lapped in the direction of the flow, as shown in Figure 10.33.

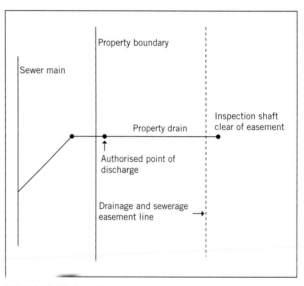

FIGURE 10.32 Crossing easements

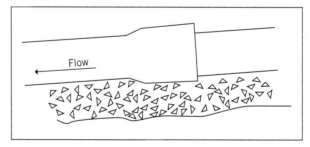

FIGURE 10.33 Pipe sockets and direction of flow

Below is a suggested sequence for installation. The first five steps are illustrated in Figure 10.34.

1 Place the required bedding material in the trench to a minimum depth of 75 mm below the bottom of the pipe for the full width of the base of the trench. (Depending on the type of bedding required – read AS/NZS 3500.2.)

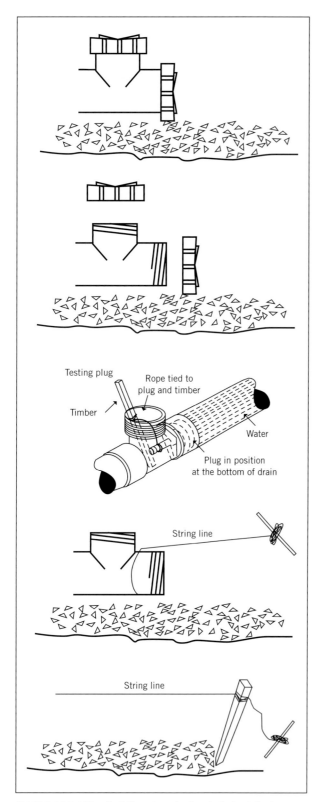

FIGURE 10.34 The first five steps in the drain installation

2   Remove the sealing cap or plug from the authorised point of discharge.

3   Insert a testing plug upstream of the inspection opening. (The purpose of the testing plug is to stop illegal entry of stormwater and soil.)

Note: It is best practice to place a large piece of timber behind the testing plug and then 'tie off' the testing plug to prevent it accidentally falling into the sewer main.

4   Attach a string line to the top of the drain and extend it along the main run of the trench. (You may have to allow for bends in the line of the drain.)

5   Fix a peg at the head (upstream end) of the drain and pull the string tight to the correct gradient. (This is to ensure there is no sag in the line.) An alternative to this would be to raise the line a suitable distance above the top of the pipe – for example, 50 mm. This prevents any fittings obstructing the line. Then ensure that the drain is straight by taking a measurement from the line to the top of the pipe after each length is laid.

6   Commence drain laying from the authorised point of discharge, ensuring that the top of the drain just touches the line, or that the distance from the string line to the top of the drain is constant.

7   Continue laying the line of the drain, including branches (junctions) where necessary, ensuring that each 45° junction is set at the correct gradient to allow for the connection of the branch. Measured sections must be cut straight using a mitre box and panel saw (or other approved cutting tools for the materials being used).

8   When you have reached the head of the drain, ensure the drain is in straight alignment. Adjust, if necessary, for correct alignment and gradient (see **Figure 10.35**).

Another method of achieving straight alignment in long lengths of drain is to open the inspection openings at each end and insert a mirror. When viewing the inside of the drain using the mirror, the soffit (top) of the drain should appear straight through to the far inspection opening.

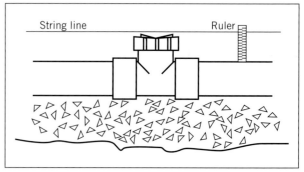

FIGURE 10.35 Maintaining alignment and gradient

9   Connect all branches, laying drains using the above procedure.
10 Check alignment and grade again.
11 Remove the string line and complete the installation of the bedding and side support materials ensuring there are no voids (spaces) in the bedding, and it is thoroughly compacted by tamping.
12 Place the concrete under the drainage traps, bends and fittings as required by AS/NZS 3500.2 and local authority requirements (see **Figure 10.36**).

Cutting in a branch

When additional fixtures are to be installed and connected to the existing sanitary drainage system, a new junction is inserted into an existing drain. This is known as 'cutting in a branch'.

FROM EXPERIENCE

When cutting in a branch, ensure the junction and repair couplings are fully supported and that there are no internal lips or projections that could trap debris and cause blockages. This will ensure that you complete the job properly and there are no call backs from the customer.

The same processes and procedures apply to cutting in a branch as for installing a new sanitary drain, except special care must be taken to minimise any internal lips and projections, particularly when cutting in a PVC branch into vitrified clay.

When cutting in a branch to an existing PVC drain, you need to allow plenty of space around the pipe to make two accurate square cuts to enable you to remove a section of drain. The piece removed can then be measured and another piece made with the same length, but incorporating a new 45° junction with inspection opening.

When inserting the new branch, the joints can be made in several ways, depending on the type of repair coupling. The most common form of repair coupling is the elastomeric adaptor, which consists of a tough type of rubber sleeve and two stainless-steel worm drive bands. Another form is the split repair coupling. The split repair coupling consists of an internal seal centrally located over adjoining pipe spigots. Two halves of a socket are clamped over the seal, using four external wedges (see **Figure 10.37**).

After the branch has been cut in and the new branch drain installed, the new drain must be either water or air tested by inserting a plug into the inspection opening at the new branch. (The existing drain is not tested.)

When cutting in a branch to an existing drain, some authorities will require you to provide for overflow relief by installing an overflow relief gully or altering the surface fittings of an existing disconnector gully.

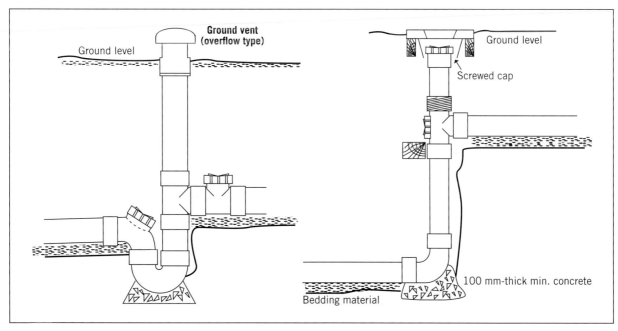

FIGURE 10.36 Boundary trap and inspection shaft arrangements with bedding material and concrete beneath drainage fittings

Source: Department of Education and Training (http://www.training.gov.au) © 2013 Commonwealth of Australia.

FIGURE 10.37 View of half-split repair coupling

Renovation of drains

Sanitary drains may be renovated (lined) using one of the following techniques:

- cured in place pipe (CIPP)
- lining with discrete pipes
- lining with continuous pipes
- lining with close-fit pipes
- lining with spirally wound pipes.

While the procedures for drain renovation are outside the scope of this unit, there are important factors for any plumber to consider when engaging a specialist contractor to perform these works, namely:

- jet cleaning of drains
- CCTV inspection after jet cleaning to determine whether the drain is in a suitable condition for the proposed method of renovation
- the effect of the liner on the hydraulic capacity of the drain
- whether the liner is suited to the nature of discharge
- hydrostatic/air testing on the completed works
- CCTV inspections during the process of drain renovation and upon completion
- compliance with the performance requirements of *NCC Volume 3, Plumbing Code of Australia*
- warranties and insurance on the completed drain renovation work.

 COMPLETE WORKSHEET 3

Testing drains

After every sewer drain has been laid, it must be tested to ensure that it complies with AS/NZS 3500.2. This testing may need to be completed in the presence of an officer of the regulatory authority. It is best practice for the plumber to test the drain prior to the regulatory authority inspection.

 **AS/NZS 3500.2 SANITARY PLUMBING AND DRAINAGE**

The traditional method of testing is the hydrostatic (water) test; however, water conservation is driving a change towards air testing, and we will discuss both methods. Vacuum testing is rarely specified for property sewerage drains, but is commonly carried out when testing new sewer main extensions. If unsure, ask the local regulatory authority which method is required.

GREEN TIP

In some areas, regulatory authorities prohibit the use of drinking water for testing of drains. Where hydrostatic testing is permitted, non-drinking water may be used.

Hydrostatic testing

Hydrostatic testing involves filling the entire drain, or sections of the drain, with water to the highest level. The procedure is as follows:

1 An expandable testing plug is fitted into the downstream end of the drain through an inspection opening and tightened to expand the plug and seal it against the walls of the pipe. This prevents the passage of water (refer to **Figure 10.34** shown earlier).

2 All other open ends of the drain to be water tested should also be plugged.

3 A large piece of timber should be placed behind the plug and tied, to ensure that if the plug does come loose, the timber will prevent it slipping away down the drain.

4 The drain is filled with water to the uppermost section of the drain to be tested – for example, the top of the disconnector gully.

5 The head pressure of water above the section being tested must be between 1 m and 3 m.

6 For PVC drains, the test will pass if there is no loss of water over a period of 15 minutes.

There are other requirements relating to water testing time and make-up water on sewerage drains. (Read AS/NZS 3500.2.)

AS/NZS 3500.2 SANITARY PLUMBING AND DRAINAGE

Air test

As indicated previously, one limitation on the water test is that the maximum allowable head pressure on any section of drain is 3 m. This limitation does not apply to air testing. Therefore, in addition to water conservation, another benefit of air testing is that vertically long sections of drain can be tested.

An air test may be applied to the completed work either in its entirety or in sections. The procedure is as follows:

1 An expandable testing plug (see **Figure 10.38**) is fitted into the drain through an open end or inspection opening and tightened to expand the plug and seal it against the walls of the pipe. This prevents the passage of air.

2 All open ends of the drain to be air tested should be plugged and tied.

3 An air pump and connection incorporating a pressure gauge is attached to the testing plug and air is pumped into the drain, or section of drain, under test to achieve a test pressure of 15 kPa. (*Note:* Some regulatory authorities may specify a different test pressure.)

4 Normally, at least three minutes is allowed for the temperature in the drain to stabilise.

5 To find any leaks, use soapy water.

FIGURE 10.38 Testing plugs available in a range of sizes

There are other requirements relating to air testing time and allowable pressure drop on sewerage drains. (Read AS/NZS 3500.2.) See also the important safety note in the 'Caution' box.

AS/NZS 3500.2 SANITARY PLUMBING AND DRAINAGE

Important safety note: Compressed air has the energy potential to cause serious injury. There have been cases of injury caused by sudden, unexpected release of testing plugs under pressure. Also, when releasing air, there could be particles in the air that can cause eye injury. Therefore, when air testing, always securely fix and tie testing plugs and wear safety glasses.

In the case of vitrified clay drains, it may be necessary to conduct a visual test and a slide test. These test methods were commonly used in the past.

Visual test

Part of the visual test includes using a mirror. Once the hydrostatic or air test has been completed, mirrors can be placed into inspection openings at each end of straight lengths of drain. After water has been run through the drain, the gradient can be verified by the width of the ribbon of water remaining in the drain. By looking into the mirrors, as the inside of the drain is illuminated by daylight/torch, the alignment and effective drainage of water within can be clearly seen.

Any lips in the pipework joints or deflections can be identified and corrected.

Other visual tests and measurements are carried out to verify that:

■ the correct fittings and approved fittings are used
■ the drain has been laid at the correct gradient
■ the bedding material is of the required depth and is compacted
■ concrete is provided where necessary.

Slide test

A 'slide' is a lead weight attached to a spring steel ribbon. In short branches or sections of property drain that cannot be visually inspected, a slide may be lowered into the drain. This is then drawn up slowly to find any pipe lips or obstructions in the drain. This is especially useful when testing the connection to the authority's authorised point of discharge, and for 'cut-ins' to existing drains.

Backfilling the excavation after testing

At the completion of a drainage installation and after the necessary tests have been passed, all inspection openings are inspected and securely sealed. The excavation must then be backfilled, as described earlier in this unit. Use only clean backfill. Do not use any material that could damage the drain. (Read AS/NZS 3500.2.)

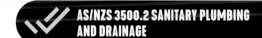
AS/NZS 3500.2 SANITARY PLUMBING AND DRAINAGE

Clean-up

After backfilling trenches and reinstating surfaces, it will be necessary to clean up. Clean up includes:

■ removing excavating equipment and carrying out service work to maintain it in good condition
■ inspection, cleaning, maintenance and storage of tools
■ disposal of unserviceable tools
■ removal and storage of usable off-cuts
■ removal, disposal and recycling of waste to correct bins
■ restoration of any soiled building surfaces to a clean condition
■ restoration of the worksite, where possible, to the original condition.

COMPLETE WORKSHEETS 4 AND 5

SUMMARY

- When installing below-ground sanitary drainage systems, it is necessary to follow the plans and specifications that indicate the location of buildings, plumbing fixtures, easements and underground services. Also, prepare by familiarising yourself with local regulations and inspection requirements, and ensure that pipes and fittings have WaterMark approval. Always be prepared to locate the approved point of discharge before laying drains, and protect it from damage and entry of stormwater, soil and bedding.

- There is safety equipment specific to trenching operations and installation of sanitary drains. When installing the below-ground sanitary drain, set out and excavate the drain in straight lines that are as short as possible. Consider the requirements for bedding, concrete support, ventilation and overflow relief.

- Generally, the sanitary drain will need to be constructed at the correct gradient and with sufficient cover. You need to know the ground conditions and, accordingly, provide lagging and waterproofing of pipes through concrete strip footings and possibly swivel and expansion joints. If you are unsure, or suspect that the ground is reactive or unstable, consult with a geotechnical engineer.

- Sanitary drains must be covered with at least 75 mm of compacted sand or fine-grained soil. Trench backfill must not contain any bricks, rocks or other hard matter that might damage the drain. In areas subject to vehicular traffic, the backfill should be compacted in layers. Backfill areas subject to vehicular traffic with crushed rock. *Ensure that you comply with authority* specifications for backfilling trenches in roadways and other public areas.

- All sewer drains must be tested to ensure they comply with AS/NZS 3500.2. This testing may need to be completed in the presence of an officer of the regulatory authority. The traditional method of testing is hydrostatic (water) testing; and air testing is increasingly being used. With the latter, it is important to understand that compressed air has the energy potential to cause serious injury. Always test drains before they are backfilled.

- When works have been completed, clean up the work area. If required by the regulatory authority, lodge the necessary documentation such as a Certificate of Compliance and as-constructed plan.

WORKSHEET 1

Student name: _____

Enrolment year: _____

Class code: _____

Competency name/Number: _____

Task: Review the section 'Preparing for work', then complete the following.

1   Explain how you would remove stormwater or groundwater from a trench for sanitary drains.

_____

_____

_____

_____

2   During excavation, you should carefully watch for other services. Why?

_____

_____

_____

_____

3   With regard to inspections, you will need to plan for completion of the drain, or section of drain, so that an inspection can be arranged with the regulatory authority. Research your local requirements and answer the following:

   a   What period of notice is required for a sanitary drain inspection appointment?

_____

   b   What tests must be carried out by you and witnessed by the inspector/auditor?

_____

_____

_____

   c   What information is required to make an inspection booking?

_____

_____

4   What is the name of the marking that identifies whether materials for sanitary drains are approved?

_____

5   What are three types of tests that can be performed on sanitary drains?

a _____

b _____

c _____

6   From Units 4 and 10, identify two pieces of information that you should obtain from a site meeting regarding the installation of drains.

a _____

b _____

WORKSHEET 2

Student name: _____

Enrolment year: _____

Class code: _____

Competency name/Number:

Task: Review the sections 'Tools and equipment' and 'Installation requirements', then complete the following.

1   What tools are used to expose the point of discharge?

_____

_____

_____

2   What instrument should be used to check the vertical alignment of shafts?

_____

_____

3   What two fluids are used to join PVC pipes and fittings?

a   _____

b   _____

4   What is used to assist in the jointing of push-fit polypropylene pipes?

_____

_____

5   What are two types of discharges or wastes where it might be necessary to install materials other than PVC?

_____

_____

_____

_____

6   Why is it important to check fittings and pipes for defects when accepting a delivery?

_____

_____

_____

7   Why do you think only approved-type pipes and fittings must be used? Discuss with your teacher or trainer if unsure.

_____

_____

8   Complete the following:

PVC pipes are available in _____m lengths.

9   You are required to provide a warranty on sanitary drainage work. Research and explain the term 'warranty'.

_____

_____

_____

WORKSHEET 3

Student name: _____

Enrolment year: _____

Class code: _____

Competency name/Number: _____

Task: Review the section 'Installing the below-ground sanitary drain', then complete the following.

1   What is the first objective of drain laying?

_____

_____

_____

2   When excavating by machine, you are only permitted to dig within a certain distance from the connection point. What is that distance?

_____

3   What is the minimum distance that excavated material should be placed away from the edge of a trench?

_____

4   Research AS/NZS 3500.2 Sanitary plumbing and drainage. What is the minimum depth of cover over a PVC drain under a private driveway with concrete paving?

_____

5   What is the minimum thickness of concrete required under the junction or bend at the base of a DN 100 PVC inspection shaft?

_____

6   On the diagram below, label all arrows and state what type of cover is used; also add in the figure for the mm minimum clearance.

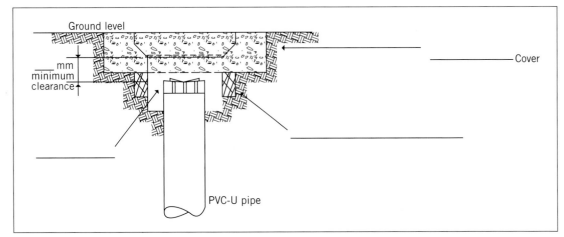

7   Where must inspection shaft assemblies be located in relation to easements?

_____

_____

_____

8   On the sketch below are letters referring to various parts of a boundary trap assembly. Research and
     name each part.

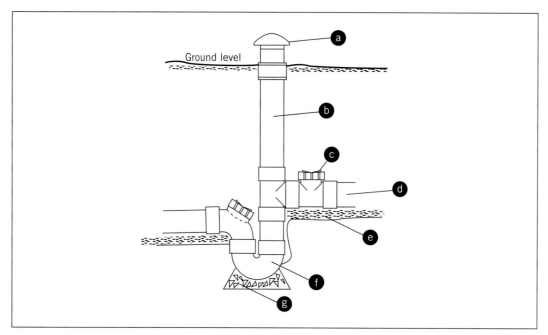

a   _____

b   _____

c   _____

d   _____

e   _____

f   _____

g   _____

9   When installing a boundary trap assembly, are you permitted to connect two lead-off drains to the
     boundary trap shaft?

_____

10 Inspection openings must be installed at the top of a jump-up. Neatly draw a jump-up, and indicate two places that inspection openings can be installed.

11 Neatly draw an overflow relief gully and indicate as follows:

a Minimum height above natural surface level.

b Minimum distance below the outlet of the lowest fixture.

12 Name three reasons why inspection openings are installed in sanitary drains.

a _____

b _____

c _____

13 When an inspection chamber exceeds a certain depth, rung-type ladders are required. Research AS/NZS 3500.2.

a Over what depth does an inspection chamber require ladders?

_____

b What type of coating is required for steel ladders?

_____

14 What is one method of sealing holes in the walls of concrete inspection chambers?

_____

_____

_____

15 What type of cement is used in the construction of concrete inspection chambers?

_____

16 Concrete is required in sanitary drainage installations. Name three locations where concrete must be placed.

a _____

b _____

c _____

17 Is it necessary to place concrete beneath DN 65 PVC bends in sanitary drains?

_____

_____

18 Research your local authority requirements. When cutting in a branch to an existing drain, what obligations do you have to provide for overflow relief?

_____

_____

_____

_____

19 Where is it necessary to install flexible joints in a sanitary drainage system?

_____

_____

_____

20 Who must determine the method of movement and support for a sanitary drain on a Type P site?

_____

 **WORKSHEET 4**

Student name: _____

Enrolment year: _____

Class code: _____

Competency name/Number: _____

Task: Review the sections 'Testing drains' and 'Clean-up', then complete the following.

1   Research AS/NZS 3500.2. When air pressure testing a DN 150 PVC sanitary drain 60 m long, what is:

   a   the minimum test period

   _____

   b   the maximum pressure drop?

   _____

2   When either hydrostatic or air testing a sanitary drain, a piece of solid timber must be placed behind the test plug. Why?

   _____

   _____

3   What is the maximum air test pressure for sanitary drains?

   _____

4   What personal protective equipment (PPE) must be worn while air testing sanitary drains?

   _____

   _____

5   What must be done with unserviceable tools, and why?

   _____

WORKSHEET 5

Student name: _____

Enrolment year: _____

Class code: _____

Competency name/Number: _____

Task: Practical exercises

All questions in the previous four worksheets must be completed and checked by your trainer/teacher before you attempt the following practical exercises.

Your trainer/teacher should provide preliminary guidance, conduct an induction and ask underpinning knowledge questions.

1   Install and test a below-ground sanitary drain to connect a bathroom, WC, kitchen, laundry, and soil or waste stack (to a minimum of 30 fixture units). The drain must be at least 10 m long and terminate at ground level.

2   After completing the above exercise, cut in a branch to the drain for a new WC and basin, and test the new branch drain.

3   Set out, assemble, install and test an approved prefabricated inspection chamber or maintenance shaft, including the connection of the inlet and outlet pipes.

4   Bench an inspection chamber for concrete, using appropriate formwork.

5   Prepare: a sub-grade; erect formwork; cut, place and tie reinforcement; place and hand screed concrete (or simulate with wet sand) for a slab of 4 square metres and a minimum depth of 100 mm to a finished level gradient of 1.65%.

11 INSTALLING STORMWATER AND SUB-SOIL DRAINAGE SYSTEMS

Learning objectives

This unit provides practical guidance on installing stormwater and sub-soil drainage systems. Areas addressed in this unit include:

- depth of cover and position of drains
- material selection
- pipe sizing and gradients
- jointing and assembly of pipes and fittings
- installation and pipe-laying procedures.

Introduction

Plumbers are required to install stormwater and sub-soil drainage systems to collect and convey roof water, surface water and sub-soil water from properties to the legal point of discharge for a stormwater drainage system. This type of work is carried out on all properties where there are stormwater discharges. It could be new work, or it could involve alterations or additions to existing systems.

As stormwater drains are normally laid underground, the focus of this unit is on the requirements for installation of underground drains. Above-ground stormwater drains must be installed in accordance with Australian Standards, authority requirements and the pipe manufacturer's installation instructions.

The stormwater drainage installation includes stormwater drains, sub-soil drains, inlet pits, stormwater pits and surface channels.

While studying this subject, you will be required to respond to questions and complete exercises referring to AS/NZS 3500.3 Plumbing and drainage. Stormwater drainage.

AS/NZS 3500.3 PLUMBING AND DRAINAGE. STORMWATER DRAINAGE

Before studying this subject, you should have completed training in work health and safety (WHS) as outlined in Chapter 3 of *Basic Plumbing Services Skills*.

Preparing for work

Plans and specifications

Plans (or design drawings) and specifications will help you to make decisions about the layout of the drainage system. Where any new stormwater drain is to be installed, the local government authority (council) must approve the building plans before work commences, and the authority may also stipulate conditions and specifications that will determine the design.

Other factors affecting the design layout include the location of existing underground services and obstructions. Proper preparation by obtaining plans and information on existing underground services *for every job* is critical.

Most large projects have specifications and drawings that will determine the drain materials to be used. In this case, the licensed plumber/drainer will have little control over the selection of materials to be used and will need to purchase the drain materials and construct the drain in accordance with the specifications and drawings. Regardless of any specification or drawing, the licensed plumber/drainer

can only construct a stormwater drainage system in accordance with plumbing regulations, codes and standards.

For small residential building projects, the specifications are unlikely to determine materials to be used, other than 'approved materials' as specified in Australian Standards. In these projects, the plumber/drainer will determine the pipe materials to be used and the layout of the stormwater drainage system.

Before commencing work, the plumber/drainer should ensure that:

- the plans have been approved and stamped by the local authority/council
- authority requirements have been met for permits and consents
- information relating to the location of all underground services is obtained (e.g. by calling Dial Before You Dig: http://www.1100.com.au; telephone: 1100)
- arrangements have been made to make the worksite safe for the duration of works (e.g. fencing to prevent public access to open trenches, etc.).

Generally, permits or consents are not required for stormwater drainage work; however, local regulations vary, and you must ascertain the local requirements before commencing work.

It may be necessary for the licensed plumber/drainer to issue a Certificate of Compliance when the work is complete. This confirms that all work is complete and meets regulatory requirements and standards, and also provides a warranty.

You will be required to research your local authority requirements in relation to applications and permits at the end of this unit.

Depth of cover

Plans and specifications may indicate that the stormwater drain requires protection against mechanical damage and deformation due to loadings from vehicles. There is different minimum cover for buried piping, and this depends upon the type of cover and vehicular loading (see Table 11.1).

TABLE 11.1 Minimum cover for external stormwater drains

Location	Minimum depth of cover in mm		
	Cast iron and ductile iron	PVC and other authorised materials	PVC and other authorised materials where there is inadequate soil cover
Light vehicle loads (except roads)	300	450	75 mm cover plus brick or unreinforced concrete*

	Minimum depth of cover in mm		
Location	Cast iron and ductile iron	PVC and other authorised materials	PVC and other authorised materials where there is inadequate soil cover
No vehicle loading or traffic (such as garden area)	Nil	100 for single-dwelling properties 300 for other properties	50 mm cover plus brick or unreinforced concrete*
All roads	300	500	N/A
*Paving to cover at least the full width of the trench			

For example, if you see that the building plan for the site indicates the location of a residential driveway where a PVC (polyvinyl chloride) stormwater drain is to be laid, and you are unsure when the brick or concrete paving will take place, you must lay the drain with at least 450 mm of cover between the top of the drain and the natural surface level. However, you must ascertain the levels on-site, and position and the manner of discharge from the local government authority (council). Where the authorised point of discharge is at a set level, there must be adequate depth to accommodate the minimum gradient for drainage by gravity.

Approved point of discharge

The location of the approved point of discharge (or legal point of discharge) is specified by the local authority. Depending on the local drainage facilities, the approved point of discharge could be:

■ street kerb and gutter
■ on-site storage tank or disposal pit
■ stormwater drain or easement
■ sub-soil distribution system or soak well.

The local authority will determine the position and manner of discharge. Generally, the licensed plumber is responsible for locating the approved point of discharge.

When planning to lay the stormwater drain, it is good practice first to locate the approved point of discharge, as there may be slight variations in the actual depth and location that will affect the layout. You should be prepared to excavate by hand to locate the approved point of discharge. Machinery must not be used to excavate within 600 mm of the connection point.

If the point of connection cannot be located, contact the authority for further direction. When the point has been located, you will need to be prepared to provide protection to the point and prevent the entry of any contaminants, sand, silt or rubbish.

The approved point of discharge provided by the local authority is usually DN 100, but may be larger for larger properties. The connection point may be located either inside or outside the property, in the road or in the footpath. If you will need to work outside the property boundary to gain access to the approved point of discharge, then you will need to seek the permission of the relevant landowner or council and obtain relevant additional permits.

Level of approved point of discharge

The local government authority will determine the available level of stormwater discharge. Generally, any floor levels within the property must be above the level of the approved point of discharge or external drainage network operating water level. Where this is not achievable, it may be necessary to install a pumped system or a reflux valve in accordance with AS/NZS 3500.3 and local government authority requirements.

AS/NZS 3500.3 PLUMBING AND DRAINAGE. STORMWATER DRAINAGE

If drainage by gravity is not achievable, then advice must be sought from the local authority.

Safety and environmental requirements

Correct preparation and planning involves obtaining all necessary equipment to comply with WHS and environmental regulations. Examples include:

■ trench support
■ preventing entry of sand, silt and rubbish to the approved point of discharge.

If trench support is required, as outlined in Unit 2, then trench shields and/or timbers must be available on-site before drain laying commences.

Contaminants, sand, silt or rubbish must not be permitted to enter the approved point of discharge, and you must be prepared to dewater trenches if necessary and filter out any such contaminants. While you are drain laying in water-charged ground, you must have pumping equipment and possibly ground spears to pump water out of the trenches.

Each state and territory has WHS regulations that require employers and self-employed people to identify hazards and assess and control risks at the workplace in consultation with their workers.

While you will have studied general safety considerations and equipment, there are some specific issues of safety when installing below-ground stormwater drains, and you should be prepared for them before commencing work on-site. Excavation work is best carried out by at least two people – one person on the ground, and another person operating the excavation machine. It is very important that you have an established communication system for stopping machine excavation.

The person on the ground must be watchful for other services that may be in the ground. Often service pipes and conduits that are exposed during excavation are in a different position from that indicated on plans.

Regardless of whether trench support is required, you must remain watchful for loose rocks and soil because, even in shallow trenches, falling material can cause serious injury.

 During excavation, keep a close watch for services and objects in the ground. This is not only for protection of underground services, but for your health and safety. Plumbers can encounter live wiring, corroded gas pipelines and chemical containers.

Quality assurance

Materials and products

All pipes and fittings used in a stormwater drainage system must comply with AS/NZS 3500.3 and bear Australian Standards marking. Also, ensure that the installation will comply with local authority requirements.

 AS/NZS 3500.3 PLUMBING AND DRAINAGE. STORMWATER DRAINAGE

GREEN TIP

Stormwater drainage may be used to transport rainwater from roof areas to storage tanks. Where rainwater is to be used for drinking or human consumption, the connecting pipe material/s must comply with AS/NZS 4020 Testing of products for use in contact with drinking water. Drinking water is water that is intended and suitable for human consumption, food preparation, utensil washing or oral hygiene.

AS/NZS 4020 TESTING OF PRODUCTS FOR USE IN CONTACT WITH DRINKING WATER

Installation work

When preparing for your work, you must have planned for the installation of the stormwater drains, and have the necessary reference numbers and phone numbers to book inspections and tests with the regulatory authority (if required).

In some areas, inspection of the stormwater drains is not required by regulatory authorities. However, if you are unfamiliar with local requirements, or are moving to a different area, you must research the local regulations and standards.

Final inspections are carried out by some regulatory authorities to ensure the stormwater drainage system has been installed correctly in accordance with local regulations and AS/NZS 3500.3. Inspection of all, or a percentage, of completed plumbing work is a form of quality assurance of your work for the regulatory authority and property owner.

 **AS/NZS 3500.3 PLUMBING AND DRAINAGE. STORMWATER DRAINAGE**

On larger building sites throughout Australia, additional requirements for quality assurance of plumbing work are common, such as sign-off on portions of final plumbing work for progress payments, and additional tests and inspections.

Later in this unit, we discuss the specific procedures for inspection and testing of stormwater drains.

Some of the issues that should be considered when preparing to install stormwater drains are:
- conformance with company operating procedures
- conformance with plumbing regulations, standards and the local authority
- conformance with project plans and specifications
- ensuring the necessary applications are lodged with the authorities
- arranging for all necessary inspections and tests
- ordering pipes, fittings and products in accordance with the job plans and specifications
- receiving and inspecting pipes, fittings and products for conformance with orders
- storage of pipes, fittings and products to prevent damage or deterioration
- correct selection, operation and maintenance of tools and equipment
- compliance with the contract work schedule
- maintaining records of job variations.

Sequencing of tasks

When preparing for work, you will need to consider the project schedule and coordination of your work with other trades. Before commencing work on-site, it is best practice to arrange for a site meeting with the construction supervisor and trade supervisors. A good site meeting should provide, among other things, the proposed construction schedule and contact details of the trade supervisors.

Many building projects have a fairly loose construction schedule, and this is when communication with other trades is more important. If a clash of trades occurs due to poor coordination, there can be significant problems. For example, your excavation work should not conflict with any work from scaffolding around the building.

Aside from the coordination of your work with other trades, you will need to plan and carry out the installation of the stormwater drain in the correct sequence with other works. For example, stormwater drains must be laid and backfilled before preparation for paving.

When preparing to lay stormwater drains, the best approach is first to establish and reveal the unknown factors, such as the precise location and depth of the approved point of discharge, and the location of other services and obstructions.

As outlined in Unit 1, before work commences, the worksite must be prepared to ensure the stormwater drainage system can be installed in an efficient and safe manner.

Tools and equipment

Safety equipment

Safety equipment specific to trenching operations and installation of sanitary drains includes:

- hard hat
- ear muffs
- safety glasses (particularly when using jackhammers and drills)
- sunscreen
- gloves
- reflective vest
- safety boots
- trench support (see Unit 2).

Tools

A range of hand and power tools and mechanical equipment is required. This may include:

- shovel, pick and crowbar to expose the approved point of discharge
- compression cutters, when working with vitrified clay pipe
- string line, to ensure the drain is laid straight and to the correct gradient
- linear measuring equipment, including a scale ruler, 8 m tape measure, and 25–30 m tape measure
- level measuring equipment, including spirit level and laser level or similar, to establish site ground levels, adjust the vertical alignment of shafts and jump-ups, and ensure drains are laid at the correct gradient
- excavator (including rock breaker, if required)
- cutting tools, including drop saw, hacksaw, mitre box and panel saw
- files, pipe de-burring tool
- angle grinder or compression cutter (to cut into older pipe materials, if required)
- jackhammer (to break rock and concrete)
- testing equipment, such as testing plugs.

Special equipment – selection and operation

Laser levels are commonly used in drain-laying operations. All laser levels must have a compliance plate attached to the body of the instrument. Among other information, this plate will give the 'class' of the laser. You may only use a Class 1 laser. To use any other class of laser, you must be fully trained, and a laser safety officer must be on-site.

Laser levels should be regularly calibrated to ensure accuracy. This instrument should always be treated with respect, and should be carefully stored when not in use. Laser levels are commonly available for hire. For more information, refer to Unit 3.

 COMPLETE WORKSHEET 1

Determining the installation requirements

Position of installation

In planning the position of the stormwater drainage system, you must follow the project plans and specifications, and only connect to the authorised point of discharge.

Stormwater drains must be located externally to the building where possible, so as to allow future access to the drain for maintenance and alterations.

The alignment of the drain should be straight, with a minimum continuous even gradient and minimum bends, and clear of buildings and easements. Where the drain is in close proximity to a building, care must be taken to ensure that trenching required for the drain does not compromise the structural strength of the building. (Read AS/NZS 3500.3.)

 **AS/NZS 3500.3 PLUMBING AND DRAINAGE. STORMWATER DRAINAGE**

Depending on soil composition, the distance from the bottom side of the trench to the base of the building footings can range from a ratio of 8:1 (stable rock) to 1:4 (silt). For example, if a drain is to be laid in an area that has stable rock, a ratio of 8:1 is to be used. In this ratio, 1 is 'L' (relative distance from the edge of the trench to the edge of the footing) and 8 is 'H' (relative vertical height from the base of the trench to the base of the footing) (i.e. rise:run). See Figure 11.1.

When excavating for the drain, it might not be possible or sensible to dig near the footing to determine the footing depth; in these cases, it is best to refer to the building drawings, which will specify the footing design depth.

In any situation where the required distance from the footing cannot be achieved, you should consult a qualified engineer to determine the necessary requirements.

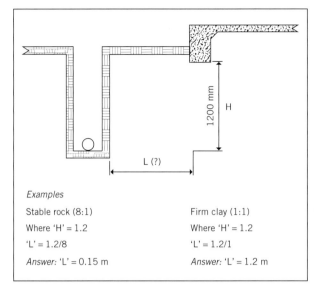

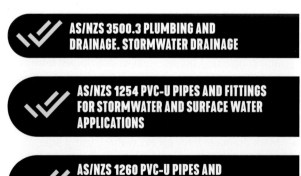

FIGURE 11.1 Calculating the required distance from trench to footing

Source: Department of Education and Training (http://www.training.gov.au) © 2013 Commonwealth of Australia.

Quantity and type of materials

The materials you are most likely to use when installing stormwater and sub-soil drainage may include:

- PVC-U pipes (commonly referred to as 'PVC')
- reinforced concrete
- cast iron
- vitreous clay
- ductile iron
- joint-making materials, such as solvents, cement mortar, rubber rings and bolted fittings
- geo-textiles
- bedding materials, such as sand, screenings, crushed rock, concrete and approved excavated materials.

The selection of pipe materials for drainage purposes is usually determined by cost and ease of handling and joining. PVC is available in long lengths, and is lightweight, easily joined and low in price. This makes PVC a preferred material in many areas, particularly in smaller pipe sizes. Reinforced concrete pipe (RCP) is generally more economical for larger sizes. Cast iron is often preferred where there may be a threat of mechanical damage, such as under roads or buildings. Vitrified clay is a very suitable material for stormwater pipes, but is generally more expensive than concrete and PVC. Other pipe materials may be useful for special situations, but are not commonly used.

Stormwater-class pipes and fittings must not be used for collection and transport of drinking water. Therefore, they should not be used to transport rainwater to rainwater tanks where it is to be used for drinking. Only pipes and fittings shown to comply with AS/NZS 4020 Testing of products for use in contact with drinking water may be used to collect drinking water. Typically, these products are pressure pipes – for

example, PVC pressure pipe complying with AS/NZS 1477 PVC Pipes and fittings for pressure applications. If you are unsure about the suitability of any product, check with the manufacturer.

From the building plan/s and specifications, determine the number and location of downpipes, pits, grates and so on that need to be connected to the stormwater drainage system. One method of doing this is by using a highlighter pen to mark all of the connections and branches.

The type of material is determined from the design drawings and specifications, and it must comply with AS/NZS 3500.3 and local authority requirements.

The types and quantities of fittings are added together to create a materials list, similar to that in Table 4.5 in Unit 4.

Ordering materials

Identifying materials

The materials for stormwater pipes and fittings must be approved for installation as listed in AS/NZS 3500.3, and bear the relevant standards marking. For example, PVC fittings are labelled as AS/NZS 1254 PVC-U Pipes and fittings for stormwater and surface water applications or AS/NZS 1260 PVC-U pipes and fittings for drain, waste and vent application.

Ordering materials

When ordering and purchasing stormwater pipes and fittings, it is good practice to plan ahead for secure storage in a place where they will not be damaged or stolen. After you have confirmed that they are approved and comply with the standards and local regulations, they can be ordered from plumbing suppliers from your materials list.

Collecting and checking materials

When receiving the delivery of stormwater pipes and fittings, you should count the number and types of items and compare them to your materials list order. Ensure you have identified all of the items as correct before accepting the order. If you have the wrong materials or damaged items, it may take some time to exchange and order the correct acceptable items.

COMPLETE WORKSHEET 2

Installing stormwater and sub-soil drains

Marking out the installation

When marking out stormwater and sub-soil drainage systems, the following must be considered.
- Inspect the project drawings for the location, size and levels of the drain.
- From the drawings, note or highlight all openings and connection points, such as downpipes from buildings, pits, sumps, traps and the connection to the approved point of discharge.
- Use pegs and string lines to set out the drainage lines.
- Mark drainage lines for excavation, using lime, paint or similar methods.
- Drainage lines should be as direct as possible to the connection points, with the minimum number of bends or changes in grade or direction.
- Keep drainage lines as far as practicable from building foundations, as outlined in AS/NZS 3500.3.
- Keep drainage lines away from other services, as outlined in AS/NZS 3500.3.

- Mark pit locations where drainage lines intersect at an upstream angle of more than 60°, or where an inlet pit is required.

AS/NZS 3500.3 PLUMBING AND DRAINAGE. STORMWATER DRAINAGE

Excavation

Excavation of trenches would normally be carried out by machine where access is not a problem. Where access is a problem, or for short or complicated locations, hand excavation may be necessary. Excavation by machine is not allowed within 600 mm of the authority point of discharge.

The excavation procedures for laying stormwater drains are similar to those for sanitary drains described in Unit 10.

Where drains are to be installed in unstable conditions, such as water-charged ground or filled ground where there may be excessive ground movement, specialist professional advice should be sought before excavation begins. This information may be obtained from the building plans or project manager. This may also become apparent after excavation commences. Filled ground may contain bricks, builder's waste and rubbish (see **Figure 11.2**).

FIGURE 11.2 Identifying filled ground

When excavating for stormwater drains, a number of precautions need to be taken, as described below.
- The plumber/drainer is responsible for the control of erosion and loss of sediment from the site caused by rainfall or stormwater, as discussed in Unit 1. Precautions to take may include covering exposed areas, grading suitable flow paths that minimise grades and limit erosion potential, and installing barriers, filters and silt or sediment traps as appropriate to the location.

- Trenches must be formed to the correct width and depth. The trench width is measured at the top of the pipe between the unsupported faces of the trench or the inside face of the trench support system. In the event of over-excavation, the excess depth of the trench must be refilled with concrete or with bedding material compacted as close as possible to the original soil density.
- Other services must not be put at risk by the excavation. Specified cross-over angles, separation from other services and access must be maintained. (Read AS/NZS 3500.3.)
- Do not excavate parallel to or below footings of buildings without taking expert advice on shoring, underpinning and backfilling, due to the risks of building movement and worker safety. Where excavating close to a single dwelling, the base of the trench should be clear of the footing, as outlined earlier in this unit.
- No excavation should be made in water-charged ground without adequate provision for dewatering. The water level must be lowered to below the trench base and maintained at that level until the trench has been backfilled. Pumps and spearheads or similar devices are the most common method of dewatering, as shown in **Figure 11.3**. Dewatering may affect surrounding buildings and services, and associated effects must be considered before excavation begins.
- Trench excavations deeper than 1.5 m (and to a lesser depth in unstable ground) must be provided with a trench support system to prevent the trench wall collapsing onto workers, as outlined in Unit 2.

Installation requirements

Surface drainage

The shape (topography) of the site is created by features such as hills and valleys, high points and low points. The topography affects the way water flows over the surface. To excavate and lay drains to the correct levels, you will need to understand site topography and the terms 'fall', 'gradient' and 'level'. Levelling is discussed in detail in Unit 3.

Paved surfaces are usually laid so that they 'fall' towards a drain – that is, they slope towards the drain so that any surface water will run downhill or fall to the drain, as shown in **Figure 11.4**.

Sub-soil drainage

Sub-soil drainage systems are designed to remove excess groundwater and reduce soil moisture levels. The sub-soil drainage system must be designed and installed in a manner that does not adversely affect adjacent buildings and the surrounding environment.

One product used for sub-soil drainage is Vinidex Draincoil®. It can be obtained in DN 50, 65, 80, 100 and 160 diameters (see **Table 11.2**), and is generally manufactured from high-density polyethylene (HDPE; see **Figure 11.5**), although DN 100 is also available in PVC.

High-capacity infiltration of water along the entire length of the pipe is achieved by rows of perforations. These are cut into the 'valleys' of the corrugations.

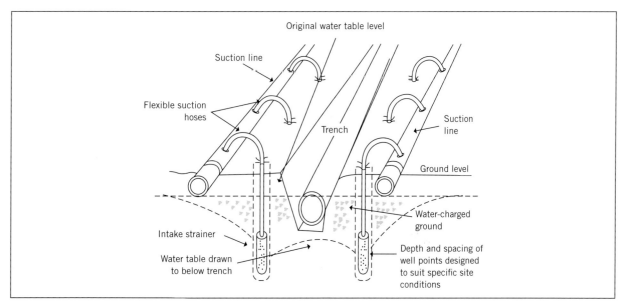

FIGURE 11.3 Dewatering trenches using well points or spearheads

Source: Department of Education and Training (http://www.training.gov.au) © 2013 Commonwealth of Australia.

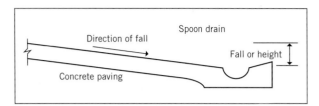

FIGURE 11.4 Paved surface drainage

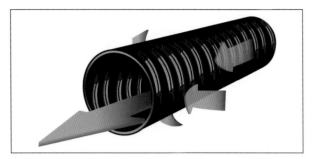

FIGURE 11.5 Draincoil® HDPE sub-soil drainpipe

Draincoil® is supplied in long coiled lengths (up to 200 m for some diameter pipes), as the pipe is both lightweight and flexible. The profiled or corrugated wall structure gives Draincoil® strength to resist external loads.

Australian Standard AS 2439.1 Perforated plastics drainage and effluent pipe and fittings – Perforated drainage pipe and associated fittings establishes classes of pipe based on stiffness – that is, the force required to achieve a particular deflection. This class rating may be used as a guide to the product application and loads to be encountered in service.

AS 2439.1 PERFORATED PLASTICS DRAINAGE AND EFFLUENT PIPE AND FITTINGS – PERFORATED DRAINAGE PIPE AND ASSOCIATED FITTINGS

Draincoil® is available in two classes, which relate to its ability to carry external loads:
1 SN 20 (PVC): for works subject to heavy vehicular traffic loads
2 SN 8 (HDPE): for land drainage, and road or civil engineering works not subject to heavy traffic loads.

TABLE 11.2 Draincoil® pipe dimensions

SN	Nominal diameter DN (mm)	Outside diameter D (mm)	Inside diameter d (mm)	No. of slotted rows	Clear water opening (mm²/m)	Coil length (m)	Coil mass approx. (kg)
8	50	50	44	6	1500	10/20/200	1/75/3.5/35
8	65	65	55	6	1500	10/20/200	2.3/4.6/46
8	80	80	68	6	1500	20/100	7.0/35
8	100	100	86	6	1500	10/20/100	4.75/9.5/48
8	160	160	138	6	1500	40	34.4
20	100	100	86	6	1500	100	70

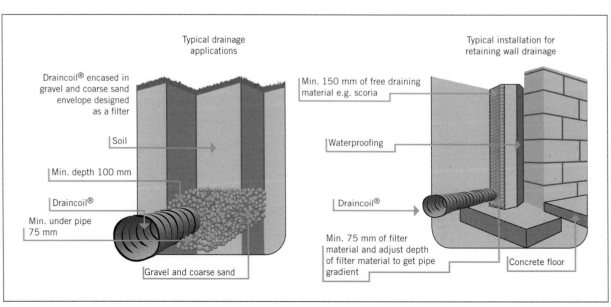

Typical drainage applications

Typical installation for retaining wall drainage

Draincoil® encased in gravel and coarse sand envelope designed as a filter

Soil

Min. depth 100 mm

Draincoil®

Min. under pipe 75 mm

Gravel and coarse sand

Min. 150 mm of free draining material e.g. scoria

Waterproofing

Draincoil®

Min. 75 mm of filter material and adjust depth of filter material to get pipe gradient

Concrete floor

FIGURE 11.6 Typical drain installations for drainage of soil and retaining walls

An envelope of granular aggregate around the pipe improves the flow of soil water into the pipe by increasing the effective diameter of the drain. In permeable soils, water flows through entry points mainly at the sides and bottom of the pipe. It is therefore important that, where used, the gravel envelope completely surrounds the pipe.

Draincoil® must be spaced correctly to achieve proper drainage. Table 11.3 gives a guide for drain spacing proportional to drain depth and soil type for agricultural applications. Figure 11.6 illustrates typical drain installations for drainage of soil and retaining walls.

TABLE 11.3 Typical depth and pacing for Draincoil® agricultural subsurface drains

Soil type	Depth (m)*	Spacing (m)
Deep light sand	1.8–2.1	27–40**
Loam	1.2	13
Clay loams and clay	0.9–1.1	7–13

*The depth of the impermeable soil layer is the critical factor in determining maximum drain depth. Spacing varies depending on the soil texture. The relevant authority in each state or territory is available to advise a drainage depth and spacing for the appropriate crop.
**Good results have been achieved at spacing up to 80 m.

Source: Vinidex Pty Limited

Specialist advice is required for design of sub-soil drainage to control soil moisture levels around building structures. Draincoil® should also be laid with a gradient. Table 11.4 provides a guide.

TABLE 11.4 Preferred and minimum gradients for Draincoil®

Application	Minimum gradient	Preferred gradient
Agricultural	0.2% or 1:500	0.4% or 1:250
Civil engineering	0.5% or 1:200	1.0% or 1:100
Recreational	0.3% or 1:330	Between 1.0% and 3.0% or 1:100 and 1:33

Source: Vinidex Pty Limited

Under some conditions, fine sands or soil particles may flow into the pipe with the water and cause blockages due to siltation in a pipe laid to a shallow gradient. Under these conditions, it may be advisable to use a woven fabric filter 'sock' to screen particles. An additional benefit is that the filter 'sock' can stretch over the corrugations of the Draincoil®, increasing the effective water access area to the pipe. Draincoil® is also available un-slotted for use where infiltration is not required or fine tree roots may cause obstruction.

Around the pipe and for a distance of 150 mm above, it is recommended that permeable, granular material, reasonably compacted, be used (see Figure 11.7). This will give the pipe the necessary support and allow drainage of groundwater. The maximum aggregate size recommended in this zone is 13 mm. In particular, impermeable clay should not be used. Organic materials such as straw, woodchips and sawdust are also not recommended.

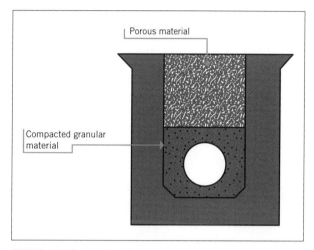

FIGURE 11.7 Sub-soil drainage

Source: Vinidex Pty Limited

The remainder of the trench should be filled with porous material, and compacted to ensure there is no subsidence at the surface. Heavy clay should not be used, as it is difficult to compact and may restrict the flow of soil water to the pipe.

Joining
Moulded couplings are available, together with a comprehensive range of simple 'push on, clip over' fittings for Draincoil®.

Bending
Draincoil® pipes can be installed continuously around curves and corners of 300 mm radius without the need for additional fittings. Flexibility also ensures that the pipe will accommodate ground movement.

Depth of cover
The depth of cover over sub-soil drains should be minimum 300 mm where there is no vehicular traffic, and up to 450 mm where there is occasional vehicle loading. For more information, refer to the manufacturer's recommendations.

The maximum burial depth for Draincoil® depends on the quality of the support the pipe receives from embedment. Draincoil® will float in water; therefore, in wet conditions, it is advisable to backfill the trench as the pipe is laid. Draincoil® is slotted in the protected valleys of the corrugations and small deflections will not cause blockages.

Stormwater drainage

Stormwater drainage systems are generally designed to collect stormwater by guiding it into appropriate channels, gutters, drains and pipes, and then conveying it safely to the point of discharge. The preferred point of discharge is the connection point to an external drainage network. Where this is not possible, stormwater should be discharged to the natural drainage system in a suitable manner, or to a rubble drain or transpiration bed. Stormwater must not be discharged in a manner that allows ponding or flow to adjoining properties, nor must it be allowed to enter buildings.

PVC stormwater pipe and fittings systems are for use above or below ground and must be manufactured in accordance with AS/NZS 1254.

PVC pipes are supplied in effective 6 m lengths. Total pipe length = effective length + socket depth. Pipes are supplied with an integral solvent weld socket. PVC stormwater pipe and fittings must be installed in accordance with AS/NZS 3500.3 and AS/NZS 2032

Installation of PVC pipe systems and local authority requirements.

Sizing and gradient

Stormwater drains must be at least DN 90 for dwellings on allotments less than 1000 m². Where allotments are larger than 1000 m², stormwater drains must be sized according to AS/NZS 3500.3, based on catchment areas and design flows.

Stormwater drains ranging in size from DN 90 to DN 150 must be laid at a minimum gradient of 1% (1:100).

Bedding

Bedding for PVC site stormwater drains must be in accordance with AS/NZS 3500.3, and as shown in Figure 11.8.

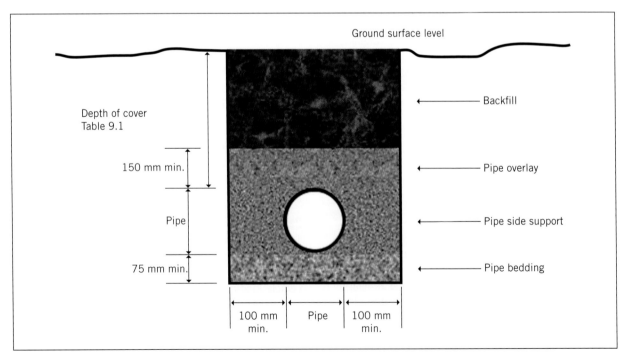

FIGURE 11.8 Bedding for PVC site stormwater drains

AS/NZS 3500.3 PLUMBING AND DRAINAGE. STORMWATER DRAINAGE

Drain support and provision for movement

Each new residential building site is classified in accordance with AS 2870 Residential slabs and footings. The building plans or geotechnical report will contain details of the site classification.

Stormwater drains require the same considerations for movement as sanitary drains. (Refer to Unit 10.) For example, if a stormwater drain must be installed through a concrete footing on a site classified as Class H2, it must be provided with at least 40 mm-thick lagging and waterproofing through the concrete. It will also need to be installed with flexible joints. *Note:* As the available range of flexible joints for DN 90 PVC pipe is limited, the drain may need to be constructed in DN 100 PVC DWV pipes and fittings.

Jointing methods

All joints and jointing materials must comply with the requirements of AS/NZS 3500.3. In general, preferred jointing methods are solvent cement for PVC pipes and rubber ring joints for concrete and vitrified clay pipes. Other materials such as stainless steel or glass-reinforced plastics may be specified for a project and approved for certain conditions. For these materials, follow the pipe jointing procedures outlined in the manufacturer's instructions or in AS/NZS 3500.3.

AS/NZS 3500.3 PLUMBING AND DRAINAGE. STORMWATER DRAINAGE

Common jointing methods and procedures are illustrated in **Figures 11.9** to **11.20**.

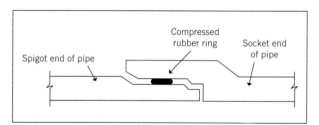

FIGURE 11.9 Reinforced concrete pipe with compressed rubber ring joint

PVC pipes

PVC pipe is by far the most common material used for stormwater drains. The pipes and fittings are joined using solvent cement.

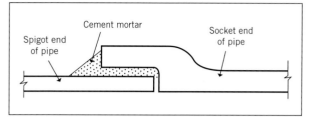

FIGURE 11.10 Vitrified clay pipe cement mortar joint

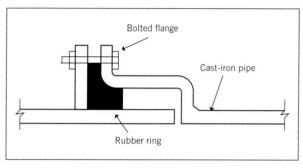

FIGURE 11.11 Cast-iron pipe bolted gland joint

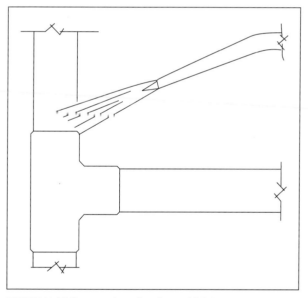

FIGURE 11.12 Copper pipe silver brazed joint

Jointing methods – solvent cement jointing of PVC pipes

To achieve strong leak-free joints, plumbers/drainers should:

- select the correct solvent cement for the application: Type P for pressure joints with an interference fit, and Type N for non-pressure joints that may have a small clearance
- select the correct pipe and fitting using the manufacturer's part list

- follow jointing steps 1–8 carefully (as outlined below). Short cuts will result in poor joints that are likely to cause system failure.

How solvent cement works

Solvent cement is a solution of resin in a mixture of solvents, which soften the surfaces when applied to PVC-U pipe and fittings. It is not a glue. A thin, uniform coat is applied to both the spigot and socket, and the joint is assembled while the surfaces are still wet and fluid. The cement layers intermingle and become one. The strength of the joint develops as the solvent permeates the PVC-U and the volatile constituents evaporate.

Iplex Pipelines' solvent cements and benzene-free priming fluids are manufactured in accordance with AS/NZS 3879 Solvent cements and priming fluids for PVC (PVC-U and PVC-M) and ABS and ASA pipes and fittings. Iplex Pipelines' solvent cement Type N is used for non-pressure applications and is formulated with the gap-filling properties needed with clearance fits.

AS/NZS 3879 SOLVENT CEMENTS AND PRIMING FLUIDS FOR PVC (PVC-U AND PVC-M) AND ABS AND ASA PIPES AND FITTINGS

The importance of priming fluid

Before applying the solvent cement, it is essential to use priming fluid for successful jointing (see Table 11.5). The fluid not only cleans and degreases, but also removes the glazed surface from PVC-U, which allows the solvent cement to permeate into the wall of the pipe or fitting.

TABLE 11.5 Average number of joints per litre of Iplex solvent cement

Nominal pipe size (mm)	Approx. joints per litre
50	60
80	60
100	48
150	40
225	16
300	8
375	6

It must be applied with a clean, lint-free cotton cloth. Brushing the priming fluid on, or simply pouring the fluid over, the pipes and fittings does not remove grease and dirt.

Jointing instructions for PVC pipes

Do not work with hot pipes or on hot windy days without protecting pipes from the wind. Keep the lid on the solvent cement to minimise evaporation. Do not use solvent that is more than 12 months old.

Step 1: Cut spigot square and de-burr

Cut the spigot as square as possible using a mitre box and hacksaw, power saw or PVC pipe cutter (see Figure 11.13). Remove all swarf and burrs from both the inside and outside edges with a de-burring tool, file, reamer or sandpaper. Swarf and burrs, if left, will wipe off the solvent cement and prevent proper jointing.

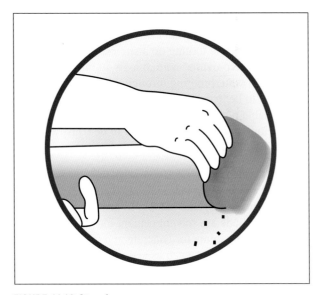

FIGURE 11.13 Step 1

Source: Vinidex Pty Limited

Step 2: Check alignment

Check the pipe and spigot or fittings for proper alignment (see Figure 11.14). The time for any adjustments is now, not later.

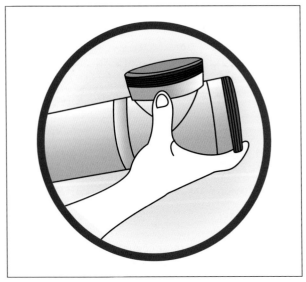

FIGURE 11.14 Step 2

Source: Vinidex Pty Limited

Step 3: Mark clearly

Mark the spigot at a distance equal to the internal depth of the socket (see Figure 11.15). Use only pencil or a marker. Do not score or damage the surface of the pipe or fitting.

FIGURE 11.15 Step 3

Source: Vinidex Pty Limited

Step 4: Clean and soften the surface

Thoroughly clean the inside of the socket and the area between the pencil mark and the spigot end with a clean, lint-free cotton cloth dipped in priming fluid (see Figure 11.16). (Do not use synthetic material.) This removes dirt and grease and softens the PVC surface. Apply priming fluid immediately prior to joining. Do not brush or pour on the priming fluid.

It is recommended that gloves be used. If contact with skin occurs, immediately wash the affected area with soap and copious quantities of water.

FIGURE 11.16 Step 4

Source: Vinidex Pty Limited

Step 5: Coat socket first, then spigot

Apply a thin, uniform coat of solvent cement to the socket. Take care to ensure that solvent build-up does not occur in the root of the socket, as a pool of cement there will severely weaken the pipe or fitting. Now apply a uniform coat of solvent cement to the external surface of the spigot up to the pencil mark (see Figure 11.17).

FIGURE 11.17 Step 5

Source: Vinidex Pty Limited

Step 6: Assemble, then hold for 30 seconds

Assemble the joint quickly, before the cement dries, by pushing the spigot firmly into the socket as far as the pencil mark, ending with a quarter turn to spread the cement evenly. Hold the joint in this position and alignment for at least 30 seconds without movement (see Figure 11.18).

FIGURE 11.18 Step 6

Source: Vinidex Pty Limited

Step 7: A vital five minutes

Wipe off the excess solvent cement from the outside of the joint and, where possible, from the inside of the joint. Do not disturb the joint for at least a further five minutes, as movement may break the initial bond and alignment (see Figure 11.19).

FIGURE 11.19 Step 7

Source: Vinidex Pty Limited

Step 8: Curing and testing

The cure time is the time needed for the joint to achieve sufficient strength to allow it to be tested by internal pressure or vacuum. The minimum cure time for solvent weld joints in stormwater pipes and fittings is 24 hours (see Figure 11.20).

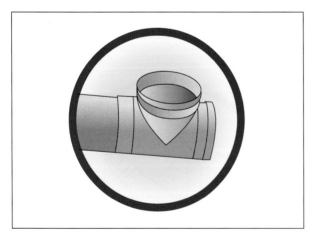

FIGURE 11.20 Step 8

Source: Vinidex Pty Limited

Storage of solvent cement and priming fluids

Consider the following when storing solvent cement and priming fluids:

- Solvent cement and priming fluids are highly flammable. In the event of fire, smother with sand or earth or use a suitable fire extinguisher.
- Store solvent cements and priming fluids in a cool place, away from heat, flames and sparks.
- Ensure bottle lids are tightly closed when not in use.
- Use solvent cements within 12 months of the date stamped on the bottom of the bottle/can. If the solvent cement has become so thick that it does not flow easily, discard it.
- Do not add any other ingredients or solvents to these products.

Safety precautions

Consider the following safety precautions:

- Do not use solvent cements or priming fluid in confined spaces without adequate ventilation, or near open flames or sparks.
- Do not smoke while using solvent cements or priming fluids.
- If spilt on skin, wash off with soap and water.
- If poisoning occurs, consult a doctor or Poisons Information Centre.
- Keep containers sealed when they are not in use.

Note: State and territory regulations generally require that, when storing any dangerous goods such as solvent cement and priming fluids in a workshop or van, material safety data sheets (MSDS) are readily available for reference.

If swallowed:

- *Solvent cement:*
  - Do not induce vomiting.
  - Call a Poisons Information Centre or a doctor immediately.
- *Priming fluid:*
  - Do not induce vomiting.
  - Call a Poisons Information Centre or a doctor immediately.

Also: avoid contact with eyes. If contact occurs, flush with copious amounts of clean water.

Inspection openings and covers

Inspection openings and enclosures must be installed in the locations indicated on the project drawings and specifications, or as indicated in AS/NZS 3500.3.

AS/NZS 3500.3 PLUMBING AND DRAINAGE. STORMWATER DRAINAGE

Installation procedure

Drainpipe installation can be carried out using the same procedures as outlined for sanitary drains in Unit 10, and for using levelling equipment as described in Unit 3 to achieve the required gradient and levels at pit connections.

GET IT RIGHT

CORRECT JOINTING

PVC pipes and fittings must be correctly made using priming fluid and solvent cement. **Figure 11.21** shows a junction in a stormwater drain that has separated at the branch. There has been no application of priming fluid and very little application of solvent cement. The drain leaked and caused damage to building foundations.

1 Could this stormwater drain have passed a hydrostatic or air test?

_____

_____

_____

2 What is the purpose of priming fluid?

_____

_____

_____

3 How would you check that the pipe is fully inserted into the socket of the fitting?

_____

_____

_____

Pre-cast or in-situ pits can be installed at intersections or changes in direction, as indicated on the project drawings. Bedding must be placed and compacted, as required, before pipe laying commences. Pipe installation can then proceed as follows:

- Commence drain installation at the upstream end and work downstream towards the point of discharge.
- Connect downpipes, either with a direct connection or via a grated sump, as shown in Figure 11.22.
- Lay DN 90 to DN 150 pipes and fittings to the gradient, as shown in Figure 11.23.
- Install inspection openings, as shown on the project drawings and specifications, or as outlined in AS/NZS 3500.3.
- Use the jointing method for PVC pipe, as described earlier in this unit, or as specified by the product manufacturer.
- Ensure that there are no projections or obstructions within the pipe that could cause blockages.
- Connect to pits or traps as shown on the project drawings. Where pipes pass through below-ground external walls, or pits more than 1 m deep, two flexible joints must be provided, as shown in Figure 11.24.
- Ensure that where pipes pass through the walls of pits that they are made watertight by caulking the annular space with stiff mortar or sealing with an epoxy-based sealant, as shown in Figure 11.25.
- Connect the site stormwater drain to the authorised point of discharge. Where the external drainage network is at a greater depth than the site stormwater drain, a jump-up may be required. See the example in Figure 11.26.
- Where the gradient of drain is 1:5 or steeper, anchor blocks will be required, as shown in Figure 11.27.
- Where no external drainage network exists, stormwater drains may connect to a rubble drain or a transpiration bed, as shown in Figures 11.28 and 11.29.
- Where roof drainage water is to be collected for domestic use, it may be directed into a storage tank, as shown in Figure 11.30. *Note:* It is best practice to install a first flush diverter to prevent unwanted debris from entering the water storage tank. The regulatory authority may not permit DN 90 PVC piping in a charged stormwater system (depicted in Figure 11.30), and may require inspection openings to enable periodic cleaning maintenance.

AS/NZS 3500.3 PLUMBING AND DRAINAGE. STORMWATER DRAINAGE.

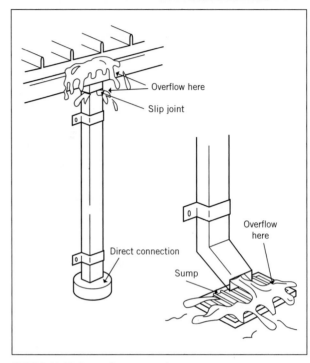

FIGURE 11.22 Downpipe connections

Source: Department of Education and Training (http://www.training.gov.au) © 2013 Commonwealth of Australia.

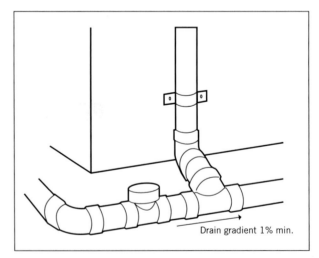

Drain gradient 1% min.

FIGURE 11.23 Drain gradient to connection point

Source: Department of Education and Training (http://www.training.gov.au) © 2013 Commonwealth of Australia.

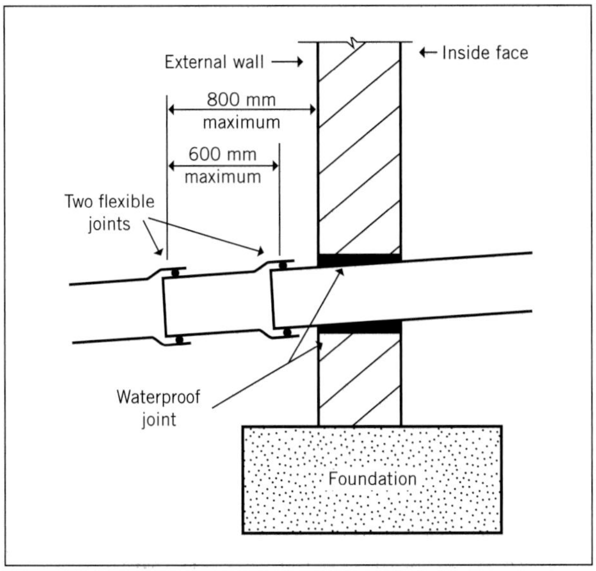

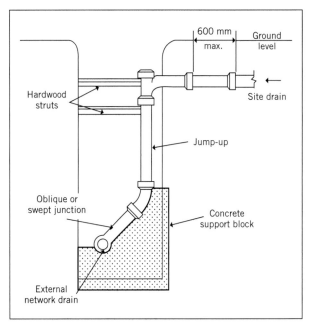

FIGURE 11.24 Providing flexible pipe joints at external walls

Source: Department of Education and Training (http://www.training.gov.au) © 2013
Commonwealth of Australia.

FIGURE 11.26 Jump-up connection to site stormwater drain

Source: Department of Education and Training (http://www.training.gov.au) © 2013
Commonwealth of Australia.

FIGURE 11.25 Caulk pipe inserts to make stormwater pits watertight.

On-site detention systems

On-site detention (OSD) systems are available in various styles and construction types. The function of OSD is to reduce flooding by providing temporary storage of stormwater during storms. After the storm, the stored water is slowly released, normally through a controlled orifice plate. A typical storage system will quickly fill, but it could take several hours to empty, with the benefit of reducing the load on the stormwater main. The local government authority (council) may specify a maximum stormwater discharge rate, which is expressed in litres per second.

The local authority may specify an on-site stormwater detention system to reduce and, where possible, eliminate additional loading from new developments on the existing public stormwater drainage system. The local authority may provide guidance and specify design requirements for OSD structures, including capacity, gradients and discharge rate. The structures require site-specific engineering design and must be installed in accordance with that design (see Figure 11.31).

Combined rainwater/detention system

Rainwater tanks may be used for rainwater detention purposes in addition to providing storage for domestic water use within the premises. A combination rainwater/detention tank will delay the discharge of excess roof water to the stormwater drainage system at a controlled rate. The design criteria for combined rainwater/detention systems are determined by the local government authority.

Testing

In general, site stormwater drains external to buildings are not tested unless required by the local regulatory authority. However, *all stormwater drains within or under buildings and rising mains must be tested and be free of leaks*. The testing procedures are specified in the section 'Site testing' in AS/NZS 3500.3.

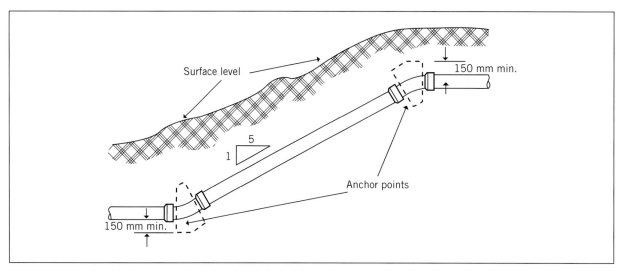

FIGURE 11.27 Anchor blocks are required when the drain is 1:5 or steeper

Source: Department of Education and Training (http://www.training.gov.au) © 2013 Commonwealth of Australia.

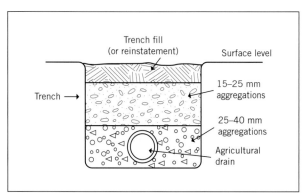

FIGURE 11.28 Rubble drain

Source: Department of Education and Training (http://www.training.gov.au) © 2013 Commonwealth of Australia.

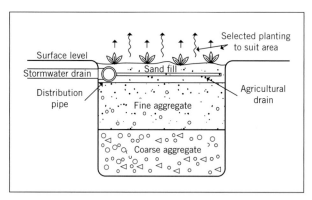

FIGURE 11.29 Transpiration bed

Source: Department of Education and Training (http://www.training.gov.au) © 2013 Commonwealth of Australia.

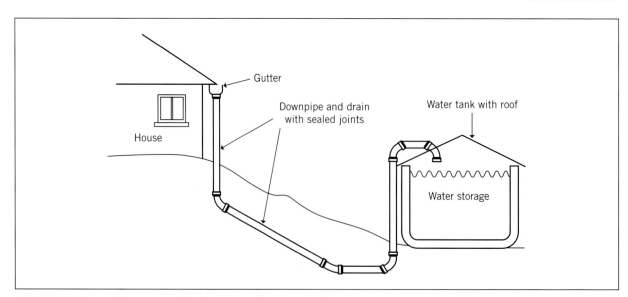

FIGURE 11.30 Domestic water supply – above-ground storage tank

Source: Department of Education and Training (http://www.training.gov.au) © 2013 Commonwealth of Australia.

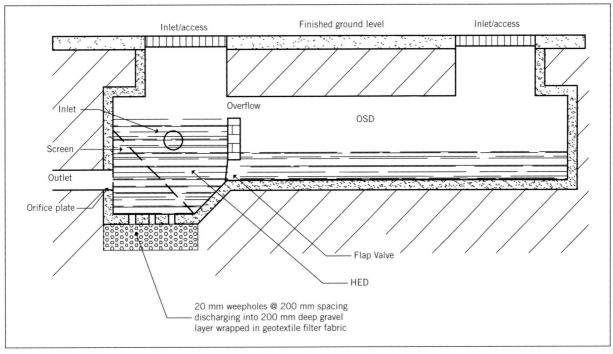

FIGURE 11.31 Example of an on-site detention system

Source: Department of Education and Training (http://www.training.gov.au) © 2013 Commonwealth of Australia.

AS/NZS 3500.3 PLUMBING AND DRAINAGE. STORMWATER DRAINAGE.

You will be required to ascertain your local testing requirements in the exercises at the end of this unit.

Backfilling the excavation

FROM EXPERIENCE

Before backfilling, ensure that the trench fill material is free from builders' waste, rocks, concrete and bricks that could damage the drain. If the drain is broken, water can escape and cause damage to buildings. Also, tree roots can enter and cause blockages.

Backfilling is the process of placing and compacting bedding material into the open excavation surrounding the pipe. Backfilling requirements will vary from job to job, depending on the nature of the site, the size of the pipes and the construction techniques. Backfilling involves placing the correct materials for bedding, haunch support, side support and overlay. Backfill must be placed without damaging or displacing the drainpipes and will follow this general procedure:

- Place and compact the side support or haunch support.
- Place and compact the overlay in layers not exceeding 200 mm of loose material per layer.
- This final step is known as reinstatement. Place the final layer of trench fill to match the original

surrounding surface and level. This may require topsoil and grass, compacted gravel, concrete, bitumen or other surface as specified.

When carrying out backfilling in a public property, such as a road or footpath, the backfill and restoration work must be carried out according to the local government authority's requirements.

Clean-up

Work area

Cleaning up the worksite is necessary for the health and safety of workers, and to ensure work can be carried out with minimum delays. The face of a building site constantly changes, and access is always required for deliveries and for workers to various areas. For these reasons, building sites are usually short of space. Stormwater drainage work usually takes up a lot of the building site, as it requires excavation and modification to the ground surface around buildings. Adverse weather conditions can delay drainage work, but drainage work is also necessary to minimise wet ground conditions.

Most large sites will have site safety procedures and a schedule for cleaning up. Small sites such as single dwellings may not have procedures, but the same principles apply for both. The procedures could include the following:

- The disposal areas for site debris should be identified at site induction. If not, ask your trade supervisor.

- Clean up as you go, if possible, but at least on a daily and weekly basis and on completion of an installation.
- Food and drink waste, wrappers and containers should be disposed of in the correct bin immediately after use, as build-up attracts vermin and creates associated health risks.
- Packaging should not be allowed to become a windblown, trip or other hazard. Dispose of it in the correct bin as soon as practicable, or take it with you from the site.
- Off-cuts and surplus material that may be used later in the job should be held in a safe storage area until required.
- Off-cuts and surplus material that are not suitable for re-use should be disposed of in the correct bin at the end of each day's work, or sooner if practicable.
- Do not use cupboards, empty rooms or corners to store rubbish and debris.
- On completion of an installation, check the relevant work area. Remove materials to the store, and tools and equipment to the correct location.
- Periodically check and maintain erosion control measures during construction to prevent siltration of the stormwater drains. Such measures include temporary or permanent silt traps, hay bales and surface-stabilising materials.

Tools and equipment

The procedure for cleaning tools and equipment could include the following:

- Tools and equipment must be cleaned, maintained and stored.
- Tools and equipment may be the responsibility of the apprentice or worker, the contractor or others such as hire companies.

- The apprentice or worker will be responsible for a toolkit and personal protective equipment (PPE).
- Keep a running check on your toolkit and your PPE. Check, clean and maintain it at the end of each day.
- Check and maintain company equipment on a regular basis according to company policy, keeping all equipment in good working order.
- Clean, maintain and return hire equipment according to the hire agreement.
- Do not overload or abuse any tools, plant or equipment.
- Dispose of unserviceable tools and equipment that could be unsafe.

Documentation

This section is relevant to documentation required after the work to install stormwater drains is complete.

Generally, councils and regulatory authorities do not require 'as-constructed' plans of stormwater drains; however, requirements vary around Australia. Therefore, you will need to research your local authority requirements. Councils or regulatory authorities in many areas often require notification when the project is complete, in order to carry out an inspection or audit.

In some areas, plumbers are required to lodge a Certificate of Compliance within a certain period of time after the job is complete.

 COMPLETE WORKSHEETS 3 AND 4

SUMMARY

- When installing stormwater drainage systems, it is necessary to follow the plans and specifications that indicate the location of buildings, downpipes, pits, easements and underground services. Also, prepare by familiarising yourself with local regulations and inspection requirements, and ensure that pipes and fittings have Australian Standards markings. Always be prepared to locate the approved point of discharge before laying drains, and protect it from damage and entry of soil, pipe bedding and rubbish.

- When installing the stormwater drain, set out and excavate the drain in straight lines that are as short as possible. Consider the requirements for bedding, support, pits and inspection openings.

- Generally, the stormwater drain will need to be constructed at the correct gradient and with sufficient cover. You need to know the ground conditions, and, if necessary, provide lagging and waterproofing of pipes passing through concrete strip footings, and swivel/expansion joints. If you are unsure, or suspect that the ground is reactive or unstable, consult with a geotechnical engineer.

- Inspection openings and enclosures must be installed as indicated on the project drawings and specifications, or in AS/NZS 3500.3. On-site detention systems (OSD) must be installed in accordance with the engineering design. Always test stormwater drains before they are backfilled.

- When works have been completed, clean up the work area. If required by the regulatory authority, lodge the necessary documentation, such as a Certificate of Compliance and an as-constructed plan.

WORKSHEET 1

Student name: _____

Enrolment year: _____

Class code: _____

Competency name/Number: _____

Task: Review the section 'Preparing for work', then complete the following questions.

1   In relation to plans and specifications, what four matters must the plumber/drainer check before commencing work?

   a   _____

   b   _____

   c   _____

   d   _____

2   Research your local authority requirements for applications and permits for stormwater drains. Complete the following:

   a   Name the local authority responsible for approving stormwater plans.

   _____

   b   What applications and permits are required to lay a stormwater drain?

   _____

   _____

3   What is the minimum depth of cover over a PVC stormwater drain in a driveway at a single-dwelling property?

   _____

4   Complete the following.

   All pipes and fittings used in a stormwater drainage system must comply with _____ and bear _____ marking.

5   Why is it best practice to have two people carrying out excavation work?

   _____

   _____

   _____

6 Before laying stormwater drains, it is best practice to establish and reveal unknown factors. Name three factors that should be investigated before laying stormwater drains.

a _____

b _____

c _____

7 Explain the method of excavating near the approved point of discharge for stormwater, and describe what tools should be used.

WORKSHEET 2

Student name: _____

Enrolment year: _____

Class code: _____

Competency name/Number: _____

Task: Review the section 'Determining the installation requirements', then complete the following.

1   Research AS/NZS 3500.3 and calculate the minimum distance 'L' that is required between the trench and footing for undisturbed firm clay conditions.

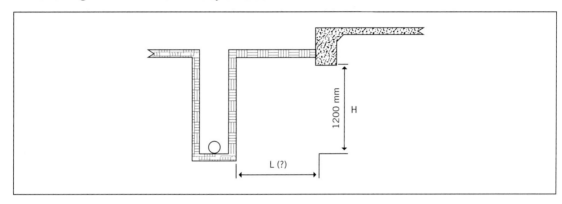

'L' =

_____

_____

_____

2   With reference to the above drawing, how would you find the depth of the footing?

_____

3   Where must stormwater drains be located in relation to buildings and easements?

_____

_____

4   What type of pipe material could be used in a residential driveway where the depth of cover for PVC cannot be achieved?

_____

5   Research AS/NZS 3500.3 and state the minimum separation distances between stormwater drains and the following underground services.

   a   Electrical cable without marker tape or mechanical protection:

_____

   b   Consumer gas pipe installed with marker tape:

_____

WORKSHEET 3

Student name: _____

Enrolment year: _____

Class code: _____

Competency name/Number: _____

Task: Review the sections 'Installing stormwater and sub-soil drains' and 'Inspection openings and covers', then complete the following.

1 What are three ways of determining whether the stormwater drain will be in filled ground?

a _____

b _____

c _____

2 If the stormwater drain will be laid in unstable, water-charged or filled ground, what action must be taken?

_____

_____

_____

3 Research Table 11.4. What is the preferred gradient for Draincoil® sub-soil drainage in civil engineering works?

_____

4 Research AS/NZS 3500.3. Where are clean-out points required for sub-soil drains?

_____

_____

5 What materials should not be used around sub-soil drains?

_____

_____

6 When sub-soil drains are installed in fine sand or silt, the drain must be protected from blockage. Explain how this is achieved.

_____

_____

_____

_____

7 What is the maximum size of aggregate, to be placed around sub-soil drains?

_____

8   State the minimum size and gradient of stormwater drain for a dwelling on an urban allotment less than 1000 m².

_____

9   Research AS/NZS 3500.3 and state three options for bedding, side support and overlay material for site stormwater drains.

a   _____

    _____

b   _____

    _____

c   _____

    _____

10   What two types of trench fill material may be used for stormwater drains?

a   _____

    _____

b   _____

    _____

11   For DN 150 PVC stormwater pipe, approximately how many joints could be made with 1 litre of solvent cement?

_____

12   List five safety precautions when working with solvent cement and priming fluid.

a   _____

b   _____

c   _____

d   _____

e   _____

13   Research AS/NZS 3500.3. List five locations where inspection openings are required on site stormwater drains (for other than single dwellings).

a   _____

b   _____

c   _____

d   _____

e   _____

14 If a stormwater drain is to be laid through a concrete footing on a site classified as H2, what is the thickness of lagging and waterproofing required around the pipe where it passes through the concrete?

_____

15 Research AS/NZS 3500.3. What are three places where square junctions are permitted in stormwater drains?

a _____

b _____

c _____

16 What is the function of an on-site detention system, and how is water released from the system?

_____

_____

_____

_____

17 Research your local regulatory requirements for inspection, testing and certification of stormwater drains, and provide details for the following:

a Inspection requirements:

_____

_____

b Testing requirements:

_____

_____

c Compliance certification:

_____

_____

18 Complete the following: When carrying out backfilling in a public property such as a road or footpath, the backfill and restoration work must be carried out according to _____ _____.

19 What are three erosion control measures to prevent silt from entering stormwater drains?

WORKSHEET 4

Student name: _____

Enrolment year: _____

Class code: _____

Competency name/Number: _____

Task: Practical exercise

All questions in the previous three worksheets must be completed and checked by your trainer/teacher before you attempt the following practical exercise.

Your trainer/teacher should provide preliminary guidance, conduct an induction and ask underpinning knowledge questions.

Install a stormwater and sub-soil drainage system. The stormwater drain must be 4 m long and connect from a downpipe to an approved point of discharge. The sub-soil drain must be 4 m long with a clean-out point at the topmost end, and connect downstream to a disposal and collection pit.

Your trainer/teacher may choose to combine the above exercise with exercises in other units.

12 INSTALLING SEWERAGE PUMP SETS

Learning objectives

This unit provides the basic principles and knowledge needed to install sewerage pump sets. Areas addressed in this unit include:

- pump performance basics
- pump location
- pump types, couplings and connections
- pump motors and controls
- description of the components, operation and installation of:
  - compressed air ejectors
  - wet wells
  - dry wells
  - holding tanks (discharge from waste fixtures)
  - pressure sewer systems
  - small bore macerator systems.

Introduction

Plumbers are required to install sewerage pump sets to convey sewage and waste to the sanitary drainage system, regulatory authorities' connection points or on-site disposal systems. This type of work is carried out in locations where it is not practical or possible to convey sewage and waste via gravity – for example, in basements where the plumbing fixtures may be lower than the sewer connection point (see **Figure 12.1**). In these situations, pumps are required to raise the waste to a higher level. This could be new work, or it could involve alterations or additions to existing systems.

Sewerage pumps can be in the form of:
■ compressed air ejectors
■ wells (wet and dry)
■ holding tanks
■ pressure sewer systems
■ small bore macerator systems.

In Units 4 and 10, we studied the requirements for planning the layout and installing a below-ground sanitary drainage system. Those requirements apply to any drain or waste pipe that conveys discharges via gravity to a sewerage pump system. Sewerage pump systems can be constructed on-site in accordance with AS/NZS 3500.2 Sanitary plumbing and drainage and local regulatory requirements. Pre-packaged systems are also available and may be used with the approval of the local regulatory authority.

While studying this subject, you will be required to respond to questions and complete exercises referring to AS/NZS 3500.2.

AS/NZS 3500.2 SANITARY PLUMBING AND DRAINAGE

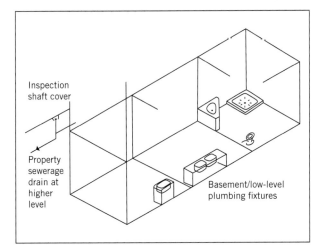

FIGURE 12.1 Sanitary fixtures in basement/lower than connection point

Source: Department of Education and Training (http://www.training.gov.au) © 2013 Commonwealth of Australia.

Before studying this subject, you should have completed training in work health and safety (WHS) as outlined in Chapter 3 of *Basic Plumbing Services Skills*.

Preparing for work

Plans and specifications

When installing sewerage pump sets, it is necessary to follow manufacturer's installation instructions, Section 12 of AS/NZS 3500.2, and any local regulatory authority requirements. Small bore macerator systems are usually packaged to connect a single fixture or small group of fixtures, and no special design is necessary.

AS/NZS 3500.2 SANITARY PLUMBING AND DRAINAGE

For projects where it is necessary to handle large volumes of sewage, a hydraulic engineer must have provided a design solution for either a compressed air ejection system or a wet well with submersible pump set. The design must limit the number of pump starts and detention time, while providing allowable flow rates. It must also accommodate the type of demand – for example, sporting facilities will have short periods of high demand and long periods of inactivity.

Generally, the following must be considered:
■ AS/NZS 3500.2, Section 12, and local sewerage authority requirements
■ pump system manufacturer's requirements
■ approval from the local sewerage authority
■ plans and specifications for large projects (for the purposes of this unit, more than one residential building may be considered a 'large project')
■ plumbing permit/consent from the sewerage authority
■ sewerage/regulatory authority plumbing inspection requirements.

AS/NZS 3500.2 SANITARY PLUMBING AND DRAINAGE

To plan the installation of a sewerage pump set, you will need to gather information from a number of sources. **Table 12.1** indicates sources of information, and the type of information each source can provide.

TABLE 12.1 Sources and types of information

Source of information	Information provided
Plan or specification	Pump details. This may be a pump that has been specified by the engineer, or information that will enable correct selection of a pump. The details of pipe sizes, connection requirements, pump position and materials required.
Job instruction	Scheduling details. Date to commence work.
Site inspection	Verify the pump installation can be successfully carried out. Useful information after visiting the site includes: ease of access to the site pump location ready for installation availability of power source the need for special equipment – e.g. scaffolding, lifting equipment, lighting, generators, etc. type of personal protective equipment needed.
Pump manufacturer	All physical dimensions of the pump and the connection details. Installation requirements for the pump.

Before installing any sewerage pump system, approval must be obtained from the relevant local authority. This may be the local council or the sewerage authority. The following is an example of the type of information required for applying to install a sewerage pump system, and should only be used as a guide:

1   Where it is practical to achieve drainage by gravity, pumping to sewer is not normally permitted.
2   An application for pumping to sewer should include:
   a   plans in duplicate, showing:
      ■   a description of the pumping system
      ■   the capacity and construction of:
      i   collection wells
      ii   pump suction wells, or
      iii   sewage ejectors
   b   details of all pumps:
      i   type
      ii   number of pumps
      iii   pumping rate
      iv   expected frequency of operation
      v   pump performance curve details
   c   the point of connection to the rising main, or the authority's sewer, and elevation above the pump
   d   the venting arrangements for pump wells
   e   size, length and materials to be used in the rising main or pump line
   f   rising main (pump line) location; and if required to be located outside the property, as directed by the local government authority.

Pump performance basics

When installing sewerage pump sets, you will be required to understand the properties of sewage and the performance measures for various sewerage pump sets. The properties of sewage and waste are:

■   pressure
■   flow rates.

There is a mathematical relationship between pressure and flow, and while this is outside the scope of this unit, it is important to understand the basics. Generally, pump performance is measured by:

■   *pressure:* in metres head of water
■   *flow rate:* in litres per second.

Pressure

The SI (Système International) system of measurement is used to measure pressure, and the unit used is pascals (symbol: Pa).

■   1000 pascals = 1 kilopascal (kPa)
■   1 000 000 pascals = 1 megapascal (mPa).

Generally, the term 'metres head' (symbol: H) is used to describe the energy possessed by a weight of fluid, and is used to measure pump performance. When we use 'metres head', we are referring to the vertical height in metres that the pump must raise water. The pressure unit 'metre' is measured rather than defined, and the actual pressure varies depending on the density of the water. When water is most dense at 4°C, one metre head of water exerts a pressure of 9.81 kPa at the base of the column of water. Hot water at approximately 100°C is less dense than cold water, and at this temperature, one metre head of water exerts a pressure of 9.4 kPa at the base of the column of water.

Generally, we use an approximate conversion: 1 (H) = 10 kPa.

The standard atmosphere (atm) is an established constant. It is approximately equal to typical air pressure at earth mean sea level and is defined as follows:

$$1 \text{ standard atmosphere (atm)} = 101\,325 \text{ Pa}$$
$$= 101.325 \text{ kPa}$$

At sea level, the air above us generally exerts the greatest pressure of approximately 101 kPa, or 1 atm, and pump performance is best at this level because this air pressure assists pump suction. As we climb in elevation above sea level, there is less air above us; the air pressure is less than 101 kPa, or 1 atm, and the pump suction capacity is less.

When we talk about pumps and pumping, the term 'head' refers to the pressure generated by the pump or the pressure the pump needs to overcome. It is important to understand that pumps must overcome pipe friction and static head on both the suction and delivery sides of the pump.

When choosing pumps for sewerage, it is best practice to consult with the pump supplier to select a suitable pump. You will need to supply information on the following:

- type of waste
- vertical height of the pump above the suction point
- length of pipe between the suction point and the pump
- vertical height of the pipe outlet above the pump
- length of pipe between the pump and the pipe outlet
- flow rate required.

Flow rate

The SI system of measurement used to measure flow rate is the cubic metre per second (m^3/s) and litres per second (L/s). For small pumps, the flow rate used is litres per second; while for very large pumps, cubic metres per second is used.

Every cubic metre contains 1000 litres. Therefore, when converting m^3/s to L/s, it is simply a matter of multiplying by 1000; and when converting L/s to m^3/s, simply divide by 1000.

Examples:

1.5 m^3/s	1.5 x 1000 = 1500 L/s
6 L/s	6/1000 = 0.006 m^3/s

Calculating design flow

When selecting and ordering a pump and sizing a pump well, you will need to estimate the flow rate. This flow rate is known as design flow. Design flow is the peak flow that will discharge into the well from all upstream fixtures and other sources of water.

The full calculations for design flow are outside the scope of this unit and must be carried out by a suitably qualified and experienced person. However, it is beneficial to understand the terminology and basics of pump well and pump sizing.

It is relatively simple to estimate the flows from residential premises because we have reliable historic information. It is not as simple to calculate combinations of flows from residential, commercial and industrial premises. AS/NZS 3500 Plumbing and drainage does not provide guidance on estimating sanitary flow.

The following sizing method is based upon information in WSA 02-2004-3.1 *Gravity Sewerage Code of Australia*, published by the Water Services Association of Australia (see 'Further reading' at the end of this unit).

Design Flow = PDWF + GWI + RDI

Where Design is the flow rate in litres per second

PDWF is Peak Dry Weather Flow in litres per second

GWI is Ground Water Infiltration in litres per second

RDI is Rainfall Dependent Inflow in litres per second

The *Gravity Sewerage Code of Australia* is used and adapted by water authorities and councils around Australia to size large pump stations and pipe networks where the design must consider GWI and RDI. However, in the case of relatively small new private sewerage pump stations, there would be little, if any, GWI and RDI, so therefore the main objective is calculating PDWF.

Average Dry Weather Flow (ADWF) is the combined average daily flow from all sources. Based upon historical research and evidence, ADWF is 180 litres per person per day, or 0.0021 litres per second (180 ÷ 24 hr ÷ 60 min ÷ 60 sec.).

Equivalent Population (EP) is the equivalent number of persons per unit/residence, etc. For example, the equivalent number of persons per single residential lot is 3.5, and the number of persons for high-density residential units is 2.5.

Example calculation of Average Dry Weather Flow (ADWF):

20 residential units with 2.5 persons per unit

ADWF = 0.0021 x EP

ADWF = 0.0021 x 20 x 2.5

ADWF = 0.105 litres per second

To calculate Peak Dry Weather Flow (PDWF), we need to multiply ADWF by peaking factor 'd'.

The peaking factor varies depending upon the area of the development.

If, in the above example, the 20 residential units were located on a 1-hectare allotment, we could look to WSA 02-2004-3.1 *Gravity Sewerage Code of Australia*, and find that the peaking factor 'd' is 7.57.

Example calculation of Peak Dry Weather Flow (PDWF):

PDWF = d x ADWF

PDWF = 7.57 x 0.105

PDWF = 0.795 litres per second

Therefore, our Design Flow will be PDWF 0.795 L/s, plus any consideration required for GWI and RDI. If there is no Ground Water Infiltration of Rainfall Dependent Inflow, our Design Flow will be the Peak Dry Weather Flow – that is:

Design Flow = PDWF + GWI + RDI

Design Flow = 0.795 + 0 + 0

Always consult an experienced specialist to calculate Design Flow.

For further information read:

- WSA 02-2014-3.1 *Gravity Sewerage Code of Australia*
- WSA 04-2005-2.1 *Sewage Pumping Station Code of Australia*.

Pump performance

When selecting a pump, it is necessary to match the performance of the pump to that needed by the system. To do that, an engineer would refer to a pump manufacturer performance curve, which indicates how a pump will perform in relation to pressure head and flow.

Total dynamic head (TDH) in a pumping system is the total amount of pressure that the pump must overcome when water is flowing in the system. It is made up of two parts: the vertical rise and pipe friction loss. This must be calculated accurately to determine the correct size and scale of pumping equipment.

To calculate TDH, we need to calculate two things:

A the vertical rise

B the friction losses of all the pipe and components the liquid passes through at the inlet and outlet of the pump.

After calculating both A and B, we add them together to determine TDH. The calculations are outside the scope of this unit; however, it is important to understand the basics of pump sizing.

LEARNING TASK 12.1

Look at Figure 12.2.
1 What is the approximate maximum head for this pump?
2 At the maximum head, what is the capacity in litres per second?
3 Reading only the 'HEAD' curve, what is the approximate maximum pumping capacity in litres per second?

Safety and environmental requirements

When installing sewerage pump sets, you will need to use appropriate personal protective equipment (PPE). The specific types of PPE for this type of work may include:

- safety glasses or goggles to protect the installer while cutting and drilling
- safety boots
- overalls
- dust masks or respirators to protect the user from breathing in dust or fine particles when cutting and drilling

EXAMPLE 12.1

Figure 12.2 shows a pump performance curve. Imagine that we need to pump 13.3 litres per second through a calculated TDH of 9.2 M HEAD. It can be seen that this pump will be operating at optimum efficiency for our requirement and is therefore suited to the application.

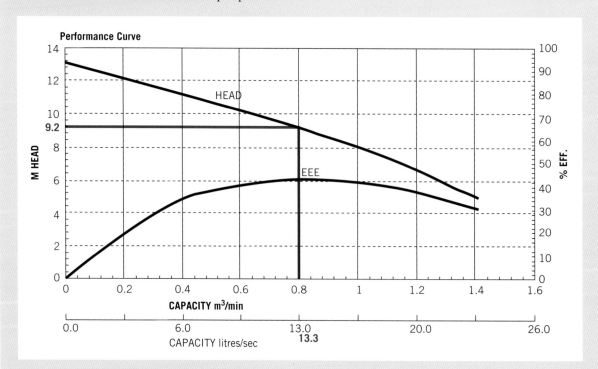

FIGURE 12.2 Example pump performance curve

- gloves, for the protection of the installer's hands when working with pipes and pump components
- ear plugs or ear muffs while drilling and cutting
- hard hat when working in a wet well or confined space below others
- fume or dust masks when exposed to bio-hazards, fumes or dust.

When replacing or repairing sewerage pump sets, you should take precautions to protect yourself from infection and viruses from waste water, and cuts/pricks from sharp objects. Cover all exposed skin by wearing appropriate PPE, and thoroughly wash, as required, with suitable disinfectant soap or solution.

> A wet well can be classified as a confined space. Before working in any confined space, you must be prepared to comply with local regulations. When working in confined spaces, and depending upon the job, a certain number of workers must have completed a course in confined space entry. It will also be necessary to have equipment available to extract workers from confined spaces.

Quality assurance

As a minimum, the quality assurance requirements for installing sewerage pump sets relate to compliance with manufacturers' installation instructions, the project plans and specifications, local regulations and AS/NZS 3500.2.

AS/NZS 3500.2 SANITARY PLUMBING AND DRAINAGE

It is important that wet wells installed below ground level are constructed to prevent the entry of stormwater (infiltration) and the escape of waste water (exfiltration). Infiltration would cause an unnecessary loading of stormwater on the sewerage system, and exfiltration would pollute the environment.

The local regulatory authority or sewerage authority may have specific inspection and testing requirements for sewerage pump sets and/or wet wells to confirm there is no infiltration or exfiltration, and this may be in the form of either an air test or a water test. (See the section later in this unit on testing.)

Sequencing of tasks

The installation, particularly of pump sets and underground wet wells, requires the correct sequencing of tasks and working to plan.

The wet well must be designed with a specific holding capacity to match the application. This holding capacity should be constructed at the *minimum* depth below ground level to *minimise* the excavation for the wet well. To achieve this, the sanitary drain upstream of the wet well should *first* be laid with the *minimum* cover and gradient.

When planning work to install sewerage pump sets and wet wells, you will need to coordinate the installation with others, including:

- the electrician, to arrange for power and controls for the pump
- equipment commissioning personnel (where applicable)
- the project manager.

Tools and equipment

The tools and equipment needed to install sewerage pump sets may include:

- hand and power tools:
  - electric drills – both conventional and hammer drills
  - adjustable shifting spanners or wrenches
  - socket sets – both imperial and metric
  - masonry saw
  - hammer
  - cold chisel
  - screwdrivers – both plain and Phillips head
- measuring and alignment tools:
  - tape measure
  - steel ruler
  - square
  - spirit level
  - straight edge
  - string line and plumb-bob
  - pressure testing equipment
- concreting tools:
  - hand trowels
  - shovel
  - bucket
  - concrete mixer
- lifting equipment (sewerage pump sets can be heavy to move):
  - hand trolley
  - hoists and jacks
  - rollers
  - forklift
  - chain blocks
  - crane.

 COMPLETE WORKSHEET 1

Pump installation requirements

Pump location

Generally, pumps must be installed in an accessible location, so that they can be removed for maintenance, repair and replacement. Pumps for waste fixtures should be located as close as possible to the fixtures.

Where wet wells are deeper than 1.2 m, a corrosion-resistant ladder must be fixed in the well to enable access for cleaning and pump maintenance.

Small bore macerator pump packages are available to connect individual fixtures and small groups of fixtures, and should be located in accordance with the manufacturer's specifications.

Pump base

The base of pump wells should be constructed of materials that resist corrosion from sewage and sewage gases and at a self-cleansing gradient towards the inlet of the pump/s. The minimum self-cleansing gradient has traditionally been regarded as 1 in 12.

Pumps should be placed on a stable base or structure that does not transmit sound from the pump to the building. The base should also be of adequate thickness and strength to attach pump bracket fixings (where required by the pump manufacturer).

Materials

All materials and equipment used in connection with sewerage pump sets and wells must be resistant to corrosion from sewage and sewage gases. Pre-cast and cast in-situ reinforced concrete should be made of sulphate-resistant cement. Other authorised materials include stainless steel, brickwork, glass-reinforced plastics and suitable plastics such as polypropylene.

There may be limitations on the depths of prefabricated plastic wells.

Before purchasing any materials in connection with the installation of sewerage pump sets, verify that the materials are approved by the manufacturer and comply with the relevant standards. Pipes connected to the outlet of pumps should be pressure-type pipe, in accordance with the provisions of AS/NZS 3500.1 Plumbing and drainage. Water services, and be capable of withstanding a hydrostatic pressure test equal to twice the shut-off head of the pump.

Further information on the installation of pumps can be found in *Basic Plumbing Services Skills – Water Supply*.

Installing sewerage pump equipment

Pump types
Centrifugal pumps

Centrifugal pumps have applications that range from small domestic water-pressure pumps and swimming pool pumps made in plastics, to high-pressure process pumps for oil refineries. Some centrifugal pumps used in sewage treatment plants are very large.

Centrifugal pumps are very simple, consisting of a casing (known as the volute), inside which an impeller is mounted on a rotating shaft that is supported by bearings. Where the drive shaft passes through the pump casing, water is prevented from leaking out by a shaft seal.

How does a centrifugal pump work? Imagine swinging a full bucket of water in a vertical circle over your head. The water is kept in the bucket by centrifugal force. The impeller in a centrifugal pump spins the water around and centrifugal force causes the water to be thrown outward, the same as in the bucket.

The casing, or volute, on the pump is spiral-shaped to channel the water around to the exit or discharge point. This means the pump must rotate in the correct direction.

Centrifugal pumps are used to apply pressure to a liquid only; for pumping raw sewage, they must have a cutter (macerator) installed to pump (transmit) solids. This is explained in more detail later in this unit.

Centrifugal pumps must be installed as close as possible to the inlet liquid level, as they have limited suction lift capability, and must not be installed with more than around 6 m of vertical suction lift.

Figure 12.3 shows a centrifugal pump without a motor.

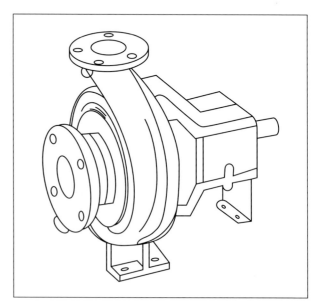

FIGURE 12.3 Centrifugal pump without a motor

Source: Department of Education and Training (http://www.training.gov.au) © 2013 Commonwealth of Australia.

Submersible pumps

A submersible pump is designed to operate fully immersed in the cooling environment of the fluid surrounding it. The electric motor is close coupled to the pump and is electrically insulated, either by housing it in a watertight casing or, in some models, encasing the motor windings in epoxy resin.

Submersible pumps have numerous advantages and are more commonly used than non-submersible pumps.

There are two types of submersible grinding pumps: 'grinder' pumps and 'mutrator' pumps. A grinder pump is a low-capacity pump for small applications used to shred the sewage into small particles prior to it passing through the impeller. A mutrator pump has a combined cutter and impeller action within the one device. It has hardened impellers and should be matched to effluent quality to ensure economic and trouble-free operation.

In submersible pumps, the pump inlet is positioned slightly above the wet well floor. Effluent is drawn in, and large particles are ground down and transferred by the outlet pipe to the point of gravity discharge.

Depending on the volume of sewage to be handled, pumps may be installed in duplicate. In these installations, they are usually installed with sophisticated controls to alternate the pump starts so as to increase the life of each pump.

Figure 12.4 shows a submersible-type pump with a motor and float switch.

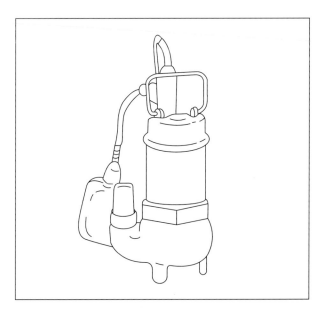

FIGURE 12.4 Submersible pump with a motor and float switch

Source: Department of Education and Training (http://www.training.gov.au) © 2013 Commonwealth of Australia.

Couplings and connections

Pumps must be connected by fittings that permit easy disconnection from the outlet pipe. Options include flanges, unions or gibault joints. Corrosion-resistant couplings and connections must be used.

Pumps may be coupled and connected to the discharge pipe (rising main) in a variety of ways, including:

- free-standing pump connected to the hose or pipeline with a disconnection coupling or flange
- suspended pump with integral non-return valve and high disconnection coupling
- wet well with pump installed on twin guide bars with automatic low-level disconnection.

The outlet pipe must incorporate a gate valve and check valve located on the outlet (delivery) side of the pump. The outlet pipe work must then discharge to the gravity sewerage system in accordance with AS/NZS 3500.2.

AS/NZS 3500.2 SANITARY PLUMBING AND DRAINAGE

Note: Use a flexible connection where it is likely that the pump might transmit vibration and stress to connecting pipework.

Pump controls

A variety of controls are used to operate pumps, depending on the pump size and application. Controls start the pump(s) when the effluent is at a high level, and turn the pump off at a low level before the pump can run dry.

Controllers include:

- float switches (electric controller)
- electrode level-sensing probe systems (electronic controller).

Small submersible pump packages usually have an attached floating pump control (float switch) that automatically switches the pump on when the effluent level meets a pre-set high level, and then switches the pump off when the water level falls to a pre-set low level. One problem with small submersible pump packages with an attached float switch is that the float switch can be fouled in a position where it cannot sense effluent level.

Float switches are of two types: mercury and ball bearing. Mercury float switches operate by making or breaking an electrical circuit through mercury (a liquid metal). The ball-bearing type operates by a ball bearing rolling over a micro-switch.

Pumps in wet wells often have electrode level-sensing probe systems with rods of various lengths to sense low, high and alarm levels. These controllers are found in larger (multiple) submersible pump installations that have separate liquid level controllers to switch the pump on and off as required. In a situation where the high-level switch fails to sense the liquid level, an alarm activates. Alarms are either a flashing light or an audible alarm, or a combination of both.

Submersible pumps must be controlled so as to limit the number of starts per hour to within the capacity of the pump, and should, as far as practicable, empty the contents of the wet well at each operation. The required pumping rate should be matched to the expected inflow, holding capacity of the well, and allowable discharge, in accordance with the pump manufacturer's specifications and AS/NZS 3500.2.

In sophisticated control systems, pumps are monitored and protected from overloading and other catastrophic failure.

Pump motors

Pump motors or engines are designed and matched to pumps based on the pump speed and power requirements.

Electric motors are the most common method of driving sewerage pumps. The electricity network throughout Australia is supplied at two voltages – 240 volts and 415 volts – with a frequency of 50 hertz. The electricity supply is often referred to as 'alternating current' (AC). You should be aware that if the motor is wired incorrectly, it could force the pump in the wrong direction. This can be dangerous, so ensure that only a qualified electrician connects wiring to the pump motor, using the pump manufacturer's instructions.

On small-duty pumps, often only a single-phase 240-volt motor is used; however, for the majority of pumps in industry, three-phase 415-volt motors are common. When there is an option to use either a single- or a three-phase motor, it is best practice to consider the power consumption and select the three-phase motor, which can be more economical in the long term. Other considerations include motor protection features such as thermistors, which shut down power to the motor on sensing excessive heat in the motor windings.

> **GREEN TIP**
>
> Pumps and motors should not be undersized or oversized. It may be necessary to consult a pump engineer to ensure that the equipment is matched to the application. This will minimise power consumption and optimise the life of the pump and motor.

Compressed air ejectors

The compressed air sewage ejector is a form of sewerage pump that has been in use for over 100 years. It is now seldom used in Australia, having been largely replaced by the wet well with submersible pump.

The two main components of the compressed air ejector are the heavy-duty pot and the air compressor.

Sewage enters the pot by gravity. The pot is a heavy-duty storage container incorporating inlet and outlet check valves, and a float mechanism. As the effluent rises in the pot, the float rises and a switch activates the air compressor. Air then enters the pot, pressurising the contents, and the sewage is ejected from the pot through the outlet check valve (see Figures 12.5, 12.6 and 12.7).

The size of ejector pots is determined by the maximum flow rates from the plumbing fixtures, and multiple pots may be installed in parallel. Each ejector pot must be fitted with a vent at least DN 40 extended to the open air, or connected to a relief or stack vent.

Compressed air ejection systems should be installed and connected in accordance with AS/NZS 3500.2.

Wet wells

A wet well is generally defined as a below-ground chamber for the collection of sewage. Either a submersible or non-submersible pump is used to transfer the contents of the well to an approved point of discharge.

Wells are widely used both within the network utility operator sewerage system and in property sewerage systems. For example, the Hoppers Crossing Pumping Station, which is operated by Melbourne Water, is the height of a 22-storey building and contains eight pumps. Each pump stands 4.5 m high and can pump 5000 litres per second. The pumping station is a massive type of wet well that lifts and transfers unscreened sewage from the Western Trunk Sewer to the Western Treatment Plant in Werribee, near Melbourne (see Figures 12.8, 12.9 and 12.10).

Within property sewerage systems, wet wells are commonly used to collect and transfer the discharge from a large number of fixtures in a building located below the gravity sewer connection point. Plumbers are required to install and maintain these smaller-scale wet well sewerage systems.

Wet wells consist of a tank and pump (or pumps) capable of handling unscreened sewage.

Wet wells are required to be of sound construction and made of materials that resist corrosion internally from the sewage and externally from aggressive soil conditions. Concrete is the most common material used for construction of wet wells, either prefabricated or cast in-situ. Corrosion-resistant metals, glass-reinforced plastics and brickwork are other materials used for wet-well construction.

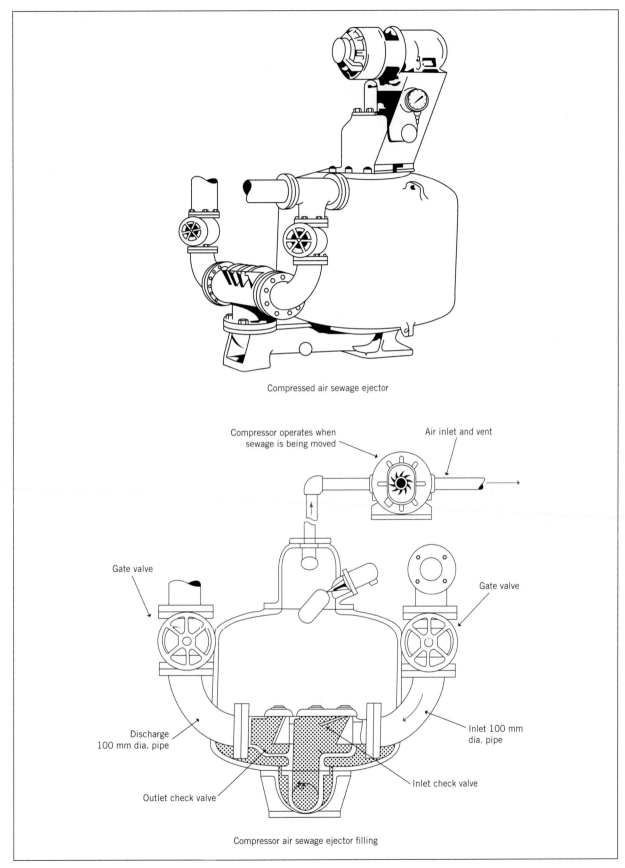

Compressed air sewage ejector

Compressor operates when
sewage is being moved

Air inlet and vent

Gate valve

Gate valve

Discharge
100 mm dia. pipe

Inlet 100 mm
dia. pipe

Outlet check valve

Inlet check valve

Compressor air sewage ejector filling

FIGURE 12.5 Compressed air sewage ejector

Source: Department of Education and Training (http://www.training.gov.au) © 2013 Commonwealth of Australia.

FIGURE 12.6 Maintenance on compressed air sewage ejector

The base and cover are constructed of materials similar to the walls of the well, and the entire weight of the well must be supported on stable ground. The base must have a self-cleansing grade towards the pump inlet (traditionally a minimum gradient of 1 in 12), and the cover is fitted with removable airtight access covers sized for maintenance purposes (see Figure 12.11).

Where wet wells exceed 1.2 m in depth, a fixed ladder must be installed. The construction requirements for wet wells are explained in greater detail in AS/NZS 3500.2.

FIGURE 12.8 Hoppers Crossing Pumping Station, during construction

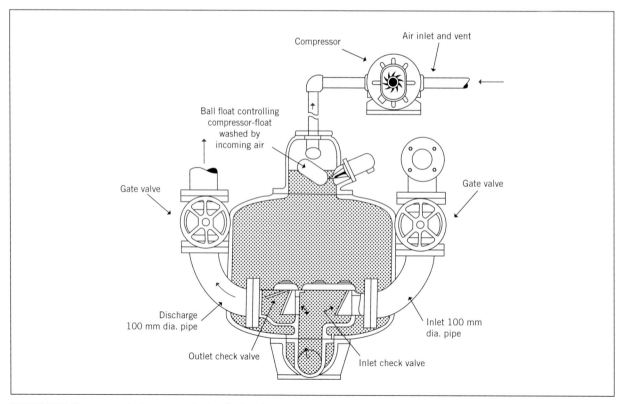

FIGURE 12.7 Compressed air sewage ejector discharging

FIGURE 12.9 Hoppers Crossing Pumping Station pump and valve, during construction

AS/NZS 3500.2 SANITARY PLUMBING AND DRAINAGE

The inlet drain transferring the discharge from all the fixtures enters the wet well and terminates with a square junction. The top of the square junction is open, while the bottom end is usually extended to near the bottom of the well above the well's lowest water level.

Wet wells are vented with a DN 80 pipe connected near the top of the well and extended to terminate as a drainage vent, in accordance with AS/NZS 3500.2 requirements.

AS/NZS 3500.2 SANITARY PLUMBING AND DRAINAGE

Dry wells

If the pump is installed in a separate chamber to the sewage, it is referred to as a *dry well system*. Dry well systems vary in size. In an authority's sewerage system, dry wells are very large in scale, sometimes having an elevator to permit access to personnel and equipment.

FIGURE 12.10 Hoppers Crossing Pumping Station valve, during construction

In smaller installations, dry wells are available prefabricated and fitted with pump equipment. Dry well systems may be utilised in small-scale installations (such as a small townhouse complex), or they could be used in a larger installation (such as a large shopping complex).

Dry well systems incorporate two tanks. One is simply a holding tank into which sewage discharges. The other is a tank within close proximity that houses (generally) two centrifugal pumps that are capable of handling unscreened sewage. Pumps are driven directly by electric motors. High-level sensors detect the effluent level in the well, which controls the pumps. The pumps start when the effluent is at a high level, and continue to operate until the low-level sensor turns the pumps off.

Sewage is pumped to its point of discharge, which would most likely be into a sewerage authority inspection chamber, where it flows again under gravitational flow.

Holding tanks (discharge from waste fixtures)

The pump discharge from a group of waste fixtures may be achieved in a similar way to a wet well with

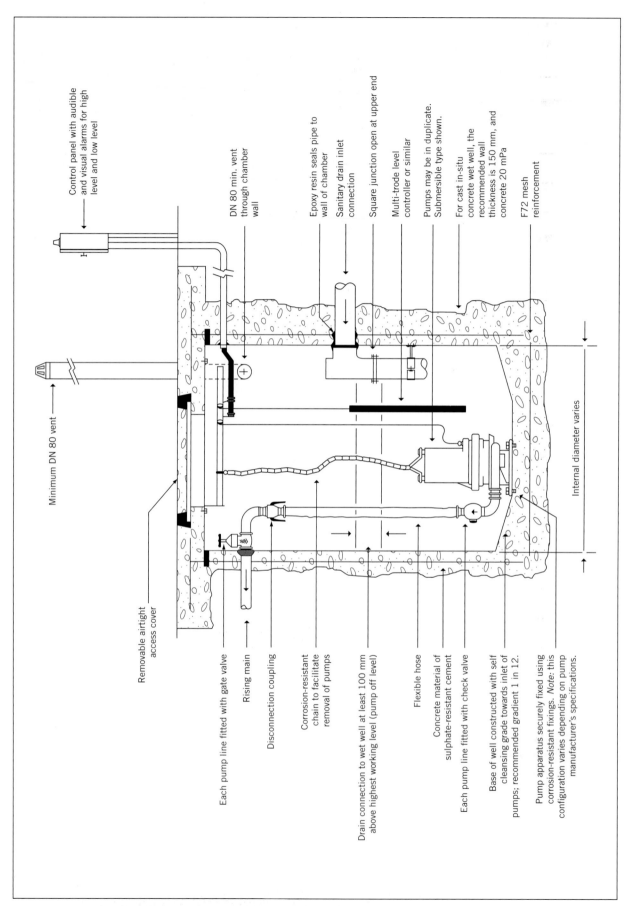

FIGURE 12.11 Section diagram of wet well

Control panel with audible and visual alarms for high level and low level

DN 80 min. vent through chamber wall

Epoxy resin seals pipe to wall of chamber

Sanitary drain inlet connection

Square junction open at upper end

Multi-trode level controller or similar

Pumps may be in duplicate. Submersible type shown.

For cast in-situ concrete wet well, the recommended wall thickness is 150 mm, and concrete 20 mPa

F72 mesh reinforcement

Internal diameter varies

Minimum DN 80 vent

Removable airtight access cover

Each pump line fitted with gate valve

Rising main

Disconnection coupling

Corrosion-resistant chain to facilitate removal of pumps

Drain connection to wet well at least 100 mm above highest working level (pump off level)

Flexible hose

Concrete material of sulphate-resistant cement

Each pump line fitted with check valve

Base of well constructed with self cleansing grade towards inlet of pumps; recommended gradient 1 in 12.

Pump apparatus securely fixed using corrosion-resistant fixings. *Note:* this configuration varies depending on pump manufacturer's specifications.

GET IT RIGHT

WET WELL CONSTRUCTION

FIGURE 12.12 Wet well construction

The pictures in Figure 12.12 show a wet well over 3 m deep containing dual submersible pumps that are continually becoming blocked. The base of the well is flat, and there is no vent or ladder. Research AS/NZS 3500.2 and this unit to answer the following questions.

1   How should the base of the wet well have been constructed?

_____

_____

_____

2   What is the minimum size of vent required?

_____

_____

_____

3   Under what circumstances is a ladder required?

_____

_____

_____

submersible pump, although commonly one or more fixtures are connected to an above-ground 'holding tank'. The holding tank system is constructed and operates similarly to the wet well, except that it is designed to cope only with the discharge from waste-type fixtures. Pumps in holding tanks are not required to handle solid matter; therefore, grinding pumps are unnecessary. Pumps in holding tanks are generally termed 'sullage' pumps.

The pumps may be submersible or non-submersible. Non-submersible pumps can be positioned either inside or outside the tank, with the inlet pipe extended to the bottom. Where the pump is located inside the tank, a non-return valve and a gate valve are required on the outlet (delivery) side of the pump.

Where the pump is located outside the tank, a gate valve is required on the inlet and outlet of the pump. A non-return valve is also required.

The tank is vented with a DN 50 pipe located 100 mm above the inlet pipe and must comply with the venting requirements for waste fixtures specified in AS/NZS 3500.2.

AS/NZS 3500.2 SANITARY PLUMBING AND DRAINAGE

The pumped outlet must be a minimum DN 25 pressure pipe and discharge according to the requirements of AS/NZS 3500.2.

AS/NZS 3500.2 SANITARY PLUMBING AND DRAINAGE

Submersible pumps require a level, stable base on which to sit.

Figure 12.13 depicts a submersible pump mounted in a holding tank. This pump is fitted with a mesh screen to prevent any solids that may have inadvertently entered the tank from entering and damaging the pump.

The base of a holding tank should ideally be constructed with a self-cleansing grade towards the pump inlet.

Pressure sewer systems

Pressure sewer systems (PSS) are usually designed to convey domestic sewage from the property drain to a sewerage treatment plant, gravity carrier, or larger pump station and rising main. They are designed to handle domestic sewage only; they are not designed to handle commercial or industrial sewage. For commercial and industrial applications, you should consult the pump manufacturer.

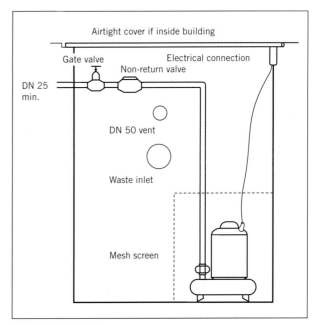

FIGURE 12.13 Holding tank with submersible pump

Source: Department of Education and Training (http://www.training.gov.au) © 2013 Commonwealth of Australia.

Authorities use these types of systems as an alternative to gravity sewerage systems. They are also used by property owners whose plumbing fixtures are lower than the authority's sewer main.

The information provided in this section is based on the installation of a proprietary system called NOV MONO™ InviziQ™, which has two tank sizes available – in 900 litres or 2200 litres.

Sewage from the home flows by gravity into the PSS holding tank. When the level rises to a set point, the pump starts automatically. For example, the NOV MONO™ InviziQ™ System 900-litre tank has pump stop and run levels preset at 120 mm and 280 mm, respectively, from the base of the unit (see **Figure 12.14**).

The self-priming pump draws the sewage into an integral macerator (cutter), turning the solids into a slurry of small particles and allowing the sewage to be discharged through a small-bore pipe (DN 32 to DN 125) into the pressurised main sewerage system.

As the sewage is transported under pressure by a positive displacement pump and not by gravity, a PSS can be installed in locations that vary in topography, such as:

■ mountainous or hilly land
■ flat land
■ clay, rocky soil or areas of shallow topsoil
■ areas of significant environmental sensitivity
■ built-up areas
■ areas of low population density.

Always refer to the manufacturer's specifications to ascertain the components of their systems.

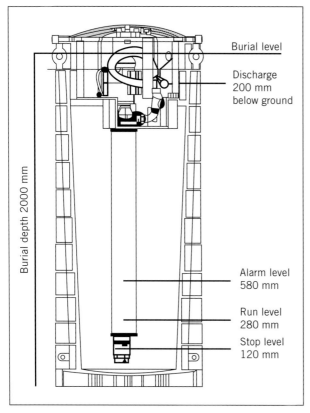

FIGURE 12.14 NOV MONO™ InviziQ™ System 900-litre tank pump stop and run levels

Source: Reproduced with permission of NOV Australia Pty Ltd.

The components listed below are included in the NOV MONO™ InviziQ™ System (see **Figure 12.14**), which comprises:

- either a 900- or 2200-litre tank with lid
- a pump with inbuilt macerator
- a control panel
- all internal tank pipework
- an advanced level sensor
- a check valve and ball valve on the pump's discharge
- 20 m of electrical cable and 20 m of comms cable.

Tools

The tools shown in **Figure 12.15** will be required to install and maintain the NOV MONO™ InviziQ™ System.

Installation instructions

Prior to installation, a site assessment must be carried out and a safe work method statement (SWMS) and job safety analysis (JSA) prepared in accordance with the local regulatory WHS requirements. The installation information shown in **Figures 12.16** and **12.17** is based on the NOV MONO™ InviziQ ™ system and should be used as a guide. For all types of systems, consult the manufacturer's installation instructions. Failure to do so may result in the cancellation of the product warranty.

The installation requirements for the PSS will vary depending on ground conditions, and should be in accordance with the design specifications determined by a suitably qualified engineer.

The PSS holding tank should be located in an area where there is good drainage of surface water away from the tank. There should be a minimum gradient of 1:4 away from the lid to ensure adequate drainage. It should be as close as possible to the plumbing fixtures and in a non-trafficable area.

Pipework connections – inlet

The connection to a NOV MONO™ InviziQ™ system 900-litre tank from the residence is made using DN 100 PVC pipe and the grommet seal provided (loose). The drainage system into the PSS is installed in the same way as a conventional sanitary drainage system, in accordance with AS/NZS 3500.2.

AS/NZS 3500.2 SANITARY PLUMBING AND DRAINAGE

Drill a hole (using a 121-mm hole saw) at the correct depth using the flats provided. Ensure that the hole is central to the flat. Place the grommet seal in the hole with the label facing out. Push the pipe through. Soapy water may assist this process.

Alternatively, use a rubber spigot connection with an internal diameter of 110 mm.

For a 900-litre tank, the maximum depth of the inlet from ground level is 1325 mm; the minimum is 675 mm. The drainage system upstream of the PSS must be installed as a conventional sanitary drainage system, in accordance with the *Plumbing Code of Australia* and AS/NZS 3500.2.

AS/NZS 3500.2 SANITARY PLUMBING AND DRAINAGE

Pipework connections – outlet

A 32-mm BSP stainless steel fitting with external thread is supplied for the outlet. The pump discharge pipework is a pressure application, not a gravity application, and so it must comply with the relevant sections of AS/NZS 3500.1.

AS/NZS 3500.1 PLUMBING AND DRAINAGE. WATER SERVICES

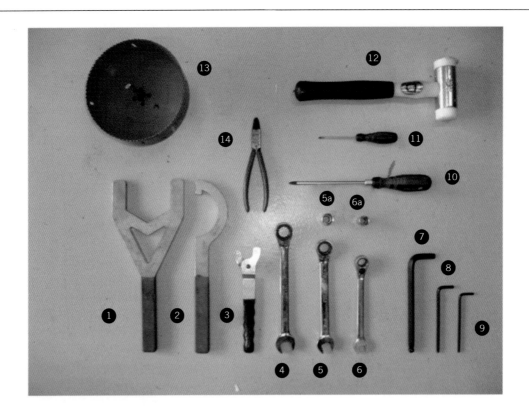

ITEM	DESCRIPTION	ITEM	DESCRIPTION
1	Pump body support tool*	7	10 mm Allen key
2	Hook tool*	8	6 mm Allen key
3	Adjustable pin wrench	9	5 mm Allen key
4	19 mm Ring spanner	10	Phillips head screwdriver
5	18 mm Ring spanner	11	Small flat head screwdriver
5A	18 mm Socket (Optional with drill)	12	Soft hammer
6	13 mm Ring spanner	13	121 mm Hole saw (Used for installation only)
6A	13 mm Socket (Optional with drill)	13a	168 mm Hole saw (Optional 2200 L tank for installation only)
		14	Bent internal circlip pliers

* 7.9.1 SPECIAL TOOL KIT AVAILABLE CONTACT NOV MONO™ (grif 556)   **GRIF 569
SERVICE PLATE SINGLE PUMP ONLY.

ITEM	DESCRIPTION
1	Pump body support tool
2	Hook tool
SEPARATE OPTION NOT PART OF KIT	
**	Service plate single pump only

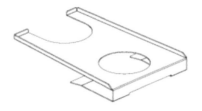

FIGURE 12.15 Tools required to install and maintain the NOV MONO™ InviziQ™ System

Source: Reproduced with permission of NOV Australia Pty Ltd.

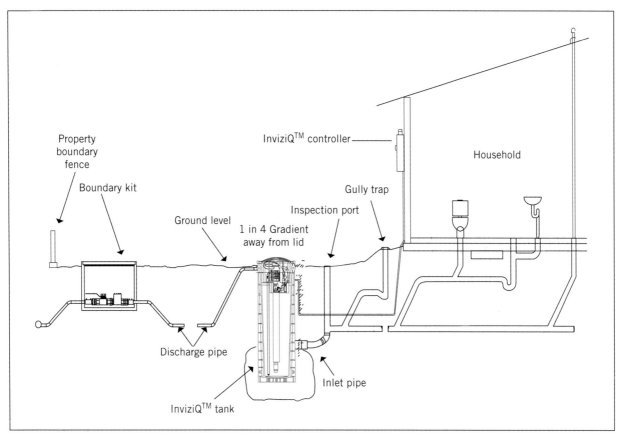

FIGURE 12.16 Typical layout of a NOV MONO™ InviziQ™ System

Source: Reproduced with permission of NOV Australia Pty Ltd.

Ensure that the discharge pipework leading away from the PSS tank is in the pattern indicated in Figure 12.16 and is well supported for backfilling.

Electrical installation

A qualified licensed electrical contractor must carry out the electrical installation.

Commissioning the system

Once the installation is completed (see Figure 12.18), and prior to turning the system on, it must be commissioned in accordance with the manufacturer's installation instructions. Failure to do so may result in the cancellation of the product warranty.

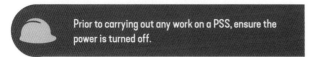

Prior to carrying out any work on a PSS, ensure the power is turned off.

FROM EXPERIENCE

The ability to install and commission sewerage pumping equipment in accordance with manufacturer's specifications and instructions is critical to reliable pump performance and durability.

Small bore macerator systems

In situations where there is one or a small group of fixtures, a small bore macerator system is more convenient and more economical than constructing a wet well system. Small bore macerator systems are primarily used for single-closet pans (see Figure 12.19), but units are also available to collect the discharge from a small group of fixtures in close proximity. The macerator unit is similar in size to a closet cistern. It is installed directly behind the water closet outlet and is connected using a standard DN 100 push-on pan adaptor fitting. Some macerator units are available with waste pipe connections for the connection of multiple waste fixtures, including basins, showers, baths and dishwashers.

When a discharge is received, the macerator automatically operates to grind the organic waste and paper into slurry. It is then pumped via a DN 20 (minimum) pipe that must be connected to the drain according to the manufacturer's specifications and AS/NZS 3500.2.

AS/NZS 3500.2 SANITARY PLUMBING AND DRAINAGE

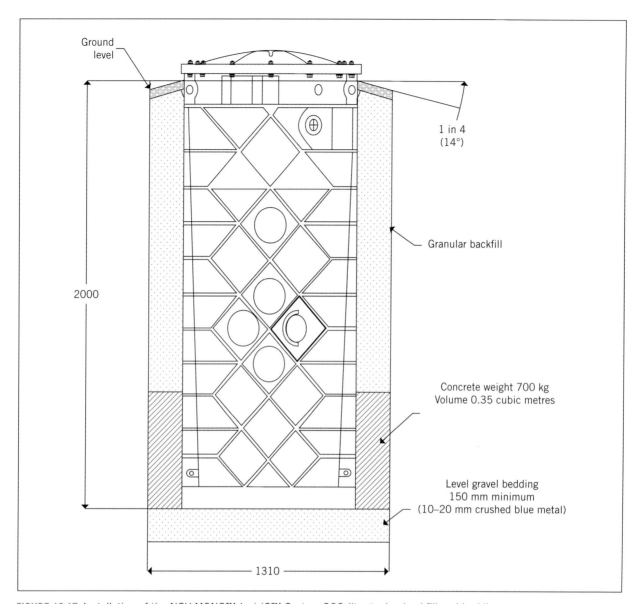

Ground level

1 in 4 (14°)

Granular backfill

2000

Concrete weight 700 kg
Volume 0.35 cubic metres

Level gravel bedding
150 mm minimum
(10–20 mm crushed blue metal)

1310

FIGURE 12.17 Installation of the NOV MONO™ InviziQ™ System 900-litre tank – backfill and bedding

Source: Reproduced with permission of NOV Australia Pty Ltd

Saniflo is one manufacturer of small bore macerator systems. Saniflo has on-line videos on different systems and installation considerations for venting, temperatures, pipe sizing, pump lines, pump location and alarms. To access the plumbing training videos visit https://saniflo.com.au/cms/51/plumberportal.

Testing pumps and piping

The local regulatory authority or sewerage authority may have specific inspection and testing requirements for sewerage pump sets and/or wet wells to confirm there is no infiltration or exfiltration, and this may be in the form of either an air test or a water test. The water test is sometimes referred to as a 'lid test', where the wet well chamber is filled with water to the access cover; however, drinking water is not used in water conservation areas. Non-drinking water such as stormwater and recycled water can be used.

The pump line (sometimes referred to as a 'rising main') between sewerage pump/s and the point of discharge must be tested to a hydrostatic pressure equal to twice the shut-off head of the pump before the pump line is concealed or buried.

The regulatory authority should be consulted in relation to local testing requirements. The pumping equipment and related piping must always be tested in accordance with the manufacturer's recommendations. Where required, test reports and results must be retained and/or submitted to the regulatory authority and/or manufacturer.

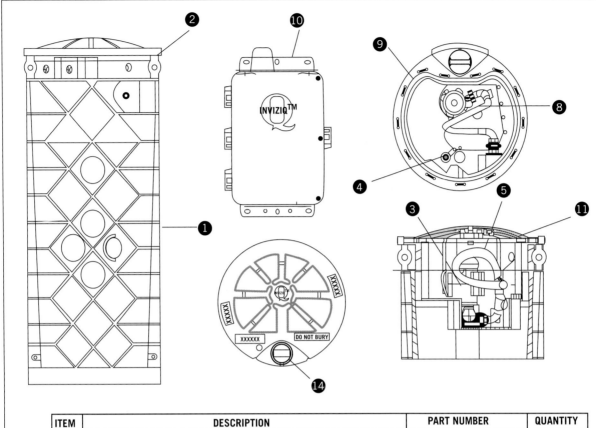

ITEM	DESCRIPTION	PART NUMBER	QUANTITY
1	Tank	CM9014XBL	1
2	Tank lid	CM9045GC	1
3	Pump and motor	SVP11CR81N	1
4	Advanced level sensor	GRIF 446	1
5	Flexible discharge assembly	GRIF 301	1
6	DWV Grommet seal (not shown)	AUX 6234	1
7	Electrical connections (not shown)	GRIF 390	1
8	2" Drain plug	GRIF 188	1
9	Lid gasket	GRIF 140A	1
10	InviziQ™ control panel	PSS-PL1MS2	1
11	Lid fastening kit	GRIF 367	13
12	3 Core orange sheath 1 mm^2 cable 30 × 0.2 mm^2 strand (not shown)	GRIF 091	20 m
13	4 Core white sheath 0.75 mm^2 cable 24 × 0.2 mm^2 strand (not shown)	GRIF 092	20 m
14	110 mm Inspection port	REFER NOV MONO™	1

FIGURE 12.18 NOV MONO™ InviziQ™ System 900-litre tank components

Source: Reproduced with permission of NOV Australia Pty Ltd.

Clean-up

Clean-up includes the following:
- Clear the worksite of debris and rubbish. Tradespeople are required to leave a safe, clean work environment that is conducive to better-quality work and cooperation with other trades.
- Sort excess materials for storage, recycling and/or disposal. Store materials in a manner that will not cause damage or deterioration to any useful excess materials.
- Recycle material by sorting similar material and disposing of it correctly.
- Dispose of rubbish in the correct manner, such that it will not attract vermin or create environmental contamination.
- Dispose of waste materials safely. Waste materials such as cement, concrete and grouting mix need to

FIGURE 12.19 Closet pan connected to 'Saniflo' small bore macerator unit

Source: Saniflo, http://www.saniflo.com.au.

be disposed of correctly. Do not pour wet cement mixtures into stormwater drains or sewers, as this will cause blockages that will be expensive to repair. Dispose of other waste such as swarf, off-cuts, rags, paper and packaging material in suitable bins.

■ Clean, maintain and store your tools and equipment. Equipment should be cleaned and oiled after each job and before packing it away. Failure to correctly maintain your tools and equipment will result in premature wear and tear, and expensive replacement. Poor maintenance of tools can also result in injury.

COMPLETE WORKSHEETS 3 AND 4

SUMMARY

- When installing sewerage pump sets, it is necessary to follow the plans and specifications that indicate the location, dimensions and layout of the equipment. Also, prepare by familiarising yourself with local regulations and inspection requirements, and ensure that pipes and fittings have WaterMark approval.
- Always read the pump manufacturer's instructions and consult with the pump manufacturer if you are unsure about the installation requirements or suitability of pumping equipment for the application. Pumps must be installed on a stable base and all equipment and materials must be resistant to corrosion, sewage and sewage gases.

- When installing sewerage pump equipment, ensure that pumps can be disconnected for servicing, and pump controls have been correctly set and fitted. The pump installation must comply with AS/NZS 3500.2, and all electrical works must be completed by a qualified electrician. The completed installation, including the rising main, must be tested and commissioned.
- When works have been completed, clean up the work area. If required by the regulatory authority, lodge the necessary documentation such as a Certificate of Compliance and as-constructed plan.

FURTHER READING

Pump manufacturers often provide useful resources for learning about pumps. Examples of pump suppliers and manufacturers well known within the plumbing industry in Australia are:

Ajax: **http://ajaxpump.com**

BKB: **http://www.bkbgroup.com.au**

Forrers/KSB: **http://www.ksb.com/ksb-au**

Grundfos: **http://au.grundfos.com**

Mono: **http://www.monopumps.com.au**

Xylem: **https://www.xylem.com/en-au**

Two well-known manufacturers of small bore macerator systems are Saniflo (**http://www.saniflo.com.au**) and Grundfos.

The Water Services Association of Australia (WSAA) publishes codes that are used and adapted by numerous water authorities and councils for construction of authority sewerage assets and pump stations. For further information read:

Water Services Association of Australia (2014). WSA 02-2014 *Gravity Sewerage Code of Australia Version 3.1.* Docklands, Australia. Gravity Sewerage Code of Australia, version 3.1, Water Services Association of Australia, © Gravity Sewerage Code of Australia

Water Services Association of Australia (2005). WSA 04-2005 *Sewage Pumping Station Code of Australia Version 2.1.* Melbourne, Australia.

WORKSHEET 1

Student name: _____

Enrolment year: _____

Class code: _____

Competency name/Number: _____

Task: Review the section 'Preparing for work', then complete the following.

1   Why is it necessary to have a special design for a large sewerage pump system?

_____

_____

_____

_____

2   What useful information is obtained from pump manufacturer documentation?

_____

_____

_____

_____

3   How is pump performance measured for:

a   Pressure? _____

b   Flow rate? _____

4   One atmosphere is the air pressure at what level?

_____

_____

5   Convert one atmosphere to kilopascals:

_____

_____

6   Research and convert the following:

a   3 atm = _____kPa

b   6 H = _____kPa (approximate conversion)

7 Complete the following:

Total dynamic head (TDH) in a pumping system is_____
_____when water is flowing in the system.

8 What two things must be known to calculate TDH?

_____

_____

9 As we climb above sea level, what is the effect on pump capacity?

_____

_____

_____

_____

10 Is a wet well 3 m deep classified as a confined space? Explain your answer.

_____

_____

_____

11 What item of PPE should be worn when working below others in a wet well?

_____

_____

12 Why is it necessary to prevent infiltration and exfiltration in a wet well?

_____

_____

_____

_____

13 Explain why it is necessary for a drain upstream of a wet well to be laid with minimum cover and gradient.

_____

_____

_____

_____

WORKSHEET 2

Student name: _____

Enrolment year: _____

Class code: _____

Competency name/Number: _____

Task: Review the section 'Pump installation requirements', then complete the following.

1 Where should the pumps for waste fixtures be located?

_____

_____

_____

_____

2 When should a fixed ladder be provided in a wet well?

_____

_____

3 What is the recommended gradient for the base of a wet well?

_____

_____

_____

_____

4 Complete the following:

Pumps should be placed on a stable base or structure that does not transmit sound from the pump to

_____ .

5 Research AS/NZS 3500.2 and state the seven locations where a pump outlet pipe may be discharged.

a _____

b _____

c _____

d _____

e _____

f _____

g _____

WORKSHEET 3

Student name: _____

Enrolment year: _____

Class code: _____

Competency name/Number: _____

Task: Review the section 'Installing sewerage pump equipment', then complete the following.

1   Explain how a centrifugal pump works.

_____

_____

_____

_____

2   What are the two types of submersible grinding pumps?

_____

_____

_____

_____

3   Research and state the name of two manufacturers of submersible grinding pumps.

_____

_____

_____

_____

4   What are three types of fittings that are used for easy disconnection of pumps?

_____

_____

_____

_____

5   State the name of two types of pump controller.

_____

_____

_____

_____

6   What is one common problem found with the small submersible pump with attached float switch?

_____

_____

_____

_____

7   What is one benefit of a pump with a three-phase motor?

_____

_____

_____

_____

8   Explain the operation of a compressed air sewage ejector.

_____

_____

_____

_____

_____

_____

_____

_____

_____

9   Research AS/NZS 3500.2. Is it necessary to provide wet well pumps in duplicate?

_____

_____

10   Research AS/NZS 3500.2. Is it necessary for wet well pumps to empty the wet well every time the pump operates?

_____

11   Research AS/NZS 3500.2. What two valves are required on the pump line from a wet well?

_____

_____

12   Research AS/NZS 3500.2. Where is the inlet to a wet well normally terminated above the highest working level?

_____

_____

13 Research AS/NZS 3500.2. Under what circumstances might a wet well vent be reduced in size?

_____

14 Research AS/NZS 3500.2 and state the venting requirements for a holding tank.

_____

_____

_____

_____

15 Research using the internet. You are required to find a suitable small bore macerator pump package for a closet pan and basin located in a basement. Name the manufacturer and the model of pump package.

_____

_____

_____

_____

16 After installing a pump line (rising main) from a wet well, you are required to test it. Explain the test pressure and when the test must be applied.

_____

_____

_____

_____

 **WORKSHEET 4**

Student name: _____

Enrolment year: _____

Class code: _____

Competency name/Number: _____

Task: Practical exercise.

All questions in the previous three worksheets must be completed and checked by your trainer/teacher before you attempt this practical exercise.

You are required to install a small bore macerator unit to a water closet pan. The point of discharge for the pump line must be at least 1 m above the macerator unit, and the total length of the pump line must be at least 4 m. All connections to the macerator unit must be in accordance with AS/NZS 3500.2 and the manufacturer's instructions.

As a minimum, you must be able to communicate the manufacturer's installation instructions to your trainer/teacher, and respond to random questioning regarding installation of the unit, and options for connection of the vent and pump line.

SEPTIC TANKS AND ON-SITE TREATMENT FACILITIES

Learning objectives

This unit provides information needed to install and maintain septic tanks and on-site treatment facilities. Areas addressed in this unit include:

- common pre-treatment (trade waste) facilities:
  - basic principles
  - methods of pre-treatment
  - pre-treatment facility materials
  - location of pre-treatment facilities
  - installation
  - ventilation
  - connection
  - control measures against rainwater/stormwater entry
  - testing and commissioning
- septic tanks and domestic treatment plants:
  - materials
  - position
  - design and capacity
  - operation and maintenance
- effluent disinfection systems:
  - operating principles of effluent disinfection systems
  - typical maintenance of chlorine disinfection systems
  - typical maintenance of ultraviolet disinfection systems
- on-site dispersal systems
  - types of on-site dispersal systems
  - distribution methods
  - installation of on-site dispersal systems.

Introduction

Plumbers are required to install systems and facilities to collect and treat waste on-site to prevent prohibited discharges from entering the environment or sanitary plumbing and drainage systems.

This unit outlines the basic requirements for:

- installation of pre-treatment facilities to retain, treat and/or prevent prohibited discharges entering the sanitary plumbing and drainage system
- installation of a domestic treatment plant (e.g. septic tank) where connection to a network sewerage system is not available
- maintenance of a chlorine disinfection system for a domestic treatment plant
- installation of an on-site effluent dispersal system for a domestic treatment plant.

Pre-treatment facilities may be required on any property where there are discharges that are prohibited from entering the network sewerage system or on-site treatment plant. Prohibited discharges are often referred to as 'trade waste'.

Domestic treatment plants and associated disinfection and on-site effluent dispersal systems are generally required where a network sewerage system connection is not available.

The pre-treatment facilities, treatment plants and systems described in this unit must be designed and installed in accordance with the regulations and guidelines of the relevant local authority. It is also essential to follow manufacturers' advice and installation instructions.

The performance requirements for on-site waste-water management systems and on-site liquid trade waste systems are specified in the *National Construction Code (NCC) Volume 3, Plumbing Code of Australia*.

Before studying this subject, you should have completed training in work health and safety (WHS) as outlined in Chapter 3 of *Basic Plumbing Services Skills*.

Pre-treatment facilities

Examples of prohibited discharges

Only domestic liquid wastes are permitted to discharge to the sewerage system. All other discharges are prohibited unless there is written approval from the local sewerage authority.

Examples of prohibited discharges are:

- acid wastes from laboratories and schools
- hot wastes from commercial laundries
- greasy waste from commercial food preparation
- plaster and similar wastes from hospitals
- sand and silt from car washing
- solvents and oils from industrial processes

- petroleum products from car and mechanical parts washing
- any material that may cause damage to the sewerage system or injury to employees or contractors of the sewerage authority
- stormwater, salt water or sea water.

Potential effects of prohibited discharges

The following are examples of the potential effects of prohibited discharges on the sewerage system.

1 Grease and oil:
   - Deposits can form in sewerage mains, reducing the flow capacity and leading to blockages.
   - Deposits can accumulate on screens at treatment facilities, causing blockages.
   - Poor effluent quality could have an adverse impact on the environment.
2 Suspended solids and sludges:
   - Deposits can form in the bottom of sewerage mains and treatment plants, reducing the flow capacity and leading to overflow.
3 High temperatures:
   - Bacterial slimes may grow.
   - Pipes may crack.
4 Low pH levels:
   - People working in the sewer system may be injured.
   - The sewer system may be damaged.
   - Toxic hydrogen sulphide gas may be released.

Utility network sewerage systems are designed to carry domestic wastes, which are of a predictable quality. Other discharges, as mentioned above, can cause problems. To ensure that prohibited discharges do not cause maintenance problems or unsafe conditions for personnel working in the main sewerage system, it is important that the owner/occupier makes an application to the relevant authority and agrees to a number of conditions. These conditions may include:

- quality of discharge
- maximum rate of discharge
- the hours during which discharge may occur
- the right of the authority to inspect equipment, and monitor discharge quality and quantity, to ensure compliance with the agreement/permit.

Prohibited wastes are commonly referred to as 'trade waste'. The definition of 'trade waste' can be found in AS/NZS 3500.0 Plumbing and drainage.

AS/NZS 3500.0 PLUMBING AND DRAINAGE

Trade waste agreements

When any trade wastes are to be discharged to the sewerage system or an on-site treatment plant, a *trade waste application* must be submitted to the relevant sewerage authority or council.

The sewerage authority or council will then specify the conditions for connection and ask the owner/occupier to enter into a trade waste agreement.

Work must not commence until the conditions for connection have been provided by the authority. The work must comply with the trade waste agreement/permit. The requirement for plans and specifications will be determined by the relevant sewerage authority and/or trade waste agreement.

The following information is usually required for a trade waste application:

■ reason for discharges
■ person responsible for managing trade waste
■ type of business
■ trading hours on-site
■ details of any unroofed areas that collect stormwater
■ details, including those listed in material safety data sheets (MSDS), of any substances or chemicals on-site that could enter the sewer
■ description of activities that generate trade waste, methods of trade waste treatment, and estimated flow rates per minute, hour and day

■ details of fixtures discharging trade waste, or other details needed to estimate flow rates
■ site plan, and process and instrument diagram, showing all fixtures connected to the pre-treatment facility. An example of a combined pre-treatment facilities plan and diagram is given in **Figures 13.1** and **13.2**.

You will be required to research the trade waste application requirements for your local sewerage/council authority at the end of this unit.

Preparing for work

Safety and environmental requirements

The main safety considerations in the installation of pre-treatment facilities systems are safe excavation and trench support, as outlined in Unit 2. In addition, it may be necessary to move and lower large tanks into excavations. Moving and lowering tanks will most likely require a crane. Crane operators must be experienced in dogging and slinging techniques, and, depending on state/territory regulations, may need to hold an appropriate qualification. Remember always to stay clear of areas beneath suspended loads.

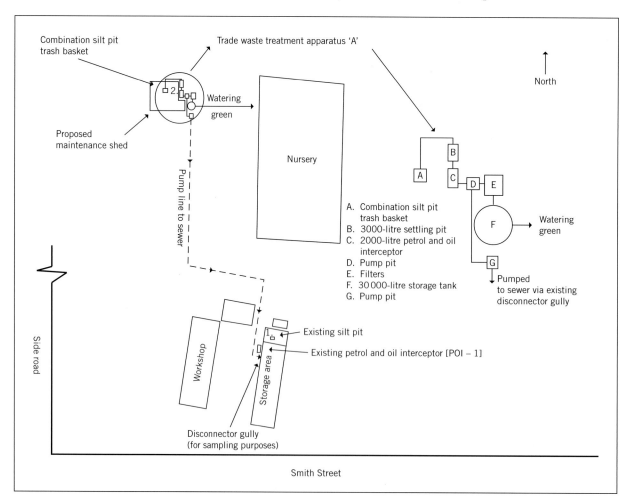

A. Combination silt pit trash basket
B. 3000-litre settling pit
C. 2000-litre petrol and oil interceptor
D. Pump pit
E. Filters
F. 30 000-litre storage tank
G. Pump pit

FIGURE 13.1 Example site plan for trade waste pre-treatment facilities

Source: Reproduced with permission of Frankston Concrete Ltd.

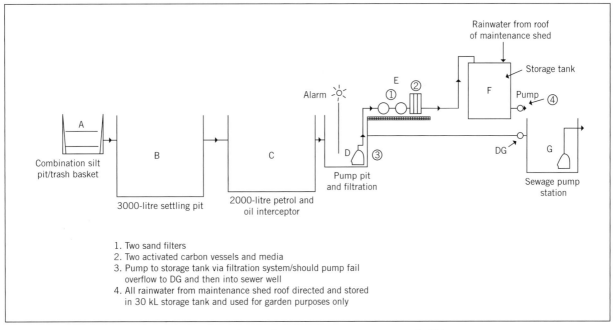

FIGURE 13.2 Example process and instrument diagram for trade waste pre-treatment facilities

Source: Reproduced with permission of Frankston Concrete Ltd.

Each state and territory has WHS regulations that require employers and self-employed people to identify hazards and assess and control risks at the workplace in consultation with their workers. All hazards must be identified before work commences.

Other examples of safety considerations specific to installing pre-treatment facilities are:

■ entry to confined spaces such as tanks and chambers

■ protection from falls into tanks and excavations

■ hazardous chemicals in pre-treatment facilities.

When preparing to handle hazardous chemicals before commissioning pre-treatment facilities, it is important to read the MSDS.

The environmental considerations in the installation of pre-treatment facilities centre on controlling the quality of discharge to the authority sewerage system or on-site dispersal system. Discharges from pre-treatment facilities must fall within the acceptable quality parameters of the relevant sewerage/local authority.

Quality assurance

Products

The regulations for pre-treatment facilities vary among sewerage authorities and local government (council) authorities. Materials and products for pre-treatment facilities must be suitable for the nature of the trade waste discharges, and comply with the sizing, dimensions and other specifications of the relevant authority.

All pipes and fittings upstream and downstream of pre-treatment facilities must comply with the *Plumbing Code of Australia* and bear WaterMark and Australian Standards certification.

Identifying, ordering and collecting materials

The materials for on-site treatment facilities must be approved and installed in accordance with AS/NZS 3500 and local regulations. For example, PVC (polyvinyl chloride) fittings for sanitary drainage must comply with AS/NZS 1260 PVC-U pipes and fittings for drain, waste and vent application and be labelled 'AS/NZS 1260'. Pre-treatment facilities such as grease arrestors must be constructed of materials suitable for the nature of the waste and for the ground conditions or environment in which they will be installed. Some authorities have a list of supplier products approved for use in their area.

AS/NZS 3500 PLUMBING AND DRAINAGE

AS/NZS 1260 PVC-U PIPES AND FITTINGS FOR DRAIN, WASTE AND VENT APPLICATION

A method for creating a materials quantities list can be found in Unit 4. When the materials list is complete, it can be used to obtain competitive prices from suppliers.

When ordering and purchasing pre-treatment facilities, pipes and fittings, it is good practice to plan ahead for secure storage to protect them from being damaged or stolen. After you have confirmed that they are approved and comply with standards and local regulations, they can be ordered from plumbing suppliers from your materials list.

When receiving the delivery of pre-treatment facilities, pipes and fittings, you should count the number and type of items and compare them to your materials list order. Ensure that you have identified all of the items as correct before accepting the order. If you have the wrong materials, or items are damaged, it may take some time to exchange and order the correct acceptable items.

Installation work

It is important to be aware of the project requirements for quality assurance. Most regulatory authorities require inspections and tests.

When preparing for your work, you must have completed applications with the relevant authorities and planned for the complete installation. You must also have the necessary reference numbers and phone numbers to book inspections and tests with the regulatory authority (where required).

In some areas, the authorities may require several inspections of pre-treatment facilities during the progress of work and when the work is complete. You must familiarise yourself with local inspection requirements before commencing work.

Inspections are carried out by sewerage authorities and/or councils to ensure pre-treatment facilities have been installed correctly, in accordance with state/territory and local regulations. Inspection is a form of quality assurance of your work for the authority and property owner.

On larger building sites throughout Australia, additional requirements for quality assurance of plumbing work are common, such as sign-off on portions of the plumbing work for progress payments, and additional tests and inspections (see Table 13.1).

TABLE 13.1 Examples of tests and inspections

Examples of tests	Examples of inspections
Hydrostatic, vacuum and air testing on trade waste drains	Drain gradient
Pressure testing pump lines from trade waste equipment	Drain materials suitable to the nature of trade waste
	Conformance to trade waste agreement
	Arrestor construction, ventilation, baffles, internal piping

Some of the issues that should be considered when preparing to install pre-treatment facilities are:

- conformance with company operating procedures
- ensuring the necessary applications are lodged with the authorities
- arranging for all necessary inspections and tests
- ordering equipment, tanks, pipes, fittings, pumps and other materials in accordance with the job plans and specifications

- receiving and inspecting equipment, tanks, pipes, fittings, pumps and other materials for conformance with orders
- storage of equipment, tanks, pipes, fittings, pumps and other materials to prevent damage or deterioration
- selection, operation and maintenance of tools and equipment
- compliance with the contract work schedule
- maintaining records of job variations.

Effluent quality

After work has been completed to install pre-treatment facilities, most authorities require a program of inspection and testing to confirm periodically that the quality of effluent is acceptable for discharge to the sewerage system. For the owner, plumber and authority, the responsibility for monitoring, testing and maintenance must be clear from the outset, including how and when it must be done.

Sequencing of tasks

When installing below-ground pre-treatment facilities, the upstream drainage systems should be completed first, as the below-ground obstructions and services may alter the final level of the inlet connection. Some tanks and chambers (particularly those made of plastic and fibreglass) may be limited in the depth at which they may be installed below the finished ground level. It is also important that the upstream drainage system is laid at the minimum depth, gradient and length to minimise the depth of excavation for tanks.

After tanks and chambers (particularly those made of plastic and fibreglass) have been installed level in the excavation, they must be immediately filled with water and backfilled in accordance with the tank manufacturer's specifications.

Tools and equipment

Safety equipment

An important safety consideration when installing pre-treatment facilities is the correct selection and use of safety equipment. Types of personal protective equipment (PPE) that may be used include:

- overalls
- boots
- safety glasses or goggles
- ear plugs or ear muffs
- gloves
- dust masks or respirators.

The MSDS for chemicals may also specify the necessary PPE for installing and commissioning pre-treatment facilities.

Tools

Tools and equipment typically used for installation of pre-treatment facilities may include:

- excavator/backhoe
- levelling equipment (see Unit 3)

- pick, shovel, crowbar, rake
- demolition saw
- scale ruler
- measuring tape
- hacksaw with a range of blades
- wood saw
- mitre box
- power drill with hammer function for masonry anchors
- socket set and ratchet spanner
- shifting spanners
- string line
- plumb-bob
- spirit level
- plugging chisel
- cold chisel
- trowel
- claw hammer.

When moving tanks and chambers, you might be required to use lifting or load-shifting equipment, such as:
- crane truck
- excavator/backhoe
- rollers
- chain blocks
- trolley
- jacks.

Special equipment – selection and operation

Special equipment is required to install some pre-treatment facilities. This special equipment must be used in accordance with the manufacturers' instructions. Examples include:
- special tools (e.g. lifting attachments) for transport of tanks and chambers
- lid keys for access covers
- special tools for installing and commissioning equipment.

Some manufacturers supply special tools, equipment and/or fixings. If these are not used, and the product is damaged, the manufacturer's product warranty may be void.

Identifying pre-treatment facility requirements

Basic principles of pre-treatment

To comply with the requirements of the relevant sewerage authority and conditions of the trade waste agreement, it may be necessary to install or construct apparatus to collect and/or treat the trade waste to an acceptable standard for discharge to the sewerage system. The basic principles of treatment are:
- balancing/averaging
- flotation
- neutralisation/dilution
- cooling
- precipitation

- screening
- settlement.

The basic principles of treatment are classified as pre-treatment, meaning preliminary treatment. This is on-site treatment of the trade waste prior to discharge to the sewerage system or an on-site dispersal system.

Methods of pre-treatment

Pre-treatment facilities are available in different types, sizes and materials, depending entirely on the waste to be treated or collected and the rate of flow of the waste. The facilities (sometimes referred to as 'arrestor' or 'interceptor') must be large enough and designed so that they slow down the flow to allow the necessary treatment to occur, such as settlement and cooling.

Arrestor pits, when installed below ground, are usually installed with access covers above the surrounding ground level to prevent the entry of stormwater. They must be placed as close as possible to the fixture/s served. Examples of descriptions and specifications of a variety of pre-treatment facilities are included in this unit.

Note that the construction and dimensional specifications for pre-treatment facilities vary around Australia. Always consult the relevant local authority when planning for work.

Size of pre-treatment facilities

The method for determining the size or capacity of pre-treatment facilities differs around Australia, but the objective is the same: to achieve a discharge quality that falls within acceptable parameters. For example, grease arrestors are required to have a minimum capacity equal to the maximum hourly discharge from all connected fixtures. This is calculated in different ways by different authorities. The capacity might be calculated on the assumption that each sink, basin, floor waste and so on discharges 50 litres per hour; or the size might be calculated based on the seating capacity of a restaurant. For example, if the total maximum hourly discharge from all greasy sinks, basins, dishwashing machines and other appliances was 2000 litres, then the grease arrestor minimum holding capacity must be 2000 litres.

Common types of pre-treatment facilities

Grease arrestors

Grease arrestors (see Figure 13.3) are also referred to as 'grease traps' or 'grease interceptors'. They are located above ground or in ground, and allow culinary waste water to cool and the grease to separate from the waste water. Grease arrestors are required in industrial/commercial kitchens and for manufacturing processes where greasy waste is a by-product of processes.

The following premises require grease arrestors if they have greasy sink, dishwasher and or floor wastes:
- restaurant kitchens
- hotel kitchens

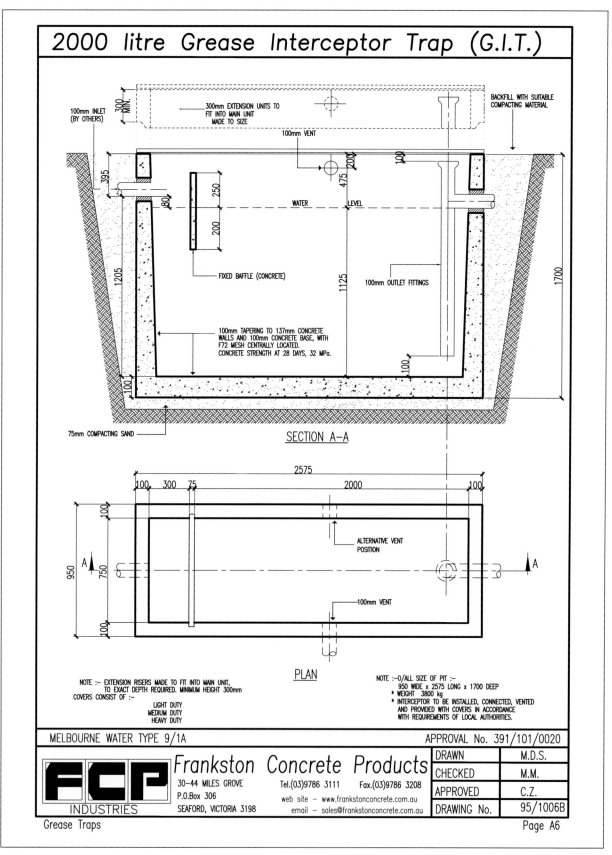

FIGURE 13.3 Grease interceptor trap

- butcher shops
- commercial meat processing plants
- smallgoods factories
- motel kitchens
- fast-food outlets
- secondary school home economics departments
- delicatessens
- bakeries
- fish shops.

Where practical, grease arrestors should be located outside buildings. Arrestors must not be installed within a building unless under the following special conditions:

- The prior approval of the sewerage authority has been obtained in each case.
- The arrestor is installed in a special chamber, aerially disconnected from the remainder of the building and ventilated to the outer air as directed by the sewerage authority.
- Independent access to the arrestor for cleaning purposes is provided, where practicable, from outside the building (suction line).

It is preferable to install grease arrestors outside a building in order to:

- reduce the cost of installation
- allow arrestors to be fitted with a loose-fitting lid
- reduce the need for two vents (inlet and outlet)
- facilitate cleaning.

Care of grease arrestors

It is essential that grease arrestors be maintained in a sanitary condition and in a state of excellent repair. Solidified grease must be removed at regular intervals, and the sludge collected and pumped out. The trade waste agreement places the obligation on the owner/occupier to have the system cleaned, and the accumulated grease and solids pumped out at regular intervals – usually every three months, depending on build-up – by an accredited contractor.

Figure 13.4 shows a typical commercial kitchen with fixtures connected via a grease interceptor trap to the sanitary drainage installation. Where there are pre-treatment facilities on-site, some authorities may specify that the sanitary drain must discharge to the sewer main via a boundary trap.

GREEN TIP

A well-maintained grease arrestor will capture fats and food solids that are washed down commercial sinks when preparing or cooking food, and when cleaning. Capturing fats and food solids (greasy waste) in a grease arrestor prevents blockages from occurring in both the sanitary drainage system and the authority mains.

Neutraliser/dilution pit

Because of the effects of acid upon the materials – particularly, cement mortar – used in sewerage work, acid wastes are prohibited from entering the sewer unless neutralised to a pH value of between 6 and 10. A neutraliser, or dilution, pit is designed to be large enough to sufficiently dilute the acid to make it harmless (see Figure 13.5).

If the acidity is too great, an alkali such as powdered lime, soda or caustic soda is added, either by hand or mechanically, according to requirements. Some authorities may specify that marble chips be added to the neutraliser. There are no standard guidelines for the use of a neutraliser. Every case is treated on its merits, and the size of the neutraliser is calculated based on the anticipated flow of the waste.

The pit should be designed to be large enough to maintain about a one-hour retention of the flow. *Note:* Some authorities require that the inlet and outlet are arranged at 90° to each other for better mixing and dilution. They may also specify the addition of a testing sump to the outlet of the neutraliser to enable periodic testing of the waste discharge by authority trade waste inspectors.

Neutralising arrestors receive the discharge from:

- chemical factories
- school science laboratories
- university laboratories
- battery charging areas
- pathology laboratories.

Cooling interceptor

Cooling interceptors are used to collect high-temperature (above 38°C) discharge (see Figure 13.6). The interceptor allows time for the discharge to cool down before entering the sewer. High-temperature water can damage the sewer pipes, traps and fittings, and cause poisonous gases to be given off.

Some authorities will specify that the cooling interceptor be fitted with a straining basket to also intercept lint from laundry washing. Lint waste can result in build-up of sludge in the interceptor.

Cooling interceptors receive the discharge from:

- hospitals
- commercial laundries
- launderettes/laundromats.

Petrol and oil interceptor/arrestor

Typically, petrol and oil arrestors, or collection pits, have been installed below ground level and take the shape of a general-purpose arrestor (see Figure 13.7). Some authorities specify that an oil skimmer be attached. This attachment collects and diverts the oil to a pit, which is periodically pumped out by a licensed liquid waste contractor. Petrol and oil interceptors/

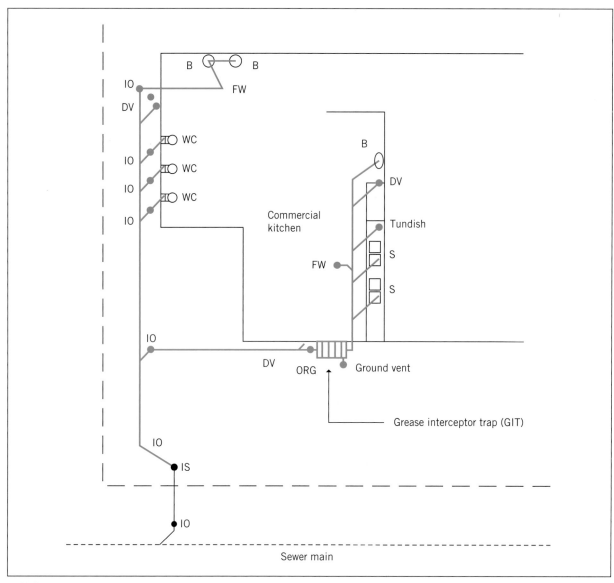

FIGURE 13.4 Property with commercial kitchen fixtures connected to grease interceptor trap

Source: Gary Cook.

arrestors are installed in premises that use petroleum-based products, such as vehicle wash-down bays, panel beaters, car detailers and motor repairers. As discussed previously, petroleum products are prohibited substances and must not be discharged into sewers.

Cross-flow oil interceptor (coalescing plate separator)

There are a number of methods of intercepting and collecting oil products. One alternative to the petrol and oil interceptor is the cross-flow oil interceptor (see Figure 13.8). An above-ground cross-flow oil interceptor called a coalescing plate separator can be installed. The oily water and sludge are collected in a wash-down pit. The heavier solids can be captured in a perforated basket and silt pit, depending on application. The oil and the remaining solids are then introduced into the cross-flow interceptor unit by a pump system.

One supplier of coalescing plate separators is KWIKFLO Group (https://www.kwikflogroup.com.au).

How it works

The KWIKFLO coalescing plate separator is an enhanced gravity separator capable of removing either oils or solids, or both (see **Figures 13.9** and **13.10**). It utilises the difference in specific gravity between two or three immiscible components of a liquid stream for separation. Oily water effluent is introduced into an inlet chamber by gravity flow or pump. Heavy solids settle out and 100% oil slugs rise immediately to the surface. The remaining oily water mixture flows through a stack of closely spaced, corrugated polypropylene plates. Both the smaller oil droplets and the fine solids are progressively separated. Downstream, a baffle or oil dam prevents the collected

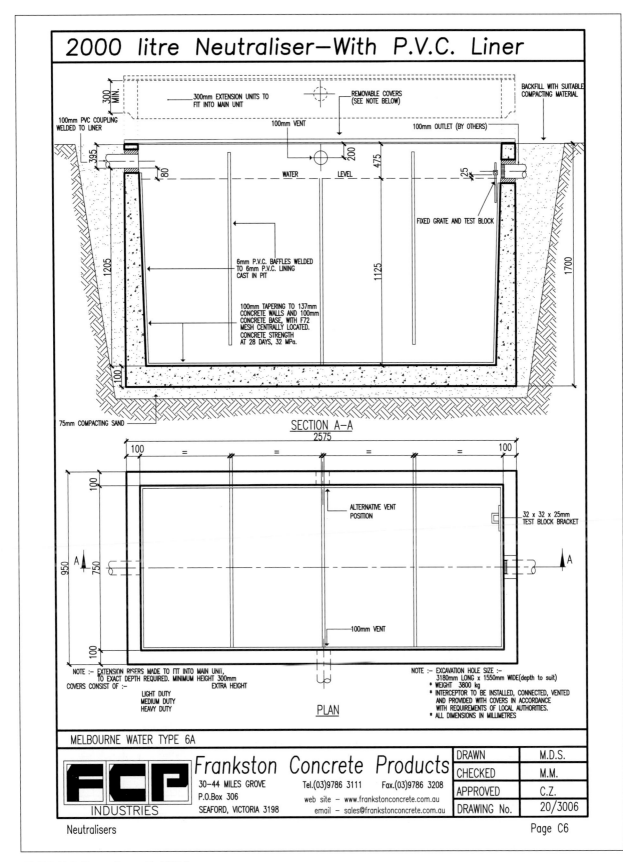

2000 litre Neutraliser—With P.V.C. Liner

300 MIN.

300mm EXTENSION UNITS TO FIT INTO MAIN UNIT

REMOVABLE COVERS (SEE NOTE BELOW)

BACKFILL WITH SUITABLE COMPACTING MATERIAL

100mm PVC COUPLING WELDED TO LINER

100mm VENT

100mm OUTLET (BY OTHERS)

395

80

WATER   LEVEL

200

475

25

FIXED GRATE AND TEST BLOCK

1205

6mm P.V.C. BAFFLES WELDED TO 6mm P.V.C. LINING CAST IN PIT

1125

1700

100mm TAPERING TO 137mm CONCRETE WALLS AND 100mm CONCRETE BASE, WITH F72 MESH CENTRALLY LOCATED. CONCRETE STRENGTH AT 28 DAYS, 32 MPa.

100

75mm COMPACTING SAND

SECTION A–A
2575

100   =   =   =   =   100

100

ALTERNATIVE VENT POSITION

32 x 32 x 25mm TEST BLOCK BRACKET

950   750

A ▲   ▲ A

100

100mm VENT

NOTE :– EXTENSION RISERS MADE TO FIT INTO MAIN UNIT, TO EXACT DEPTH REQUIRED. MINIMUM HEIGHT 300mm EXTRA HEIGHT
COVERS CONSIST OF :–
LIGHT DUTY
MEDIUM DUTY
HEAVY DUTY

PLAN

NOTE :– EXCAVATION HOLE SIZE :–
3180mm LONG x 1550mm WIDE(depth to suit)
* WEIGHT 3800 kg
* INTERCEPTOR TO BE INSTALLED, CONNECTED, VENTED AND PROVIDED WITH COVERS IN ACCORDANCE WITH REQUIREMENTS OF LOCAL AUTHORITIES.
* ALL DIMENSIONS IN MILLIMETRES

MELBOURNE WATER TYPE 6A

FCP INDUSTRIES

Frankston Concrete Products
30–44 MILES GROVE
P.O.Box 306
SEAFORD, VICTORIA 3198

Tel.(03)9786 3111   Fax.(03)9786 3208
web site – www.frankstonconcrete.com.au
email – sales@frankstonconcrete.com.au

DRAWN	M.D.S.
CHECKED	M.M.
APPROVED	C.Z.
DRAWING No.	20/3006

Neutralisers

Page C6

FIGURE 13.5 Neutraliser with PVC liner

Source: Reproduced with permission of Frankston Concrete Ltd.

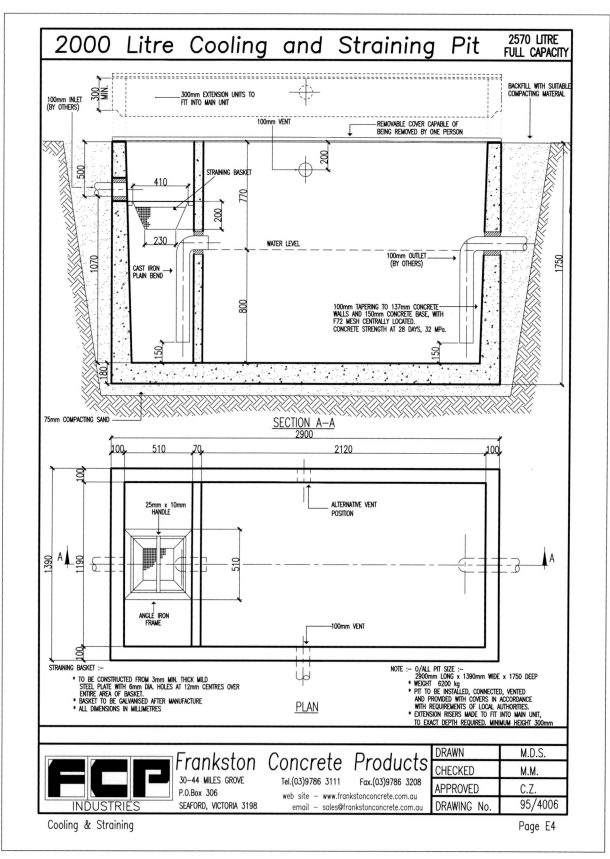

FIGURE 13.6 Cooling and straining interceptor pit

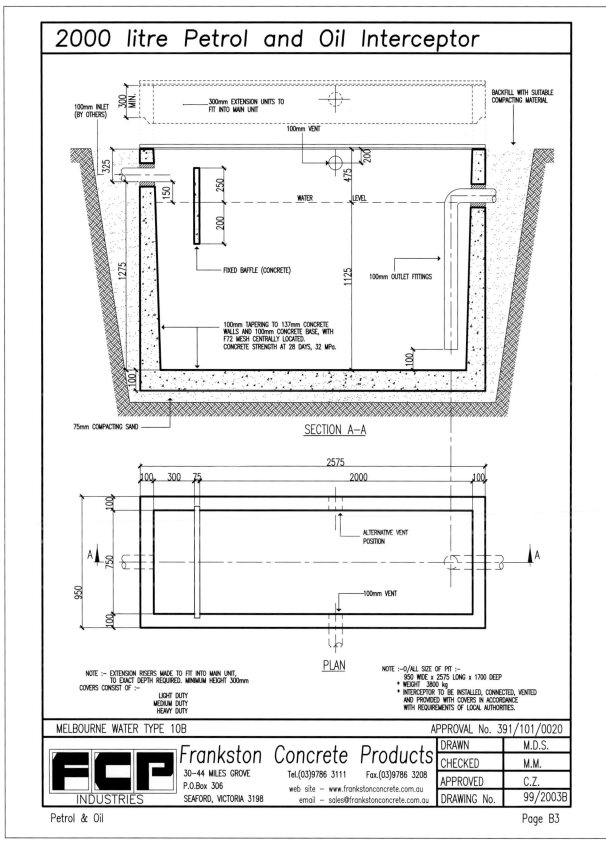

FIGURE 13.7 Petrol and oil interceptor

FIGURE 13.8 KWIKFLO cross-flow oil interceptor (coalescing plate separator)

oil from entering the outlet weir. Drains are provided in the sludge hopper for the removal of solids. Oil skimmers are provided for the manual or automatic removal of oil.

A fast, easy, regular cleaning of the plate packs with hot water is the only maintenance requirement under normal operating conditions.

Figure 13.11 shows a plan of a coalescing plate separator installation in a car maintenance centre where oil may drip from the vehicles being serviced. The oily waste discharge is collected and transferred to the sanitary drainage installation through a pre-treatment cross-flow oil interceptor. Where there are pre-treatment facilities on-site, some authorities may specify that the sanitary drain must discharge to the sewer main via a boundary trap.

Silt pit (arrestor)

A silt pit (see Figure 13.12) is generally located in a wash-down area with the concrete floor graded towards it. When the area is hosed down, the water and silt flows into the arrestor through a perforated plate or similar-type lid. The silt settles out and is trapped in the bottom section of the apparatus, while the waste water flows out to the drainage system. These arrestors may also be in a location remote from the wash-down area, and may be installed upstream of a pre-treatment facility, as shown in Figure 13.11.

Where silt pits are connected directly to the sanitary drainage system and not through a pre-treatment facility, they must be connected through a DN 100 drainage trap.

Trash basket arrestors

A trash basket arrestor (see Figure 13.13) is generally used in areas where there are large solids discharged in the waste water that may cause blockages in the sanitary drainage or authority sewerage system. Trash basket arrestors may be used in premises such as bottling plants, where bottles and bottle caps are washed into draining channels, or in fruit processing plants, where parings and fruit stones may find their way into channels. Their discharge into the sewer must be prevented.

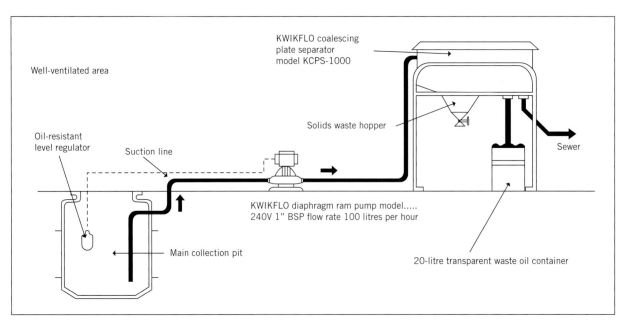

FIGURE 13.9 Typical installation of KWIKFLO coalescing plate separator Model KCPS-1000

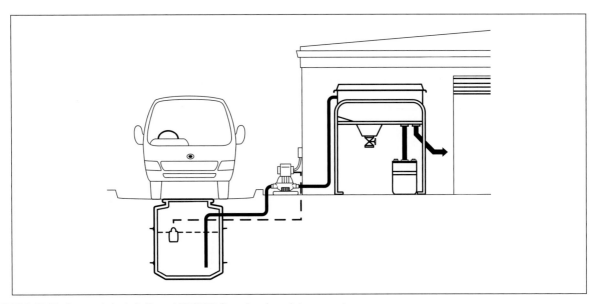

FIGURE 13.10 Car wash installation of KWIKFLO coalescing plate separator

Source: Reproduced with the permission of All Pumps.

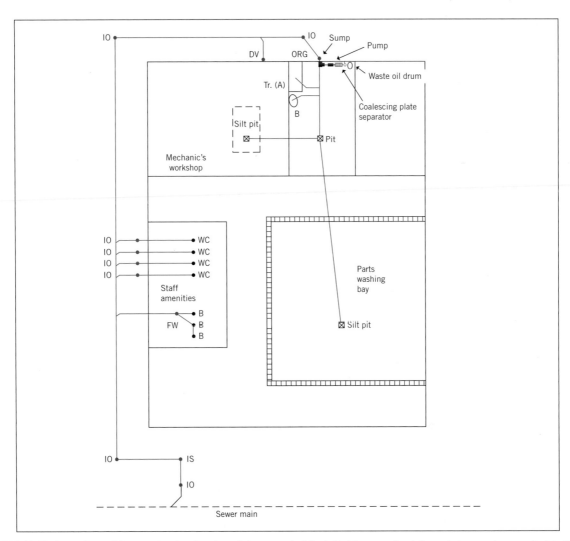

FIGURE 13.11 Cross-flow oil interceptor (coalescing plate separator) installed in a mechanic's workshop and connected to the sanitary drainage system through an overflow relief gully

Source: Gary Cook.

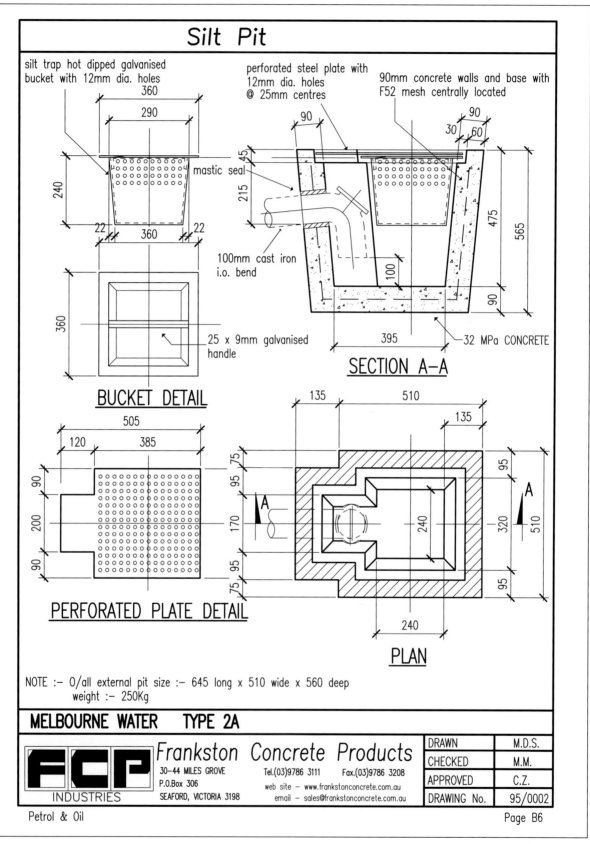

Silt Pit

silt trap hot dipped galvanised bucket with 12mm dia. holes

perforated steel plate with 12mm dia. holes @ 25mm centres

90mm concrete walls and base with F52 mesh centrally located

mastic seal

100mm cast iron i.o. bend

32 MPa CONCRETE

SECTION A–A

BUCKET DETAIL

25 x 9mm galvanised handle

PERFORATED PLATE DETAIL

PLAN

NOTE :– O/all external pit size :– 645 long x 510 wide x 560 deep
weight :– 250Kg

MELBOURNE WATER   TYPE 2A

Frankston Concrete Products
INDUSTRIES
30–44 MILES GROVE
P.O.Box 306
SEAFORD, VICTORIA 3198
Tel.(03)9786 3111   Fax.(03)9786 3208
web site – www.frankstonconcrete.com.au
email – sales@frankstonconcrete.com.au

DRAWN	M.D.S.
CHECKED	M.M.
APPROVED	C.Z.
DRAWING No.	95/0002

Petrol & Oil

Page B6

FIGURE 13.12 Silt pit

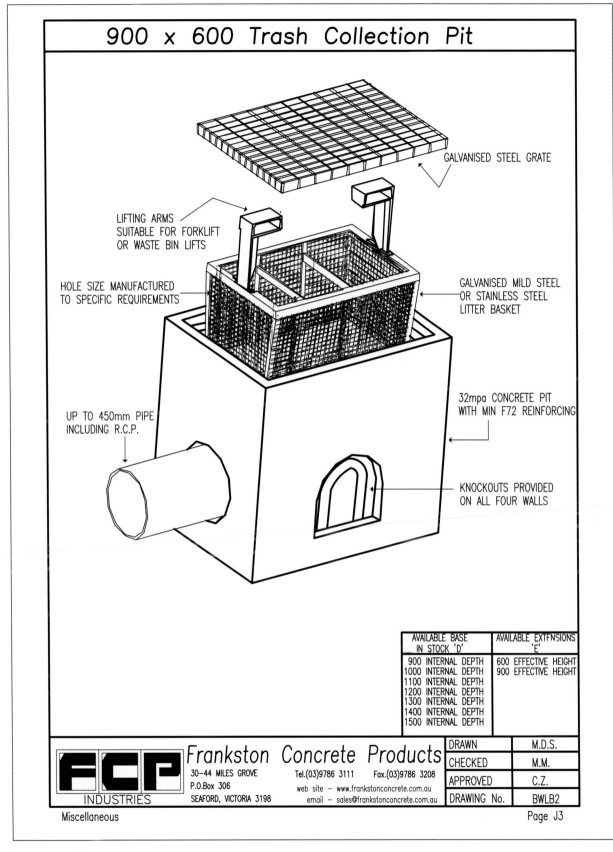

900 x 600 Trash Collection Pit

GALVANISED STEEL GRATE

LIFTING ARMS
SUITABLE FOR FORKLIFT
OR WASTE BIN LIFTS

HOLE SIZE MANUFACTURED
TO SPECIFIC REQUIREMENTS

GALVANISED MILD STEEL
OR STAINLESS STEEL
LITTER BASKET

32mpa CONCRETE PIT
WITH MIN F72 REINFORCING

UP TO 450mm PIPE
INCLUDING R.C.P.

KNOCKOUTS PROVIDED
ON ALL FOUR WALLS

AVAILABLE BASE IN STOCK 'D'	AVAILABLE EXTENSIONS 'E'
900 INTERNAL DEPTH	600 EFFECTIVE HEIGHT
1000 INTERNAL DEPTH	900 EFFECTIVE HEIGHT
1100 INTERNAL DEPTH	
1200 INTERNAL DEPTH	
1300 INTERNAL DEPTH	
1400 INTERNAL DEPTH	
1500 INTERNAL DEPTH	

Frankston Concrete Products

FCP
INDUSTRIES

30–44 MILES GROVE
P.O.Box 306
SEAFORD, VICTORIA 3198

Tel.(03)9786 3111   Fax.(03)9786 3208
web site – www.frankstonconcrete.com.au
email – sales@frankstonconcrete.com.au

DRAWN	M.D.S.
CHECKED	M.M.
APPROVED	C.Z.
DRAWING No.	BWLB2

Miscellaneous

Page J3

FIGURE 13.13 Trash basket arrestor

Basket arrestors are used in the following premises/locations:

- fish wholesalers (no cooking on-site)
- oyster/shellfish processing
- stables
- fruit canneries
- bottling plants
- markets.

A trash basket arrestor is a pit with a removable wire basket that can be withdrawn and cleaned periodically.

Garbage can wash-down area arrestor (or silt trap)

Wash-down areas for domestic garbage containers must be provided with a concrete floor that is graded to a dry basket arrestor (see Figure 13.14). Any stormwater must be diverted away from the area. The requirements for garbage can wash area arrestors (or silt traps) vary around Australia.

FIGURE 13.14 High-density polyethylene (HDPE) silt trap

Plaster trap (arrestor)

Plaster arrestors are used in premises such as dental technicians, hospitals and universities. They are usually compact (to make removal easier) and are generally installed below the sink used for the washing of the plaster moulds, etc.

Pre-treatment facility materials

The materials forming part of the pre-treatment facilities and upstream pipework and ventilation system must be suitable for the nature of the waste. For example, PVC is not generally suitable for petrol- and oil-type waste, and copper is not suitable for acidic-type waste. Also, the temperature of the wastes may affect the material's suitability.

If you are unsure about which materials are suitable, consult the local sewerage authority or refer to the pipe/material manufacturer's chemical resistance charts.

Location of pre-treatment facilities

External location

Arrestors and special units in an external location:

- must be installed in the open, wherever practicable
- where located in the open, must be installed so as to prevent the entry of surface or roof water
- must be provided with a removable cover that will be able to withstand vehicular or pedestrian traffic or other loads likely to be imposed on them, and as specified by the relevant sewerage authority.

Internal location

Where required, arrestors and special units installed in an internal location must:

- be installed in a separate chamber
- be atmospherically disconnected from the remainder of the building and ventilated to the open air above the roof level, as required by the relevant sewerage authority
- be located above the finished floor level, and have internal measurements that will allow ready access to the arrestor or special unit for cleaning or maintenance, with clearance above the arrestor equal to its depth
- where practicable, be provided with separate access from outside the building or a gas-tight door, as required by the relevant sewerage authority.

Installation and testing of pre-treatment facilities

Installation

Factors to consider for the installation of pre-treatment facilities include:

- the type and size of facility, as specified and approved by the authority
- dimensions and material construction, as specified
- allowance for correct levels between inlet and outlet to allow for drain connections
- a stable and level base to construct or install the facility
- connection in accordance with the manufacturer's instructions, providing flexible connections if required by AS/NZS 3500.2 Sanitary plumbing and drainage to allow for differential settlement
- where pumps are required, installation according to the manufacturer's requirements.

AS/NZS 3500.2 SANITARY PLUMBING AND DRAINAGE

Generally, the piping and ventilation for the plumbing/drainage *upstream* of the pre-treatment facilities must be installed, as outlined in Units 4 and 10, in accordance with AS/NZS 3500.2 with regard to gradient, size, venting, vent termination, etc. However, in some cases, the relevant sewerage authority may specify different requirements.

AS/NZS 3500.2 SANITARY PLUMBING AND DRAINAGE

The sanitary plumbing/drainage and vents *downstream* of the pre-treatment facilities must be installed, as outlined in Units 4 and 10, in accordance with AS/NZS 3500.2.

AS/NZS 3500.2 SANITARY PLUMBING AND DRAINAGE

Ventilation requirements

The vent pipes from trade waste installations and arrestor chambers may contain fumes and gases that are not compatible with the general sanitary plumbing and drainage system. For this reason, trade waste vents from trade waste installations and arrestor chambers must not be interconnected with any other sanitary plumbing or drainage vent.

Sewerage authorities may not require ventilation of some pre-treatment facilities. However, it is generally necessary to provide ventilation to grease arrestors.

The requirements and sizing for grease arrestor vents are specified by the relevant sewerage authority. Grease arrestor vent pipes must be of the high-level type, except that the arrestor chamber vent may be of the low-level type, subject to the requirements for low-level vents in AS/NZS 3500.2. The requirements for the location of vent terminals from grease arrestor chambers and vents in relation to building openings are the same as those for sanitary drain vents in AS/NZS 3500.2.

AS/NZS 3500.2 SANITARY PLUMBING AND DRAINAGE

Where the sewerage authority specifies a chamber vent for a petrol and oil arrestor or other type of arrestor where flammable gases are present, the vent must be a high-level type.

Connection requirements

Generally, authorities require traps to be fitted to the outlet of above-ground (portable) arrestors. This provides trap seal disconnection from the sanitary plumbing/drainage system. The trap riser also provides a point of access for trade waste inspectors to obtain a grab sample of the flowing waste and measure whether the quality is within the required parameters. Where an arrestor or special unit is installed below ground level, some authorities may require the installation of a disconnector gully at the outlet for the same reasons.

Premises in industrial areas may be connected to the authority sewer via a boundary trap. Boundary traps are required in these areas for several reasons. The boundary trap allows the utility to take a grab sample of the discharge to the sewer main to measure whether the quality is within the required parameters.

The nature of the discharge in industrial areas is such that authorities will vent the sewer main at a considerably higher level than in non-boundary trap areas. If a boundary trap were not installed, the property upstream vent would be venting the main at a relatively low level, and sewer gases may be offensive.

Control measures against rainwater/stormwater entry

Where trade waste processes are outside the building, control measures must be in place to prevent entry of surface rainwater and stormwater into the system (see Figure 13.16). The requirements and allowances vary around Australia, but the principle is the same. Some sewerage authorities specify a maximum unroofed area of 20 m². One way of achieving this is to construct a roof to prevent the entry

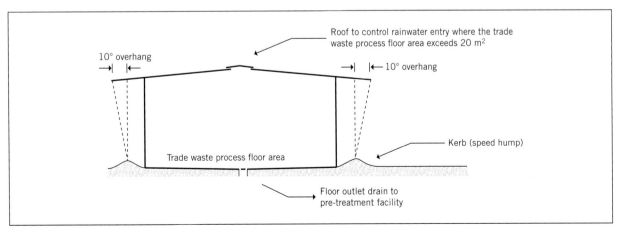

FIGURE 13.16 Controlling rainwater and stormwater entry

CHAMBER VENTILATION

FIGURE 13.15 Incorrectly installed grease arrestor

Figure 13.15 shows a grease arrestor that has been incorrectly installed. The disconnector gully vent has been combined with the chamber vent. If the grease arrestor outlet pipe becomes blocked, greasy waste will rise in the chamber and flow out of the chamber vent into the disconnector gully. Untreated waste can then flow into the sanitary drain.

1   How should the vents have been installed?

_____

_____

_____

2   Is it permitted to combine these vents?

_____

_____

_____

3   Research AS/NZS 3500.2. What clause clarifies the venting requirements in this situation?

_____

_____

_____

of rainwater into the pre-treatment facilities and sewer. For a structure where one or more sides are open to the weather, a 10° overhang of the roof is usually the minimum acceptable cover.

It must be considered that, in extreme storm conditions, wind-driven rain can enter the area below the roof. To control the entry of wind-driven rain, the roof should overhang by an amount not less than 10°. Alternatively, wall sheeting can be fitted to prevent the rainwater entering the roofed area.

To ensure that no surface stormwater can flow on to the trade waste process area, a kerb (speed hump) must be constructed at least 150 mm high around the perimeter. The overall surface water flow across the site must be considered, and the height of the kerb (speed hump) may have to be increased to effectively prevent stormwater flow on to the process area.

Where it is not practical to install a roof over the trade waste process floor area, there are other methods of controlling rainwater/stormwater entry. A diversion valve and rain sensor with timer control can be used to capture a rainfall first flush to the pre-treatment facility, and then divert rainfall stormwater discharge to the stormwater drainage system.

Testing and commissioning

The requirements for testing and commissioning a pre-treatment facility will be regulated by the sewerage authority and specified by the equipment manufacturer. These regulations and specifications are to ensure that the pre-treatment facility operates correctly and produces waste of an acceptable quality.

The authorities provide general rules about the acceptable parameters of the quality of waste. Manufacturers' requirements and specifications supplement those of the local authority and are sometimes site-specific to the facility being installed.

As with many tasks performed by workers in the plumbing industry, there are rules and regulations that apply – from designing and planning through to testing and commissioning. The rules and regulations apply to ensure the safety, health and welfare of the community. Therefore, the correct operation of a pre-treatment facility is of paramount importance.

Each pre-treatment facility must be tested and commissioned as specified by the authority and manufacturer. However, as a general rule, the following procedures may apply:

1 The procedure for testing may include:
   – filling the facility with water
   – making sure that the facility is watertight (i.e. no leaks)
   – checking for correct inlet and outlet levels
   – checking that all fittings and baffles are fitted
   – checking that the facility complies with AS/NZS 3500.2 and relevant local regulations.

2 The procedure for commissioning may include:
   – checking liquid levels in the tank
   – checking flow through the tank
   – checking that covers are fitted flush with the surface level (no trip hazards)
   – checking the operation of pumps or other ancillary equipment (if installed)
   – checking waste quality (e.g. pH levels or the temperature of the waste flowing from the outlet).

AS/NZS 3500.2 SANITARY PLUMBING AND DRAINAGE

The sewerage authority may specify that the waste quality sampling be carried out by a laboratory with NATA (National Association of Testing Authorities) accreditation.

COMPLETE WORKSHEET 1

Domestic treatment plants

Most Australians living in major towns and cities have access to sewerage systems to dispose of effluent. Where there is a sewer main connection available, properties must be connected to the sewer main. However, many rural and developing areas remain without sewerage systems and require on-site waste-water management and treatment. These areas are termed 'unsewered'.

Domestic waste water can be divided into two categories:

1 sewage (all waste water, including greywater and toilet waste – also known as 'blackwater')

2 greywater (waste water from the shower, bath, basins, washing machine, laundry troughs and kitchen – also referred to as 'sullage').

For properties, whether 'sewered' or 'unsewered', there is a growing emphasis on saving and reusing water. This development has led to advances in on-site waste-water treatment technology and increased the use of greywater treatment systems.

It is a common misconception that greywater does not contain pathogens and that it is only sewage (blackwater) that requires treatment prior to dispersal or recycling. Greywater can contain pathogens that, if poorly managed, could present a risk to human health.

Therefore, *all* on-site waste-water systems must be installed safely and in accordance with the relevant state/territory policies, local government (council) requirements and Australian Standards.

On-site waste-water treatment systems use a number of the following processes.

1 Primary treatment:
  – aerobic – physical treatment (e.g. screening, filtration, sedimentation, flocculation, flotation) and digestion under aerobic conditions (e.g. aerobic bio-filter, wet composting)
  – anaerobic – physical treatment (e.g. filtration, sedimentation, flocculation, flotation) and digestion under anaerobic conditions (e.g. septic tank).

2 Secondary treatment (sometimes followed by disinfection of the treated waste water):
  – biological treatment (aerobic digestion supported by aeration or trickling filters; controlled cycles of aerobic/anaerobic digestion)
  – filtration (through a variety of differently shaped and sized media and membranes)
  – chemical treatment (flocculation, coagulation, precipitation)
  – other biological treatment (e.g. in wetlands, reed beds, sand filters).

3 Tertiary treatment:
  – filtration (reverse osmosis)
  – nutrient reduction.

4 Other:
  – incineration.

The treatment processes available are numerous and the numbers and types of methods are increasing.

On-site treatment of sewage in unsewered areas is regulated by the relevant state/territory health departments/environmental protection agencies (EPAs) and local councils. The installation of on-site waste-water treatment systems must be carried out in accordance with AS/NZS 3500, AS/NZS 1547 On-site domestic wastewater management, local regulations and the *Plumbing Code of Australia*.

AS/NZS 3500 PLUMBING AND DRAINAGE

AS/NZS 1547 ON-SITE DOMESTIC WASTEWATER MANAGEMENT

On-site waste-water treatment systems must be located within the property boundaries of the premises producing the waste water. The responsibility for control, maintenance and management of the plant is with the owner of the premises. Different types of waste-water system require different degrees of control, depending on the complexity of the system.

Table 13.2 provides an overview of on-site waste-water management options for sewered and unsewered areas. Note that the use of treatment systems, and approvals for dispersal and recycling, vary for each state and territory. For more detailed information on required water quality standards and on-site waste-water management options, refer to the relevant local regulations.

An example of primary treatment is the septic tank. See the definition of 'septic tank' in AS/NZS 3500. A septic tank is a solids-reduction unit (SRU). It is a chamber where anaerobic *primary* treatment of sewage can occur. Its function is to reduce solid matter to a liquid, and decompose the solid matter, using anaerobic bacterial digestion. After this primary treatment, the effluent leaves the unit, to undergo *secondary* treatment, usually by aerobic bacterial digestion. In summary, anaerobic treatment takes place without oxygen being present, and aerobic treatment takes place in the presence of oxygen.

AS/NZS 3500 PLUMBING AND DRAINAGE

The effluent from septic tanks may be dispersed on-site or pumped to sewer, or may be pumped out and removed in a method approved by the local authority. The local authority (council) provides guidance on and options for the treatment and dispersal of waste water. It must be satisfied that treatment and dispersal meets its requirements, regulations and standards. Prior to the commencement of any work in an unsewered area, an application must be made on the appropriate form to the relevant local authority (council).

Each local government authority will have its own form, and you will need to obtain a copy from the authority relevant to the area where you are working. A typical example of an application form is shown in Figure 13.17.

Plans of the proposed installation must be submitted to the local government authority in the format required by that authority. The plans must be approved prior to the commencement of work.

As a minimum, plans of the on-site treatment system should indicate the proposed position of:
- all buildings
- surcharge and disconnector gullies
- below-ground sanitary drain, inspection shafts and openings, vents, fixtures and fittings, as discussed in Unit 4
- all components of the on-site waste-water treatment system
- the setback and buffer distances between the on-site treatment system and buildings, allotment boundaries, services, surface waters and bores. These distances vary in each state/territory.

At the end of this unit, you will need to research your local government authority requirements for lodgement of applications and plans.

An example block plan is shown in Figure 13.18.

TABLE 13.2 On-site waste-water management options for sewered and unsewered areas

On-site waste-water treatment systems[1, 7]	For sewered or unsewered areas	Effluent recycling options[2, 3, 6, 7]	Effluent dispersal options
Primary treatment Dry composting toilets	All areas	N/A	Excess liquid discharged to sewer, or to a soil absorption trench in unsewered areas
Incineration toilets	Unsewered only		
Primary treatment Anaerobic (septic tank) Aerobic biological filter (wet composting, vermiculture)	Unsewered areas only	N/A	Absorption trenches/beds Evapo-transpiration absorption (ETA) beds Low pressure effluent distribution (LPED) Mounds Wick trench and beds
Secondary treatment Sewage and greywater AWTS (aerated waste-water treatment systems) Biological filters (wet composting, vermiculture) Membrane filtration Ozonation Reedbeds Sand filters Textile filters Trickling aerobic filters (foam, plastic, mixture of media)	All-waste sewage treatment systems in unsewered areas only[10]	Sub-surface irrigation Surface irrigation[11]	Absorption trenches/beds Evapo-transpiration Absorption (ETA) beds Low pressure effluent distribution (LPED) Mounds Wick trench and beds
	Greywater systems in all areas	Single domestic households[5,8,9] 10/10/10: Toilet flushing Cold water supply to washing machines Surface irrigation Sub-surface irrigation Hand-held purple hose[12] 10/10, 20/30/10, 20/30: Sub-surface Surface irrigation	
		Multi-dwelling residential, business and community[4] 10/10, 10/10/10 20/30/10, 20/30: Sub-surface irrigation[13]	

1. It is recommended that onsite sewerage systems used by patients with transplants or on dialysis or chemotherapy are more frequently serviced and/or pumped-out as the drugs are likely to kill the beneficial microbes in the treatment system.

2. Sub-surface irrigation is the dispersal of water from pipes laid 100 mm to 150 mm below the ground surface (i.e. in the unsaturated biologically-active topsoil layer) (see AS/NZS 1547). The minimum water quality required is 20/30 standard.

3. Treated sewage or greywater must not come in contact with the edible parts of herbs, fruit or vegetables.

4. Treated greywater from multi-dwellings, schools, business or commercial premises must not be used for toilet flushing or used in the washing machine.

5. The use of treated greywater for clothes washing may not always result in the desired outcome, especially when washing light-coloured clothes. Householders should discuss the risks with the system manufacturer or supplier and be careful of the colours of cleaning and personal care products used.

6. No uses other than those stated are permitted.

7. See the relevant Certificate(s) of Approval for conditions of installation, performance and management.

8. Only a purple-coloured hose with a left-hand thread which screws into the recycled greywater tap (coloured purple) is permitted to be used. The tap must have a removable child-proof handle and clear signage with words and symbols which indicate 'Recycled Water – Do Not Drink'.

9. Recycled water pipes must not be connected to the drinking water supply pipes. All recycled water plumbing works must be undertaken in accordance with the most recent version of AS/NZS 3500 [set]: *Plumbing and Drainage*.

10. The exception is a water business which may install all-waste sewage treatment systems as part of a new reticulated sewerage system or an up-grade to existing onsite waste-water systems within a sewerage district or as an asset management retrofit to an existing reticulated sewerage system.

11. The exceptions are multi-dwelling residences, schools, child care centres, medical centres, hospitals, nursing homes and premises for other sensitive populations which must use sub-surface irrigation.

12. Only permitted in sewered areas, where the excess greywater is discharged to sewer.

13. The exception is a commercial premises (other than multi-dwelling residences or a school, medical centre, hospital, child care centre, nursing home or premises for other sensitive populations) in an unsewered area which has a technical requirement for surface irrigation with 20/30/20 or 10/10/10 effluent (e.g. a vineyard using drip irrigation; an existing golf course with pop-up sprinklers provided it has a 4-hour withholding period after each irrigation event and a service contract with a professional service technician).

Source: *Code of Practice: Onsite Wastewater Management*, (Publication 891.3, Feb 2015, Environment Protection Authority (EPA) Victoria).

Preparing for work

Safety and environmental requirements

The main safety considerations in the installation of septic tanks and domestic treatment plants relate to safe excavation and trench support, as outlined in Unit 2. In addition, it may be necessary to move and lower large tanks into excavations. Moving and lowering tanks will most likely require a crane. Crane operators must be experienced in dogging and slinging techniques, and, depending on state/territory regulations, may need to hold an appropriate qualification. Remember always to stay clear of areas beneath suspended loads.

Each state and territory has WHS regulations that require employers and self-employed people to identify hazards and assess and control risks at the workplace in consultation with their workers. All hazards must be identified before work commences.

Other examples of safety considerations specific to septic tanks and domestic treatment plants are:

- entry to confined spaces such as tanks and chambers
- protection from falls into tanks and excavations
- biological hazards in domestic treatment plants and on-site dispersal systems.

The environmental considerations in the installation of domestic treatment plants centre on controlling the quality of discharge to the environment. Discharges from domestic treatment plants and on-site dispersal systems must be contained on-site and prevented from entering adjoining properties, stormwater drains, streams and waterways.

Quality assurance

Products

The regulations for on-site treatment facilities vary from state/territory to state/territory. Generally, domestic treatment plants and on-site dispersal systems must comply with AS/NZS 1547.

AS/NZS 1547 ON-SITE DOMESTIC WASTEWATER MANAGEMENT

Domestic treatment plants must be sized correctly, and comply with the dimensions and other specifications of the relevant local government authority (council).

Identifying, ordering and collecting materials

The materials for septic tanks and domestic treatment plants must be approved for installation in accordance with AS/NZS 1547 and must bear the relevant standards marking.

AS/NZS 1547 ON-SITE DOMESTIC WASTEWATER MANAGEMENT

As well, the design and installation of septic tanks must be in accordance with AS/NZS 1546.1; the design and installation of waterless composting toilets must be in accordance with AS/NZS 1546.2; and the design and installation of aerated wastewater treatment systems must be in accordance with AS/NZS 1546.3.

Packaged treatment plants and systems must comply with the requirements of the relevant local authority. Some authorities have a list of systems approved for use in their area.

A method for creating a materials quantities list can be found in Unit 4. When the materials list is complete, it can be used to obtain competitive prices from suppliers.

After you have confirmed that the products and materials are approved and comply with the standards and local regulations, they can be ordered from plumbing suppliers from your materials list. When ordering and purchasing products, pipes and fittings, it is good practice to plan ahead for secure storage in a place where they will not be damaged or stolen.

When receiving the delivery of pipes and fittings, you should count the number and type of items and compare them to your materials list order. Ensure that you have identified all of the items as correct before accepting the order. If you have the wrong materials, or items are damaged, it may take some time to exchange and order the correct acceptable items.

Installation work

It is important to be aware of the project requirements for quality assurance. Most local government authorities (councils) require inspections and tests.

When preparing for your work, you must have completed applications with the relevant authorities and planned for the complete installation. You must also have the necessary reference numbers and phone numbers to book inspections and tests with the regulatory authority (where required).

In some areas, the authorities may require several inspections of on-site domestic treatment systems during the progress of work and when the work is complete. You must familiarise yourself with local inspection requirements before commencing work.

Inspections are carried out by the authorities to ensure that the on-site treatment systems have been installed correctly, in accordance with state/territory and local regulations. Inspection is a form of quality assurance of your work for the authority and property owner.

EXAMPLE COUNCIL

METHOD OF EFFLUENT TREATMENT/DISPERSAL

❏ Septic tank – sub-soil absorption

❏ Treatment plant – surface irrigation

❏ Treatment plant – sub-soil absorption

❏ Other (please specify)

NUMBER OF PLUMBING FIXTURES TO BE CONNECTED TO THE SYSTEM

_____ Water closets

_____ Sinks

_____ Showers

_____ Baths

_____ Basins

_____

_____ Waste disposal units

_____ Other (please specify)

CAPACITY OF TANK/TREATMENT PLANT (in litres): _____

SIGNATURE OF APPLICANT: _____ **DATE:** _____

PLANS AND SPECIFICATIONS: Please submit three copies of the plans and specifications of the proposed septic tank system. These plans should contain the following information:
1. Location of property including street number and/or lot number.
2. Dimension of all property boundaries and location of all other streets adjoining the property.
3. Location and dimensions of all buildings or proposed buildings, streams, water tanks, swimming pools, excavations, driveways, stormwater drains and other service pipes.
4. Location of proposed septic tank system including sewer drains.
5. House floor plan showing the location of all plumbing fixtures.
6. Fall of land.

SOIL PERCOLATION TESTS: The Environmental Health Officer may require a soil percolation test to be undertaken to assist in calculating the size of the absorption drains.

INSPECTION: A minimum of 24 hours notice is required for all inspections.

AUTHORISATION BY PROPERTY OWNER:

I hereby authorise _____ (insert applicant's name) to apply for a permit to install/alter the septic tank system at this property.

Name: _____

Address: _____

_____

Phone: _____

Signature of property owner _____ Date _____

FIGURE 13.17 Example application form to construct an effluent treatment/dispersal system

Source: Gary Cook.

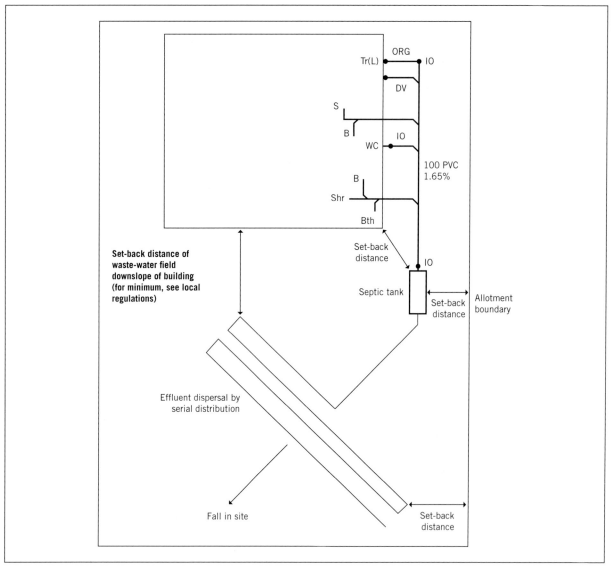

FIGURE 13.18 Example block plan of a waste-water system

Some of the issues that should be considered when preparing to install on-site domestic treatment facilities are:

- conformance with company operating procedures
- ensuring the necessary applications are lodged with the authorities
- arranging for all necessary inspections and tests
- ordering equipment, tanks, pipes, fittings, pumps and other materials in accordance with the job plans and specifications
- receiving and inspecting equipment, tanks, pipes, fittings, pumps and other materials for conformance with orders
- storage of equipment, tanks, pipes, fittings, pumps and other materials to prevent damage or deterioration
- selection, operation and maintenance of tools and equipment

- compliance with the contract work schedule
- maintaining records of job variations.

Effluent quality

After work has been completed to install domestic treatment plants, most authorities require a program of inspection and testing to periodically confirm that the quality of effluent is acceptable for discharge to the environment. For the owner, plumber and authority, the responsibility for monitoring, testing and maintenance must be clear from the outset, including how and when it must be done.

Sequencing of tasks

When installing below-ground domestic treatment plants, the upstream drainage systems should be completed first, as the below-ground obstructions and services may alter the final level of the inlet connection. Some tanks and chambers (particularly those made of plastic and fibreglass) may be limited in the depth at which they may be installed below the surface level. It is also important that the upstream drainage system is laid at the minimum gradient and length to minimise the total excavation.

After tanks and chambers (particularly plastic and fibreglass) have been installed level in the excavation, they must be immediately filled with water and backfilled in accordance with the tank manufacturer's specifications. Note that lightweight tanks (such as plastic and fibreglass) may need to have anchor blocks installed to prevent flotation, where specified by the tank manufacturer.

When installing treatment plants in conjunction with on-site dispersal systems, any other major works in the dispersal field (e.g. landscape levelling) should be completed first, and this will eliminate the possibility of damage to the system and unnecessary exposure to effluent.

FROM EXPERIENCE

Correct installation and backfilling of lightweight plastic and fibreglass chambers/tanks in the ground is essential, to ensure that they do not deform or collapse.

Tools and equipment

Safety equipment

An important safety consideration when installing septic tanks and domestic treatment plants is the correct selection and use of safety equipment. Types of PPE that may be used include:

- overalls
- boots
- safety glasses or goggles
- ear plugs or ear muffs
- gloves
- dust masks or respirators.

Tools

Tools and equipment typically used for installation of domestic treatment plants may include:

- excavator/backhoe
- levelling equipment (see Unit 3)
- pick, shovel, crowbar, rake
- demolition saw
- scale ruler
- measuring tape
- hacksaw with a range of blades
- wood saw
- mitre box
- string line
- plumb-bob

- spirit level
- cold chisel
- trowel
- claw hammer.

When moving tanks and chambers, you might be required to use lifting or load-shifting equipment such as:

- crane truck
- excavator/backhoe
- rollers
- chain blocks
- trolley
- jacks.

Special equipment – selection and operation

Special equipment is required to install some septic tanks and domestic treatment plants. This equipment must be used in accordance with the manufacturer's instructions. Examples include:

- special tools (e.g. lifting attachments) for transport of tanks and chambers
- lid keys for access covers
- special tools for installing and commissioning equipment.

Some manufacturers supply special tools and equipment, including fixings. If these are not used for the installation, the product warranty may be void.

Identifying domestic treatment plant installation requirements

Materials

Waste-water systems must be constructed of durable materials. They must be watertight and capable of withstanding vertical, lateral and uplift loads imposed by the ground and any other environmental conditions in which they are installed. The design and performance requirements for septic tanks are outlined in AS/NZS 1546.1.

✓✓ AS/NZS 1546.1 ON-SITE DOMESTIC WASTEWATER TREATMENT UNITS

Part 1: Septic tanks. Septic tanks can be manufactured from a variety of materials, such as:

- concrete cast in-situ
- pre-cast concrete
- fibre-reinforced concrete
- fibreglass
- bricks, cement rendered
- plastic.

There are many different types of domestic treatment plants, and each type has advantages and disadvantages, which must be considered when planning an installation.

The most common types of domestic treatment plants and their features are:

- *pre-cast concrete:* constructed in one piece; no underground joints; heavy; strong; and available in standard sizes
- *glass-reinforced plastic (GRP, fibreglass):* reasonably lightweight; requires anchorage; made in sections; and available in limited sizes
- *plastic septic tanks (polyethylene and polypropylene):* lightweight; strong; durable; requires anchorage; and available in a variety of sizes.

The construction and dimensions of a proprietary pre-cast concrete septic tank are shown in Figure 13.19.

Position

The position of the domestic treatment plant must be determined, based on the plans and specifications, and must be clear of building structures and other services. The material of the treatment plant and method of on-site dispersal may also determine the position of the domestic treatment plant. The manufacturers of some tanks specify a maximum depth, and in that case the trench for the sanitary drain may need to be kept short to minimise the fall in the sanitary drain and depth of the plant. If the method of on-site dispersal is via gravity away from the plant, the tank will need to be uphill of the dispersal field and the outlet of the tank may need to be at a certain level.

Design capacity

The design capacity requirements for septic tanks are outlined in AS/NZS 1547.

AS/NZS 1547 ON-SITE DOMESTIC WASTEWATER MANAGEMENT

Domestic treatment plants must have adequate capacity for treating the flow of waste water and have sufficient storage for the solids that build up before removal. Waste-water systems must be able to:

- hold the normal flow of domestic waste from up to 10 persons
- allow for normal peak flows
- allow for short-term unusual/overload production of wastes
- allow for sludge and scum to build up.

The following criteria may differ with local authority requirements. The criteria used to determine capacity are as follows:

- A daily allowance of 200 litres (L) per person for all-waste units. *Note:* This is a conservative volume to ensure that the unit has sufficient capacity to cope with peak discharge rates or large temporary overloads.
- The flow of waste water from up to 10 persons – being 14 000 L/week for all-waste waste-water flow. *Note:*

The flow limit of 14 000 L/week represents an average daily flow for design sizing purposes of 2000 L from up to 10 persons in a single residence.

- Sludge and scum accumulation from:
  - all-waste: 80 L/person/year
  - greywater: 40 L/person/year
  - blackwater: 50 L/person/year.
    Note: These accumulation rates may vary.

Also, the capacity must be designed for:

- retention of the average daily flow for at least 24 hours
- sludge removal when sludge accumulation reduces settling volume below 24 hours' retention; usually at three- to five-year intervals.

Operation and maintenance

The operation and maintenance of any domestic treatment plant must be in accordance with the manufacturer's specifications and instructions.

Septic tank

The combined discharge (soil and waste) may discharge into a septic tank. Anaerobic bacteria act on the organic matter, breaking it down. Sludge is formed, which settles to the bottom of the tank, and a scum is formed on the surface. The excess treated waste is then disposed of (depending on the type of installation) in the approved manner. Dispersal system methods will be dealt with later in this unit.

Important considerations to note:

1 Care must be taken not to discharge large amounts of antiseptic, caustic material, trade waste and/or rainwater/stormwater, as they will destroy the bacteria, causing the system to stop functioning.
2 Sludge must be removed periodically, as this, combined with the scum, will clog the tank and reduce its operating capacity.
3 The installation may be of a single tank with on-site dispersal, or two tanks for off-site disposal.
4 Some local government authorities require grease arrestors on the kitchen sink lines to prevent grease from being discharged into the septic tank or clogging the on-site dispersal system. Before installing the system, you will need to check with the relevant authority as to whether a grease arrestor is required.

Installation of domestic treatment plants

All domestic treatment plants must be installed in accordance with the manufacturer's installation instructions. A solids-reduction unit (septic tank) must be installed level. It is essential to ensure that 75 mm is maintained between the inlet and outlet. A levelling instrument (such as a laser level) is required, and should be used as outlined in Unit 3.

Rectangular concrete septic tanks
5000 litre and above

TANK CAPACITY litres	DIMENSIONS				REINFORCEMENT	CONCRETE THICKNESS		
	a	b	c	d		walls	roof	floor
5,000 LITRES	2400	1200	1200	1200	F72	100	150	150
10,000 LITRES	3000	1500	1500	1500	F82	150	150	150
15,000 LITRES	3400	1700	1700	1750	F92	150	150	150
20,000 LITRES	3800	1900	1900	1900	F92	150	150	150

REFER AS/NZS 1547

Note: all dimensions in millimetres

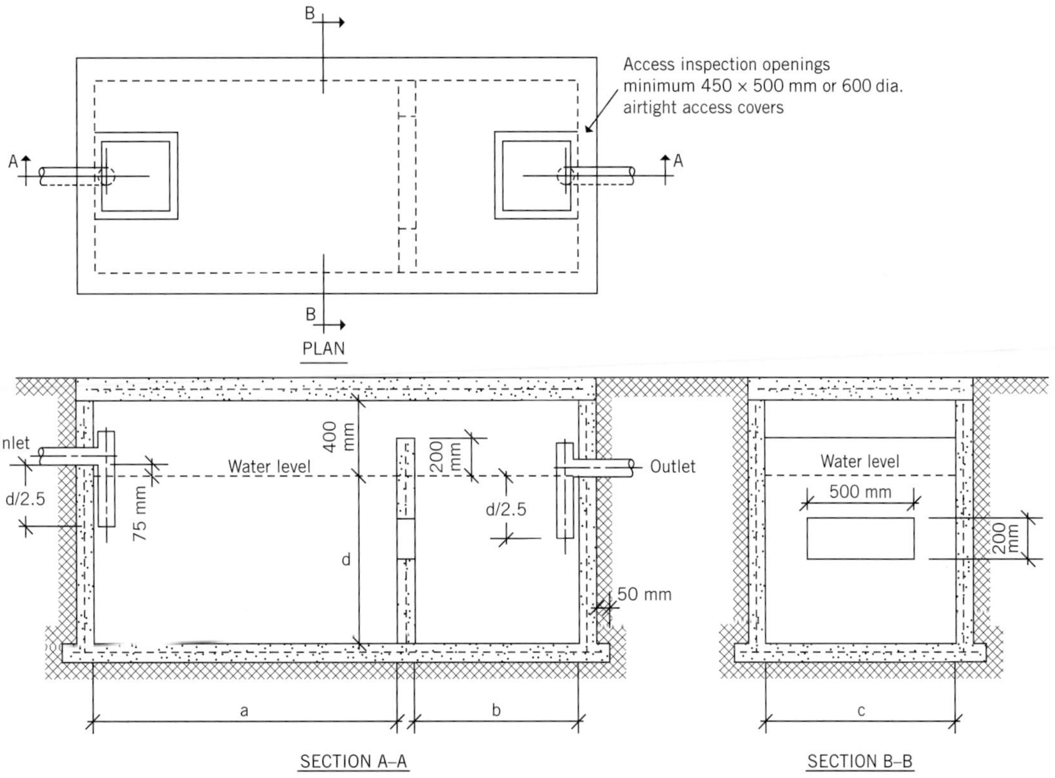

Access inspection openings minimum 450 × 500 mm or 600 dia. airtight access covers

B

A

PLAN

Inlet

Water level

d/2.5

75 mm

400 mm

200 mm

d

a

b

d/2.5

Outlet

50 mm

SECTION A–A

Water level

500 mm

200 mm

c

SECTION B–B

Frankston Concrete Products

FCP INDUSTRIES

30–44 MILES GROVE   Tel.(03)9786 3111 Fax.(03)9786 3208
P.O.Box 306
SEAFORD, VICTORIA 3198

Web site – www.frankstonconcrete.com.au
email – sales@frankstonconcrete.com.au

DRAWN	M.D.S.
CHECKED	M.M.
APPROVED	C.Z.
DRAWING NO.	95/6004

FIGURE 13.19 Rectangular pre-cast concrete septic tanks

Source: Reproduced with permission of Frankston Concrete Ltd.

The septic tank must be installed on a firm base, normally on a bed of sand. It is then backfilled and compacted in layers. Fibreglass and plastic tanks must be backfilled with sand, and it is generally recommended that these tanks be progressively filled with water during backfilling. All tanks must be filled with water to operating level as soon as possible after installation to achieve settlement and prevent flotation.

On completion, the system must be commissioned as per the manufacturer's specifications. When commissioning, no additives are required because the bacterial action will occur naturally.

The sanitary drains and vents *upstream* of domestic treatment plants must be installed as outlined in Units 4 and 10, in accordance with AS/NZS 3500.2 (see Figures 13.20 and 13.21).

AS/NZS 3500.2 SANITARY PLUMBING AND DRAINAGE

COMPLETE WORKSHEET 2

Effluent disinfection systems

When there is no reticulated sewer system, an on-site waste-water system is required to collect, treat and dispose of all household wastes. Some waste-water systems require the effluent to be treated by disinfection prior to discharge on-site.

Effluent disinfection systems must be maintained in accordance with AS/NZS 1547. The local regulatory

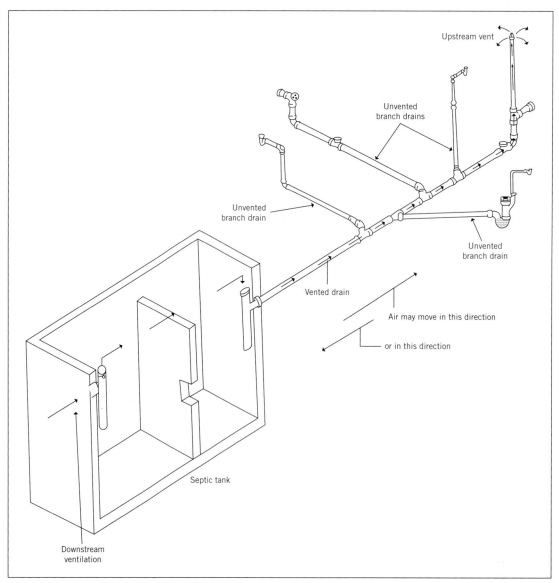

FIGURE 13.20 Septic tank and upstream drainage system

Source: Department of Education and Training (http://www.training.gov.au) © 2013 Commonwealth of Australia.

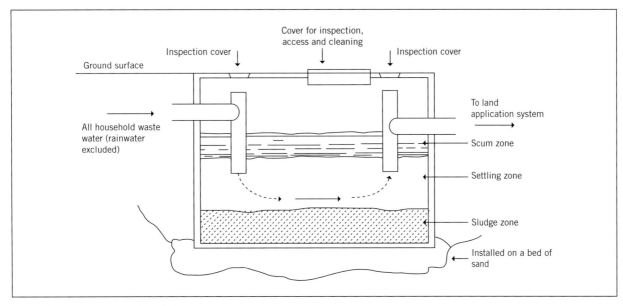

FIGURE 13.21 Septic tank installation

Source: Gary Cook.

authority (or council) may have laws that require on-site treatment plant disinfection systems to be serviced at regular intervals, with service reports provided to the authority at specified intervals.

AS/NZS 1547 ON-SITE DOMESTIC WASTEWATER MANAGEMENT

The type of waste-water system and exposure to effluent discharge may determine whether disinfection of effluent is necessary. Some examples include:
- where effluent is discharged over the surface of the block
- where effluent is discharged into soil, which could result in contamination of groundwater
- where effluent is to be re-used on-site – for example, for toilet flushing or watering vegetation.

Processes and methods of disinfection

The process of disinfection

The process of disinfection involves the destruction of pathogenic and other microorganisms in waste water. A number of important water-borne pathogens are found in Australia, including some bacteria species, protozoan cysts and viruses. All pre-treatment processes used in on-site waste-water management remove some pathogens. Chlorination and ultraviolet (UV) irradiation are two methods that are commonly used in disinfection of effluent in on-site treatment plants.

The effectiveness of disinfection of waste water is measured by the use of indicator bacteria, usually faecal coliforms. These organisms are excreted by all warm-blooded animals and are present in waste water in high numbers. They also can survive in the natural environment as long as or longer than many pathogenic bacteria, and are easy to detect and quantify.

Methods of disinfection

A number of methods can be used to disinfect waste water. These include chemical agents, physical agents and irradiation. For on-site applications, only a few of these methods have proven to be practical (i.e. simple, safe, reliable and cost-effective). Although ozone can be, and has been, used for disinfection, it is less likely to be employed due to economic reasons.

Effectiveness of disinfection

The effectiveness of chemical and UV disinfection depends on several factors:
- the nature and concentration of the disinfecting agent
- the type of microorganisms present
- temperature and pH (how acid or alkaline the water is)
- contact time (see **Figure 13.22**)
- the turbidity and quality of the water being disinfected
- the concentration and type of organic matter in the water.

Disinfectants work more effectively if the water is clear. Turbidity in the water indicates the presence of particles and chemicals. The pathogens often attach themselves to these particles and are thus shielded from the disinfectant. Other chemicals compete with the disinfectant, and this reduces the amount that can come in contact with the pathogens.

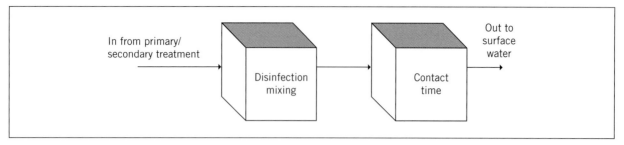

FIGURE 13.22 Generic process of disinfection

Source: Department of Education and Training (http://www.training.gov.au) © 2013 Commonwealth of Australia.

Figure 13.23 shows the different stages of treatment of an aerated waste-water treatment system and the positioning of a chlorine disinfection chamber to ensure the best possible quality prior to disinfection.

Processes in an aerated treatment plant

Initially, the waste water enters the primary or septic compartment, where the age-old principle of separating solids from liquids occurs. Under still conditions, heavy solids undergo anaerobic dissection. The septic compartment is a critical component of the treatment process. Up to 30% of biological oxygen demand (BOD) removal is achieved in this stage. Therefore, sludge build-up in the primary compartment should be monitored.

The liquid overflow from the septic compartment passes into the secondary treatment stage – the aeration or aerobic compartment/s. The waste water is aerated by means of a submerged diffuser system and air blower. The water is constantly being circulated around the contact media. This media serves:

- as an attachment point for microorganisms
- to increase the amount and range of these microorganisms
- to enable an even flow of water containing BOD and oxygen over the microorganisms.

From the aeration compartment, the effluent enters the clarifier chamber. This compartment removes any bits and pieces of matter that may come through the treatment process, such as:

- microorganisms that may have fallen or broken off the media
- particles such as hair or human-made clothing fibres not trapped in the media
- fruit seeds, etc.

It is important to note that particles that fall off the media do not always flow through to this chamber, but instead gather at the bottom of the aeration compartment (sludge). Only the suspended material finds its way through to the sedimentation, clarifier compartment. It is important to remove any sediment in the effluent, because when chlorine is used as a disinfectant, it will tend to react with any particles in the water in preference to bacteria.

Use of chlorine

Faecal coliform enter the septic tank in the tens of millions. By the end of the treatment process, the numbers are down to thousands. This is still far too

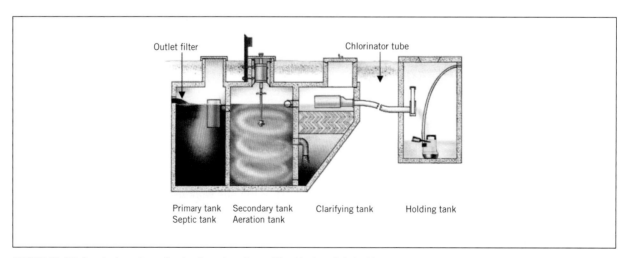

FIGURE 13.23 Aerated waste-water treatment system with chlorine disinfection

Source: Department of Education and Training (http://www.training.gov.au) © 2013 Commonwealth of Australia.

high to be discharged to a land application area. The final effluent is therefore disinfected by adding chlorine.

Chlorine is a powerful oxidising agent and has been used as an effective disinfectant in water and waste-water treatment for many years. Chlorine may be added to water as a gas (Cl_2), or as a liquid or solid in the form of sodium or calcium hypochlorite. Because chlorine gas can present a significant safety hazard and is highly corrosive, it is not recommended for on-site applications.

Typically, the solid form (calcium hypochlorite) has been most favoured for on-site applications. Different manufacturers use varying forms of solid chlorine. One such example is slow-dissolving chlorine tablets, Trichloro-S-Triazinetrione (Tri-Chlor).

Chlorine is dosed to waste water in an on-site treatment system using a simple tablet feeder device. Waste water passes through the feeder and then flows to a contact tank designed to hold the mixture for a period to ensure appropriate disinfection prior to discharge to a land application area.

Chlorination units must ensure that sufficient chlorine release occurs (depending on pre-treatment) from the tablet chlorinator. These units have historically provided erratic dosage, so regular maintenance is important. Performance is dependent on pre-treatment. At the point of chlorine addition, a suitable mixing time in a purpose-built contact chamber is necessary to ensure maximum disinfection. Tablets are usually suspended in open tubes that are housed in a plastic assembly designed to increase flow depth (and tablet exposure) in proportion to effluent flow.

Effectiveness of chlorine

The effectiveness of chlorine as a disinfectant is considerably affected by the pH of the water – pH is a scale of measurement (see **Figure 13.24**). For convenience, pH may be considered a measurement of 'power of Hydrogen', although it is actually a scale for the measurement of concentration of hydrogen ions.

If the water is alkaline, with a pH of 8.5 or above, the disinfection properties are greatly reduced. Chlorine provides residual protection for the water that would otherwise support bacterial regrowth.

An important reason for reducing the turbidity and colour of effluent before it is disinfected is to minimise the formation of undesirable by-products. These are produced when disinfectants react with organic chemicals in the water. Chlorine has the disadvantage that it can produce by-products called trihalomethanes (e.g. chloroform), which have the potential to damage human health.

Disinfection with an electro-chlorinator

Some modern treatment plants use an electro-chlorinator for the disinfection of effluent prior to

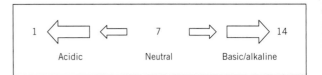

FIGURE 13.24 pH scale

dispersal to a land application area (see **Figure 13.25**). The electro-chlorinator is an electrolytic means of producing chlorine from water that contains chloride ions in excess of 150 milligrams per litre (mg/L). This type of water would originate from a town's water supply or a bore water supply.

In cases where the water supply origin is rainwater from a roof (i.e. water low in chloride), the electro-chlorinator would need to be replaced by an alternative means of disinfection. The electro-chlorinator requires an electrical power supply. However, this is generally reduced to 24 volts DC for operating and maintenance safety. Some units utilise power supply electronics to change electrode polarity as a means of keeping the electrode plates in a non-fouled condition.

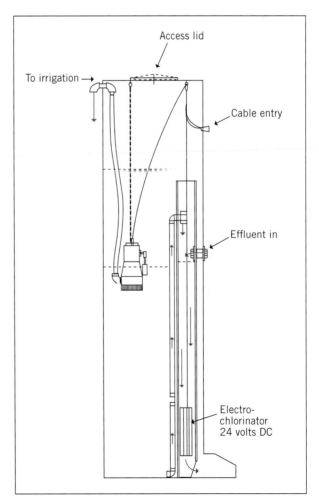

FIGURE 13.25 Electro-chlorinator

Source: Department of Education and Training (http://www.training.gov.au) © 2013 Commonwealth of Australia.

Electro-chlorination produces chlorine and hydroxide ions. This high-alkaline region around the plates causes scale formation from the water hardness. The electrode polarity reversal is intended to remove this fouling matter.

The electro-chlorinator is composed of metal plates, which are constructed from platinum-coated titanium. The plates are purposely placed using inert plastic supports.

Position of electro-chlorinator in relation to effluent flow

The total amount of chlorine (free and combined chlorine) required to react with organic and inorganic substances, and to kill the pathogens in effluent, is called the 'chlorine demand' and is dependent upon the level of pre-treatment offered to the effluent prior to disinfection. The chlorine dose is the amount of chlorine applied to waste water and is typically expressed in milligrams per litre. The dose of chlorine that may be required to disinfect effluent from a biological treatment process would typically be higher (around 10 to 25 mg/L) than, say, effluent that had passed through a sand filter (possibly 1 to 5 mg/L) over the same detention period, depending on the chlorinator design. (The chlorine doses stated are examples only and are not intended to be indicative of industry standards.)

Chlorine residual (sometimes called 'free chlorine') is that chlorine existing in water as hypochlorous and hypochlorite ions and is the amount of chlorine left over after the disinfection process. Chlorine residual should be between 0.5 mg/L and 1.0 mg/L and is required to prevent bacteria regrowth. Higher levels of residual chlorine are not desirable, and could be harmful to the environment.

Chemical-free effluent disinfection with UV irradiation

As biological control standards tighten, UV disinfection has gained in popularity as a method of disinfecting effluent water. Domestic aerated waste-water treatment plants are increasingly using UV for disinfection.

UV radiation occurs naturally in sunlight. It is also generated by mercury vapour lamps. It is a potent disinfectant and can kill bacteria and viruses. Normal doses of UV radiation have little effect on cryptosporidium and Giardia, but can inactivate them if used at a sufficiently high dosage.

Like all methods of disinfection, UV must be applied correctly to achieve the desired outcomes in disinfection. Simply put, UV disinfection is the process of allowing suitably pre-treated effluent to pass through a tube that houses mercury vapour-filled UV lamps (see **Figure 13.26**).

The effectiveness of UV disinfection depends upon the:

- UV power (expressed in micro Watts, or 'mW')
- time the effluent is in contact with the UV lamps (time in seconds)
- liquid film thickness (area in cm²) or mW sec/cm².

UV radiation penetrates the cell wall of the organism and is absorbed by cellular materials, which either prevents the organism from reproducing or causes the death of the cell. Because the only UV radiation effective in destroying the organism is that which actually reaches it, the water must be relatively

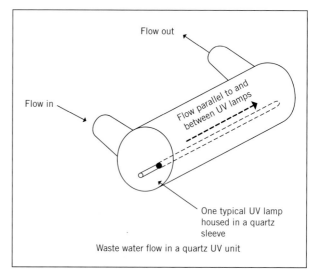

FIGURE 13.26 UV disinfection unit

Source: Department of Education and Training (http://www.training.gov.au) © 2013 Commonwealth of Australia.

free of turbidity. (Turbidity is muddiness or cloudiness caused by particles in the water.)

'Dirty' water attenuates irradiation more than 'clean' water because it affects the ability to inactivate microorganisms (bacteria, viruses, protozoa, fungi, algae and yeast). The shading effect of particles in water results in bacteria and viruses being protected from a lethal dose of UV radiation in low-quality or poorly mixed effluent water. Because the distance over which UV light can travel is very limited, the most effective disinfection occurs when a thin film of the water to be treated is exposed to the radiation.

Since effective disinfection using UV technology depends upon waste-water quality as measured by turbidity, it is important that pre-treatment provides a high degree of suspended and colloidal solids removal. The outflow effluent from a septic tank cannot be effectively disinfected by UV.

UV units used for on-site applications are mostly self-contained and provide low-pressure mercury arc lamps encased by quartz glass tubes. The unit should be installed downstream of the final treatment process and protected from weather exposure.

The temperature of effluent has a significant influence on the choice of UV lamp. Medium-pressure UV lamps are not as influenced by water temperature as low-pressure UV lamps, which have an optimum operating temperature of approximately 20°C.

UV is more sensitive to extremes in temperature than chlorination, and must be housed appropriately for the climate. UV units must be located near a power source and should be readily accessible for maintenance and inspection. Controls for the unit must be resistant to corrosion and enclosed in accordance with electrical regulations.

If the UV disinfection process is not properly managed, it may not deliver the level of pathogen destruction that is anticipated and could present health risks to downstream users.

Preparing for work

Safety and environmental requirements

The main safety consideration in the maintenance of effluent disinfection systems is the safe use of chemicals.

Each state and territory has WHS regulations that require employers and self-employed people to identify hazards and assess and control risks at the workplace in consultation with their workers. All hazards must be identified before work commences.

Other examples of safety considerations specific to effluent disinfection systems are:

- entry to confined spaces such as tanks and chambers
- biological hazards in domestic treatment plants and on-site dispersal systems.

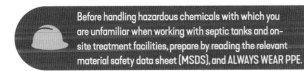

Before handling hazardous chemicals with which you are unfamiliar when working with septic tanks and on-site treatment facilities, prepare by reading the relevant material safety data sheet (MSDS), and ALWAYS WEAR PPE.

The environmental considerations in the maintenance of effluent disinfection systems centre on controlling the quality of discharge to the environment. Discharges from domestic treatment plants and on-site dispersal systems must be contained on-site and prevented from entering adjoining properties, stormwater drains, streams and waterways. The discharges must fall within the acceptable quality parameters of the relevant local government (council) authority.

Quality assurance

Products

The requirements for maintenance of effluent disinfection systems vary from state/territory to state/territory, and depend on the type of equipment. Generally, effluent disinfection systems must comply with AS/NZS 1547. Materials and products used in the maintenance of effluent disinfection systems must

be compatible and suited to the equipment. If you are unsure, refer to the equipment manufacturer for assistance and advice.

AS/NZS 1547 ON-SITE DOMESTIC WASTEWATER MANAGEMENT

Identifying, ordering and collecting materials

The materials for effluent disinfection systems must be approved for installation by the equipment manufacturer and local government (council) authority.

A method for creating a materials quantities list can be found in Unit 4. When the materials list is complete, it can be used to obtain competitive prices from suppliers.

When ordering and purchasing chemicals, it is good practice to plan ahead for safe dry storage in a place where they will not be subject to dampness or reaction with any other substances. This is critical, as some chlorine-based chemicals are highly reactive with some substances.

After you have confirmed that the materials and products are approved and comply with the equipment manufacturer's and authority's requirements, they can be ordered from suppliers on your materials list.

When receiving the delivery of materials and products, you should count the number and type of items and compare them to your materials list order. Ensure that you have identified all of the items as correct before accepting the order. The incorrect use of materials, particularly chemicals, could result in a malfunctioning system or, worse, serious injury.

Installation work

It is important to be aware of the project requirements for quality assurance. Most regulatory authorities require inspections and tests. You must familiarise yourself with local inspection requirements before commencing work.

When preparing for your work, you must comply with the requirements of the relevant authorities. You must also have the necessary reference numbers and phone numbers to book inspections and tests with the authority (where required).

Inspections are carried out by sewerage authorities and/or councils to ensure that the effluent disinfection systems have been maintained correctly in accordance with state/territory and local regulations. Inspection is a form of quality assurance for the authority and property owner.

Some of the issues that should be considered when preparing to maintain effluent disinfection systems are:

- conformance with company operating procedures
- ensuring the necessary applications are lodged with the authorities
- arranging for all necessary inspections and tests

- ordering materials in accordance with the job plans and specifications
- receiving and inspecting materials for conformance with orders
- safe and secure storage of materials and chemicals in accordance with the MSDS
- selection, operation and maintenance of tools and equipment
- compliance with the contract work schedule
- maintaining records of job variations.

Effluent quality

Most authorities require a program of inspection and testing to periodically confirm that the quality of effluent is acceptable for discharge to the environment. The correct maintenance of effluent disinfection systems is essential to ensure that the effluent quality is acceptable.

For the owner, plumber and authority, the responsibility for monitoring, testing and maintenance must be clear from the outset, including how and when these must be done.

Sequencing of tasks

When maintaining effluent disinfection systems, it will be necessary to shut the system down for a period of time; therefore, you will need to coordinate the shutdown with the owner/occupier.

Often when systems are shut down, some sections of pipe may be under pressure. You must therefore be prepared to relieve pressure in any section of pipe that is to be removed. Prepare by covering areas of exposed skin and wearing safety goggles.

Tools and equipment

Safety equipment

An important safety consideration when maintaining effluent disinfection systems is the correct selection and use of safety equipment. Types of PPE that are used include:
- overalls
- boots
- safety goggles
- ear plugs or ear muffs
- rubber gloves
- dust masks or respirators.

The MSDS for chemicals may also specify the necessary PPE for commissioning and maintaining effluent disinfection systems.

Tools

Tools and equipment typically used for maintenance of effluent disinfection systems may include:
- access cover keys
- cordless drill with a range of bit tools for self-drilling screws
- socket set and ratchet spanner
- range of screwdrivers
- shifting spanners
- range of round and flat files.

Special equipment – selection and operation

Special equipment is required to maintain some effluent disinfection systems. Also, some manufacturers provide special equipment that must be used in accordance with their installation instructions. Examples include:
- keys for access to control panels and switches
- lid keys for access covers
- special tools for disassembly and assembly of equipment.

Identifying effluent disinfection system maintenance requirements

The effluent disinfection system must comply with AS/NZS 1547 and local authority requirements. It is necessary to obtain any maintenance instructions and specifications, and to familiarise yourself with the type of system, as described earlier in this unit. Then, to identify the effluent disinfection system maintenance requirements, it will be necessary to inspect the site.

AS/NZS 1547 ON-SITE DOMESTIC WASTEWATER MANAGEMENT

When inspecting the site, it is necessary to be aware of certain land features, soil properties and application methods. These issues can determine the quality of effluent required. This may involve:
- determining whether the waste-water treatment system and disinfection system components are accessible for maintenance
- determining any repairs that may be necessary, such as replacement of components of the effluent disinfection system
- checking the detention capacity and effluent flow rate for compliance with regulatory requirements for effluent disinfection
- checking the type of land application – surface or sub-surface irrigation
- checking that the slope of the land and buffer distances comply with regulatory requirements for effluent disinfection
- checking the condition of existing vegetation.

After inspecting the site and effluent disinfection system, you will be prepared to maintain the system to ensure that the:
- system will perform in accordance with AS/NZS 1547 and local regulatory requirements
- effluent is thoroughly mixed, and the discharge meets the treatment requirements
- effluent can be tested for total chlorine or other specified measures, and the dosage rates are adjusted to achieve the levels and stability specified by Australian Standards and local regulatory requirements.

AS/NZS 1547 ON-SITE DOMESTIC WASTEWATER MANAGEMENT

Maintenance of effluent disinfection systems

Typical maintenance of chlorine disinfection systems

Maintenance of the chlorine addition must include inspection of all the pre-treatment processes.

A typical maintenance schedule could include the following:

- Inspect the primary tank: measure the scum layer and sludge depth.
- Remove, clean and refit the outlet filter.
- Check any associated media filter bed vents for damage and mosquito proofing:
  - Check that the drain-back to the final effluent chamber is running freely.
  - Do a visual inspection for ponding around the filter.
- Remove scum or any other solids build-up that may have formed in the aeration or clarification chamber and pump it to the inlet of the primary tank, if necessary.
- Service the tablet feeder equipment, as caking of tablet feeders may occur.
- Replace chlorine tablets as required.
- Adjust flow rates, collecting and analysing effluent samples for chlorine residuals.
- Check any effluent and recirculating pressure pumps for operation, including any associated float switches, valves and pipework.
- Have a grab sample from the outflow effluent tested for faecal coliform, to check the effectiveness of the disinfection.
- Check irrigation areas to ensure that heavy droplet spray outlets are working satisfactorily, with even distribution.
- Prepare reports for the property owner and/or local authority as required.
- Ensure all access covers are in place and the system is returned to normal operation.

Typical maintenance of UV disinfection systems

Routine operation and maintenance of UV systems involves semi-skilled technician support. Tasks include cleaning and replacing the UV lamps and sleeves, checking and maintaining mechanical equipment and controls, and monitoring the UV intensity. Monitoring should include routine indicator organism analysis – that is, the levels of faecal coliform present. The mercury lamps in the UV units will need to be replaced (usually annually). The exact life of the lamps will depend upon the type of equipment that is installed for the particular installation.

Continuous UV bulb operation is recommended for maximum bulb service life, as frequent on/off sequences in response to flow variability will shorten the bulb life. Generally, lamp life may range from 7500 to 13 000 hours.

The quartz sleeves that house the lamps will require cleaning with an alcohol or other mildly acidic solution at each inspection. These should be inspected and cleaned four times a year.

Maintenance of the UV system must include inspection of all the pre-treatment processes. A typical maintenance schedule should include the following:

- Primary tank inspection: measure the scum layer and sludge depth.
- Remove, clean and refit the outlet filter.
- Check any associated media filter bed vents for damage and mosquito proofing.
- Check that the drain-back to the final effluent chamber is running freely.
- Do a visual inspection for ponding around the filter.
- Remove scum or any other solids build-up that may have formed in the aeration or clarification chamber and pump it to the inlet of the primary tank, if necessary.
- Clean the quartz tubes in the UV unit in accordance with the manufacturer's instructions.
- Measure the temperature of the effluent and compare it with the UV unit operating specifications.
- Check any effluent and recirculating pressure pumps for operation, including any associated float switches, valves and pipework.
- Have a grab sample from the outflow effluent tested for faecal coliform, to check the effectiveness of the disinfection.
- Check irrigation areas to ensure that heavy droplet spray outlets are working satisfactorily, with even distribution.
- Prepare reports for the owner and/or local authority as required.
- Ensure all access covers are in place and the system is returned to normal operation.

 COMPLETE WORKSHEET 3

On-site dispersal systems

On-site dispersal systems should have a minimum working life of 15 years and are divided into six categories:

1. absorption trenches and absorption beds (see AS/NZS 1547, Appendix L)
2. evapotranspiration/absorption/seepage beds/ trenches (see AS/NZS 1547, Appendix L)
3. sub-surface and covered surface irrigation – drip irrigation (see AS/NZS 1547, Appendix M)
4. surface irrigation – spray irrigation (see AS/NZS 1547, Appendix M)

5 sub-surface irrigation – LPED (low pressure effluent distribution) irrigation (see AS/NZS 1547, Appendix M)

6 mounds (see AS/NZS 1547, Appendix N).

AS/NZS 1547 ON-SITE DOMESTIC WASTEWATER MANAGEMENT

As discussed previously in this unit, a septic tank is a solids-reduction unit (SRU). It is a chamber where anaerobic *primary* treatment of sewage can occur. Its function is to reduce solid matter to a liquid, and decompose the solid matter, using anaerobic bacterial digestion. After this primary treatment, the effluent leaves the unit to undergo *secondary* treatment, usually by aerobic bacterial digestion.

Trench-based dispersal systems accept the anaerobically treated effluent discharged from the septic tank and treat it aerobically. They further purify it and then allow the final purified product to evaporate or disperse through the soil.

The aerobic treatment takes place in the aggregate bed and soil surrounding the distribution pipes. The effluent flows around and through the aggregate where bacterial action takes place, treating the effluent and further purifying it.

A traditional on-site dispersal system consists of a series of level-graded trenches cut into the land adjacent to a septic system. These trenches have perforated pipes or arch drains in the bottom, which are then covered to a depth of about 250 mm with 20–40 mm of aggregate. The effluent is allowed to flow into these trenches where aerobic bacteria can process it, and further purify it, before it percolates away through the soil or evaporates into the atmosphere.

More conventional aerated waste-water treatment systems (AWTS) incorporate an aerobic treatment phase into the process, which treats the waste water to a better quality. This improved-quality effluent can also be disposed of in trenches. For single residential properties, the treated effluent can be used to irrigate lawns or gardens.

The length (or size) and type of the on-site dispersal system is determined by two factors:

1 The rate of inflow into the system.
   a For a domestic property:
      i   300 L/bedroom (per day) on mains water supply
      ii 200 L/bedroom (per day) on tank water supply.
   b WC pan only: 60 L/bedroom (per day).
   c Sullage only, with an adequate perennial (lasting through the year) water supply: 240 L/bedroom (per day).
2 The depth and permeability of the soil.

The length of an absorption trench may be calculated by considering the above factors and using a formula

that is found in AS/NZS 1547. The local government authority (council) will often specify a required length of absorption trench, based on factors such as treatment plant discharge rates and known site soil permeability. If the site soil conditions are unknown, or for some other reason the site conditions present difficulties, the council may require an engineer to carry out a land capability assessment. Some authorities may permit the plumber to calculate the length of the trench based on site soil absorption capability.

AS/NZS 1547 ON-SITE DOMESTIC WASTEWATER MANAGEMENT

It should be noted that, in some cases, councils might not permit the use of absorption trenches or transpiration beds if there is a possibility of contamination of the groundwater, and may insist on the use of an on-site treatment and irrigation system (covered later in this unit).

Greywater systems

As mentioned earlier in this unit, it is a common misconception that greywater does not contain pathogens, and that it is only sewage (blackwater) that requires treatment prior to dispersal or recycling. Greywater can contain pathogens that, if poorly managed, could present a risk to human health.

The trend towards recycling and reuse of domestic greywater is creating a market for greywater treatment systems. These systems must only be installed with the approval of the relevant local government (council) authority.

Untreated greywater must only be used for sub-surface irrigation of gardens. Untreated greywater must not be stored, used in a manner that allows direct human contact, or used in a manner other than by sub-surface soil dispersal. Also, it must not be used where it could contaminate food sources, children's play areas or other domestic recreation areas.

LEARNING TASK 13.1

Research AS/NZS 3500.2 and read the requirements for installing greywater plumbing and drainage systems.

1 Where a property has a greywater system and is also connected to the authority sewer, is the greywater system required to have a permanent connection to the sewer system?

2 How must a greywater system be connected to the sanitary drain?

3 Where a below-ground greywater diversion device is installed and connected to the sanitary drain, what must be installed?

Preparing for work

Safety and environmental requirements

The main safety considerations in the installation of on-site dispersal systems relate to safe excavation and trench support, as outlined in Unit 2. In addition, it may be necessary to move and lower large tanks into excavations. Moving and lowering tanks will most likely require a crane. Crane operators must be experienced in dogging and slinging techniques, and, depending on state/territory regulations, may need to hold an appropriate qualification. Remember always to stay clear of areas beneath suspended loads.

Each state and territory has WHS regulations that require employers and self-employed people to identify hazards and assess and control risks at the workplace in consultation with their workers. All hazards must be identified before work commences.

Other examples of safety considerations specific to on-site dispersal systems are:

- entry to confined spaces such as tanks and chambers
- protection from falls into tanks and excavations
- biological hazards in domestic treatment plants and on-site dispersal systems.

The environmental considerations in the installation of on-site dispersal systems centre on controlling the quality of discharge to the environment. Discharges from domestic treatment plants and on-site dispersal systems must be contained on-site and prevented from entering adjoining properties, stormwater drains, streams and waterways.

Quality assurance

Products

The regulations for on-site dispersal systems vary from state/territory to state/territory. Generally, on-site dispersal systems must comply with AS/NZS 1547. Materials and products for on-site dispersal systems must be suitable for the nature of waste discharges, and comply with the dimensions and other specifications of the relevant local authority.

AS/NZS 1547 ON-SITE DOMESTIC WASTEWATER MANAGEMENT

Identifying, ordering and collecting materials

The materials for on-site systems must be approved for installation in accordance with local authority requirements. Packaged treatment plants and systems must comply with the requirements of the relevant local authority, and some authorities have a list of systems approved for use in their area.

A method for creating a materials quantities list can be found in Unit 4. When the materials list is complete, it can be used to obtain competitive prices from suppliers.

When ordering and purchasing products, pipes and fittings, it is good practice to plan ahead for secure storage in a place where they will not be damaged or stolen. After you have confirmed that they are approved and comply with standards and local regulations, they can be ordered from plumbing suppliers on your materials list.

When receiving the delivery of pipes and fittings, you should count the number and type of items and compare them to your materials list order. Ensure that you have identified all of the items as correct before accepting the order. If you have the wrong materials, or items are damaged, it may take some time to exchange and order the correct acceptable items.

Installation work

It is important to be aware of the project requirements for quality assurance. Most local authorities require inspections and tests.

When preparing for your work, you must have completed applications with the relevant authorities and planned for the complete installation. You must also have the necessary reference numbers and phone numbers to book inspections and tests with the regulatory authority (where required).

In some areas, the authorities may require several inspections of on-site domestic treatment and dispersal systems during the progress of work and when the work is complete. You must familiarise yourself with the local inspection requirements before commencing work.

Inspections are carried out by authorities to ensure that the on-site dispersal systems have been installed correctly in accordance with state/territory and local regulations. Inspection is a form of quality assurance of the work for the authority and property owner.

Some of the issues that should be considered when preparing to install on-site domestic treatment facilities and dispersal systems are:

- conformance with company operating procedures
- ensuring the necessary applications are lodged with the authorities
- arranging for all necessary inspections and tests
- ordering equipment, tanks, pipes, fittings, pumps and other materials in accordance with the job plans and specifications
- receiving and inspecting equipment, tanks, pipes, fittings, pumps and other materials for conformance with orders
- storage of equipment, tanks, pipes, fittings, pumps and other materials to prevent damage or deterioration
- selection, operation and maintenance of tools and equipment
- compliance with the contract work schedule
- maintaining records of job variations.

Effluent quality

After the installation of domestic treatment plants and on-site dispersal systems, most authorities require a program of inspection and testing to periodically confirm that the system is operating correctly and the quality of effluent is acceptable for discharge to the environment. For the owner, plumber and authority, the responsibility for monitoring, testing and maintenance must be clear from the outset, including how and when these must be done.

Sequencing of tasks

When installing on-site dispersal systems, the upstream treatment plant and drainage systems should be completed first, as below-ground obstructions and services may alter the final level of the inlet connection. Some upstream tanks and chambers (particularly those made of plastic and fibreglass) may be limited in the depth at which they may be installed below the surface level.

After tanks and chambers (particularly those made of plastic and fibreglass) have been installed level in the excavation, they must be immediately filled with water and backfilled in accordance with the tank manufacturer's instructions. Note that lightweight tanks, such as plastic and fibreglass, may need to have anchor blocks installed to prevent flotation, where specified by the tank manufacturer. Also, it is important that the upstream system is laid at the minimum gradient and length to minimise the total excavation.

When installing on-site dispersal systems, any other works in the dispersal field (e.g. landscaping, levelling) should be completed first, to eliminate the possibility of damage to the system and unnecessary exposure to effluent.

Tools and equipment

Safety equipment

An important safety consideration when installing on-site dispersal systems is the correct selection and use of safety equipment. Types of PPE that may be used include:

- overalls
- boots
- safety glasses or goggles
- ear plugs or ear muffs
- gloves
- dust masks or respirators.

Tools

Tools and equipment typically used for installation of on-site dispersal systems may include:

- excavator/backhoe
- levelling equipment (see Unit 3)
- pick, shovel, crowbar, rake

- scale ruler
- measuring tape
- hacksaw with a range of blades
- wood saw
- mitre box
- string line
- spirit level
- trowel.

When moving tanks and chambers, you might be required to use lifting or load-shifting equipment, such as:

- crane truck
- excavator/backhoe
- rollers
- chain blocks
- trolley
- jacks.

Special equipment – selection and operation

Special equipment is required to install some on-site dispersal systems. This equipment must be used in accordance with the manufacturer's instructions. Examples include:

- special tools (e.g. lifting attachments) for transport of tanks and chambers
- lid keys for access covers
- special tools for installing and commissioning equipment.

Some manufacturers supply special tools, equipment and/or fixings. If these are not used for the installation, the product warranty may be void.

Identifying on-site dispersal system requirements

After determining the type of on-site dispersal system to be installed, the position of the system is determined from the plans and specifications approved by the local authority. The position must take into account the local authority requirements for setback and buffer distances.

You will be required to research your local and regulatory requirements for setback and buffer distances at the end of this unit.

While there are currently six types of on-site dispersal systems, the following common types are described in detail as follows (*Note*: authorities may specify alternative designs):

- absorption trenches and absorption beds (see AS/NZS 1547, Appendix L)
- evapotranspiration/absorption/seepage beds/ trenches (see AS/NZS 1547, Appendix L)
- mounds (see AS/NZS 1547, Appendix N)
- surface irrigation – spray irrigation (see AS/NZS 1547, Appendix M).

Absorption trenches and absorption beds

A cross-section of an absorption trench is illustrated in **Figure 13.27**. In an absorption trench system, the effluent is discharged through a series of trenches in deep, permeable soil (see AS/NZS 1547, Appendix L). A large proportion of the effluent percolates down beyond the root zone of the vegetation.

Absorption trenches should be constructed with a distribution system of agricultural, slotted PVC pipe or arch tunnel material. This is placed in the bottom of a level-graded trench approximately 500 mm deep and between 300 and 1000 mm wide. The distribution system is covered to a depth of 250 mm with an aggregate of about 20–40 mm. The aggregate is covered with a geotextile biodegradable material, such as paper

or hessian, to prevent soil from permeating into and clogging the aggregate material. The geotextile material is covered with topsoil to form a mound about 75 mm higher than the surrounding surface level to prevent the local groundwater from flowing over the trench area and flooding the system.

Methods of effluent distribution through absorption trenches are discussed in the following section.

Evapotranspiration/absorption/seepage beds/trenches

This system takes the form of a trench, bed or mound that is excavated into the original soil and raised above the original soil level (see AS/NZS 1547, Appendix L). **Figure 13.28** shows a cross-section of an absorption–transpiration trench.

Absorption–transpiration bed

These types of trenches are constructed in a similar manner to absorption trenches and must be located in a well-drained area or have a cut-off drain installed up

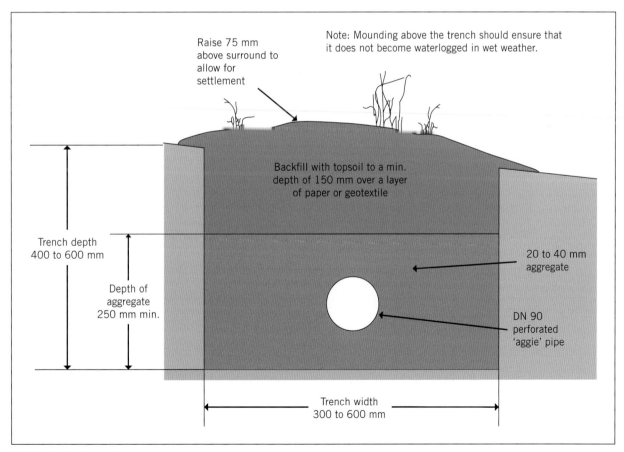

FIGURE 13.27 Absorption trench

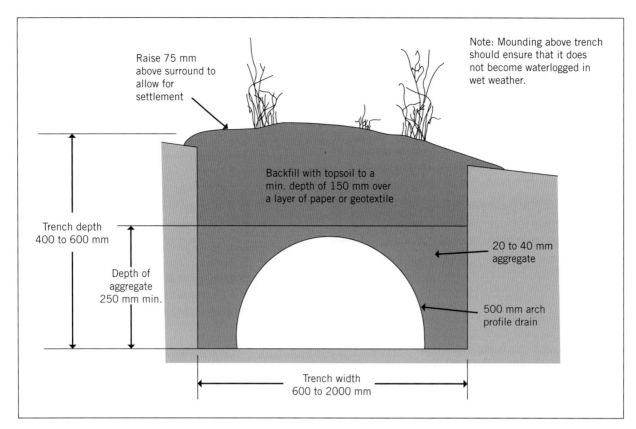

FIGURE 13.28 Absorption–transpiration trench

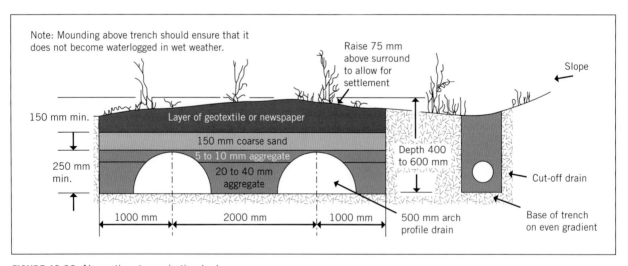

FIGURE 13.29 Absorption–transpiration bed

the slope to divert groundwater away from the trench. They are excavated to the required depth and must have a level-graded floor. The distribution rows are set out at 2 m intervals. When slotted pipes are used, they should be placed on a bed of 50 mm-deep, 20–40 mm aggregate. If arching is used instead of slotted pipe, the arch material is placed directly onto the trench floor.

Once the distribution rows are in place, the trench is filled in the following manner:

- 150 mm of 20–40 mm aggregate
- 50 mm of 5–10 mm aggregate
- 150 mm of coarse washed sand
- 150 mm minimum of soil suitable for planting grasses and small shrubs.

The finished surface of the trench must be mounded at least 75 mm above the surrounding finished surface of the ground to prevent surface water from flowing on to the area.

Figure 13.29 shows a cross-section of an absorption-transpiration bed.

Effluent distribution through absorption and absorption–transpiration systems

The distribution of effluent using absorption or absorption–transpiration, where multiple trenches are used, can be carried out using either parallel or serial methods. These are connected to a distribution box, as shown in **Figure 13.30**.

Parallel distribution

Parallel distribution requires a relatively flat area. When installing parallel systems, there must be 2 m of undisturbed soil left between each trench (see **Figure 13.31**). This allows for replacement of the dispersal system when it reaches the end of its life.

The trenches in parallel distribution are each connected to the distribution box and are all working at the same time.

Serial distribution

This is a common method of effluent dispersal (see **Figures 13.32** and **13.33**). Serial systems suit almost all site conditions, especially sloping sites, and are therefore more common than parallel distribution systems.

Serial distribution means that the effluent must fill the first trench to a depth of 250 mm before it can flow over into the second trench, which must in turn fill to 250 mm before flowing over into the third trench, and so on, to the final trench. Individual

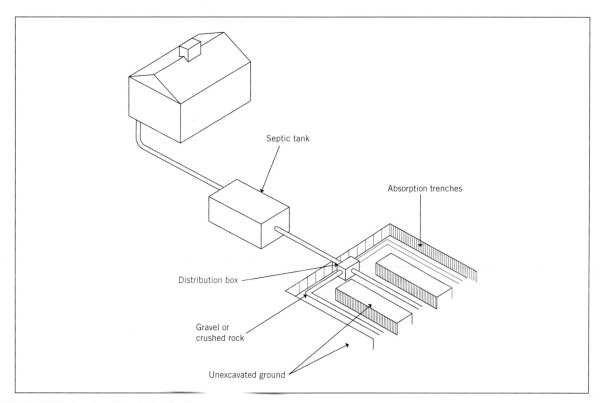

FIGURE 13.30 On-site effluent distribution

Source: Gary Cook.

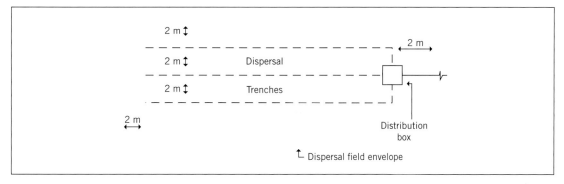

FIGURE 13.31 Effluent dispersal by parallel distribution

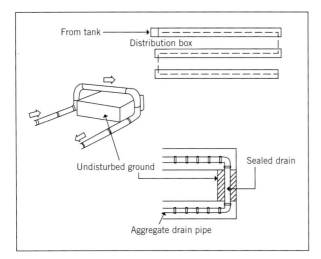

FIGURE 13.32 Effluent dispersal by serial distribution (plan view)

Source: Gary Cook.

trenches are constructed in the same manner as for parallel trenches; however, each trench is separated by a barrier of undisturbed ground about 250 mm higher than the floor of the trench. The distribution pipes are connected by a joining section that passes over the raised section of undisturbed ground.

Mounds

Raised mounds are used where the soil is not suitable due to its slow permeability (see AS/NZS 1547, Appendix N). The mounds are constructed of medium clay-free sand on top of the natural ground surface. The ground surface is ripped beforehand with a distribution bed of 20–25 mm aggregate containing the distribution system. **Figure 13.34** shows a cross-section of a mound.

AS/NZS 1547 ON-SITE DOMESTIC WASTEWATER MANAGEMENT

Surface irrigation – spray irrigation

Where the site soil has limited long-term acceptance rates for effluent dispersal, surface irrigation systems may be used as an alternative to trenches and beds (see AS/NZS 1547, Appendix M). The effluent quality requirements are such that these systems include solids reduction units and secondary aeration treatment.

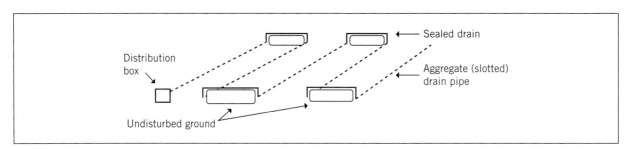

FIGURE 13.33 Effluent dispersal by serial distribution (layout)

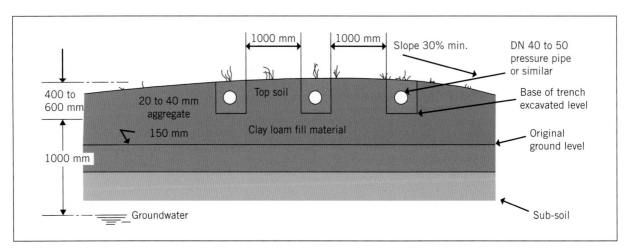

FIGURE 13.34 Mound system

AS/NZS 1547 ON-SITE DOMESTIC WASTEWATER MANAGEMENT

Sprinkler surface irrigation systems must be installed in accordance with AS/NZS 1547, and should generally be installed taking into account the following:

- Coarse droplet sprays will minimise misting and limit the distance that the effluent can be carried by the wind.
- They should have a low plume height. (See the regulations applicable to your state/territory.)
- A filter should be used to prevent blockages.
- They should be capable of irrigating the entire land application area.

AS/NZS 1547 ON-SITE DOMESTIC WASTEWATER MANAGEMENT

Aerated waste-water treatment systems (AWTS) are subject to the approval of various state, territory and local government authorities. The Aqua-nova® treatment system by Everhard Industries is an example of an AWTS (see Figure 13.35). Other suppliers of this type of system include Bio Treat, Bio-cycle, Super Treat and Envirocycle.

AWTS mechanically treat waste water to the quality required for irrigation. It is generally the responsibility of the provider/manufacturer of the treatment system to certify and guarantee that the correct effluent quality will be produced.

If an AWTS is installed, it must be serviced and maintained by a qualified person and in accordance with the manufacturer's specifications and relevant state/territory requirements and local government approvals.

Common requirements are:

- servicing of the treatment system four times per year
- testing of effluent water by a NATA laboratory once a year
- desludging of the system every three years
- submitting test and maintenance reports to the local authority.

Unless specifically approved, *surface irrigation systems must not be used for growing fruit and vegetables*. Also, stock must not enter the irrigated area, and the local authority may require that warning signs be placed near it.

Installation of on-site dispersal systems

Once the plans and specifications for the on-site dispersal system have been approved by the relevant local authority, and the position and components of the system have been determined, the system can be set out and installed.

From plans and specifications, and site inspection, the position of any underground services must be identified and marked, to avoid damaging them during excavation work. Excavation for the system should be carried out, as described in Units 2 and 10. Trenches for absorption and transpiration systems for serial and parallel distribution methods are excavated level. See Unit 3 for levelling procedures.

After the system has been installed, it must be checked for compliance with standards and regulatory authority requirements, and correct operation. This may involve following the manufacturer's commissioning instructions and procedures.

Off-site disposal

Pre-treatment facilities

As described earlier in this unit, some local authorities and/or treatment system suppliers may specify pre-treatment of wastes – for example, wastes from the kitchen sink may be required to discharge through a grease arrestor. This may apply to large residential or commercial premises. Greasy wastes can cause absorption trenches or transpiration beds to become clogged, rendering them useless. The purpose of the grease arrestor is to slow the flow rate to enable the grease and fats to cool and rise to the surface. Grease arrestors require regular cleaning, and the frequency of cleaning depends upon the amount of greasy discharge. Always check local requirements to determine whether grease arrestors are required. Greasy wastes must be pumped out and disposed of at an authorised location by an accredited contractor.

Collection wells

Collection wells are usually only required where the nature of the property is such that treated waste water from primary treatment (septic tank) cannot be disposed of on-site because of soil type or proximity to sensitive areas. Collection wells are intended for temporary storage of treated waste water from the septic tank before its regular removal to an authorised disposal area by accredited contractors, or before it is pumped to an authority's sewerage system.

The collection well *does not* contain biologically safe water and *must not* be used as a source of household water for any purpose, including garden irrigation.

Areas around septic tanks and collection wells should not be disturbed or used for play areas. The installation requirements for prefabricated collection wells are the same as for septic tanks, and the manufacturer's installation instructions must be followed.

Tanker removal

Two tanks are required in this type of installation. A collection well is installed adjacent to a septic tank.

Aqua-nova®: An effective environmentally friendly wastewater treatment process

- Separates
- Treats
- Clarifies
- Sanitises
- Irrigates

Primary anaerobic treatment process

Aerobic digestion process

Clarification disinfection

Safe, clean, water delivered automatically to your garden

Superior treatment process working for you to safeguard your family's health

Polymer Tank Systems
Tested and certified to comply with the requirements of the following authorities

QLD	10EP	Department of Natural Resources	22/08/1996
NSW	10EP	NSW Health Department - Environmental Health Branch C of A AWTS 006	14/03/2008
VIC	8EP	Environmental Protection Agency CA 68/2000	22/08/2000
SA/NT	6EP	Department of Human Services - Public and Environmental Health Services	30/05/2000
WA	8EP	Department of Health - Government of Western Australia Reg. No. 173	12/11/2004

Concrete Tank Systems (QLD & Northern NSW ONLY)
Tested and certified to comply with the requirements of the following authorities

QLD	10EP	Department of Natural Resources	21/07/1995
NSW	10EP	NSW Health Department - Environmental Health Branch C of A AWTS 006	14/03/2008

How does the Aqua-nova® Aerated Wastewater Treatment System work for you?

Aqua-nova® is a two tank system which provides the very latest in aerated wastewater treatment through a multi staged digestion process using naturally occurring bacteria and enzymes.

In the primary stage, treatment is anaerobic, where bacteria thrive in an oxygen free zone, and the breakdown of solid waste is performed. From here the waste is sent to the second stage where air is continually supplied to bacteria providing aerobic treatment to complete the total digestion process. At this point the wastewater passes through a clarifying process into the disinfection chamber where any remaining pathogenic bacteria are destroyed. Finally the now clean, clear and disinfected water is delivered by an integral submersible pump with automatic level control to a selected irrigation system in landscaped garden beds or other dispersion areas.

FIGURE 13.35 Aqua-nova® waste-water treatment system by Everhard Industries

Source: Everhard Industries, http://www.everhard.com.au.

A DN 50 suction line from the collection well should be installed so that the connection is convenient for tanker access. If there is an excessive lift (vertical rise) from the collection well to the tanker connection point, a pump must be installed and operated from the tanker pick-up connection by key (see **Figure 13.36**).

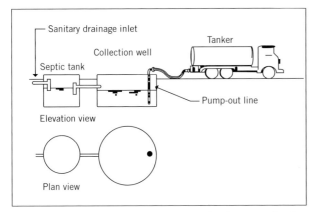

FIGURE 13.36 Removal of effluent by tanker from a collection well

Source: Gary Cook.

Pumping to sewer

Where pumping from an on-site system is necessary, refer to Unit 12 and the relevant local authority requirements (see **Figure 13.37**).

COMPLETE WORKSHEET 4

Clean-up and finalisation of documentation

Work area

After installing pre-treatment facilities, domestic treatment plants or on-site dispersal systems, it may be necessary to reinstate surfaces and clean up the work area. Clean-up includes:

- ensuring all access covers are correctly fitted and trip hazards are removed
- checking (where applicable) that soil over dispersal trenches is 75 mm higher than the surrounding ground and is seeded/planted as required
- cleaning up any chemical spills, where applicable
- removal and storage of usable materials
- removal, disposal and recycling of waste to correct bins
- restoration of any soiled building surfaces to a clean condition
- restoration of the worksite, where possible, to its original condition.

Tools and equipment

Excavating equipment should be removed and serviced to maintain it in good condition. Tools and equipment must be inspected, cleaned, maintained and stored correctly to prevent corrosion. Unserviceable and unsafe tools and equipment should either be repaired or disposed of.

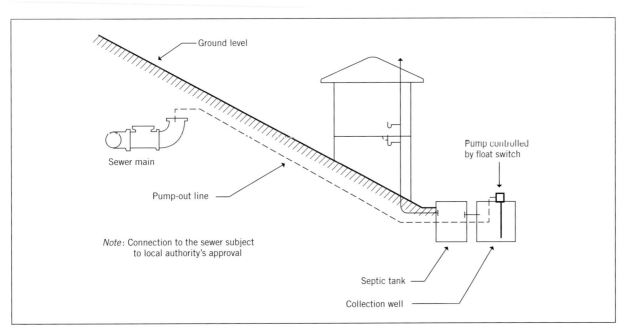

FIGURE 13.37 Pumping from a collection well

Source: Gary Cook.

Documentation

The documentation for installation of pre-treatment facilities, domestic treatment plants, on-site dispersal systems and effluent disinfection systems is discussed earlier in this unit and varies considerably around Australia.

It is important not only to lodge applications, but also to finalise documentation such as lodgement of 'as-constructed' plans where required by the relevant authorities. Some states/territories also require a Certificate of Compliance to be lodged with the regulatory authority and property owner.

 COMPLETE WORKSHEETS 5 AND 6

SUMMARY

- When preparing to install pre-treatment facilities, you will need to familiarise yourself with the nature of discharges and what pipe materials and facilities will be required. Normally, it will be necessary to enter into a trade waste agreement with the sewerage authority. Pre-treatment facilities must be installed, tested and commissioned to ensure that the installation complies with the requirements of both the sewerage authority and plumbing regulations.

- When preparing to install domestic treatment plants, you will need to familiarise yourself with the site conditions and options for on-site treatment and dispersal. Normally, it will be necessary to lodge an application with the local municipal authority and comply with their guidelines for installation, maintenance and inspections. Domestic treatment plants must be installed, tested and commissioned to ensure that the installation complies with the requirements of both the local municipal authority (council) and plumbing regulations.

- When preparing to install and maintain effluent disinfection systems, you will need to familiarise yourself with the requirements for the quality of discharge on-site, which must fall within certain parameters. Normally, the local authority will have specific guidelines for maintenance, testing and regular reporting. Effluent disinfection systems will need to be maintained in accordance with manufacturers' specifications, and PPE should always be worn while performing such maintenance.

- When preparing to install on-site dispersal systems and greywater systems, you will need to familiarise yourself with the limitations regarding where effluent and greywater can be safely discharged or used. The local authority and/or EPA will have regulations and criteria for systems that can be used.

- When works have been completed, clean up the work area. If required by the regulatory authority, lodge the necessary documentation, such as a Certificate of Compliance and an 'as-constructed' plan.

WORKSHEET 1

Student name: _____

Enrolment year: _____

Class code: _____

Competency name/Number: _____

Task: Review the section 'Pre-treatment facilities', then complete the following.

1   Research AS/NZS 3500 Plumbing and drainage and define the term 'trade waste'.

   _____

   _____

   _____

2   Provide four potential effects of prohibited discharges.

   a   _____

   b   _____

   c   _____

   d   _____

3   Explain why it is important for the owner/occupier to make a trade waste application to the relevant authority.

   _____

   _____

   _____

4   Obtain a trade waste application and any guidelines from your local sewerage/council authority. In some areas these will be available on the internet. What is the name of your local sewerage authority?

   _____

5   The following plan represents a commercial kitchen with greasy discharges connected to an external grease arrestor at 1 Main Street, Springfield. Food is not eaten on the premises. Using the trade waste application obtained earlier, complete the application and determine the holding capacity of the grease arrestor.

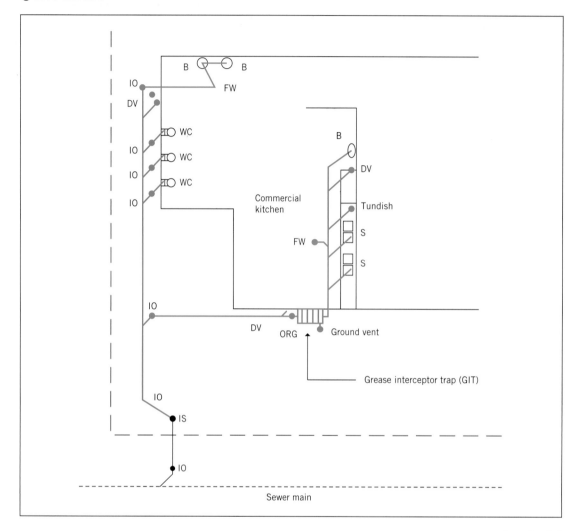

Source: Gary Cook.

Holding capacity of grease arrestor: _____

6   Under what circumstances might a petrol and oil interceptor be required?

_____

7   Provide three examples of premises where plaster arrestors are installed.

a   _____

b   _____

c   _____

8 Circle the correct response to each of the following.

   a PVC is a suitable material for the internal construction of a petrol and oil arrestor.

      YES / NO

   b Concrete is a suitable material for the internal construction of a neutraliser/dilution pit.

      YES / NO

   c Trade waste vents may be interconnected with other sanitary plumbing/drainage vents.

      YES / NO

9 What are two methods of restricting the entry of rainwater and surface stormwater to a trade waste process area?

   a _____

   b _____

10 Provide five things that can be checked when commissioning a trade waste facility.

   a _____

   b _____

   c _____

   d _____

   e _____

11 Research your local authority requirements and state who is responsible for:

   a Testing the quality of trade waste at the outlet of a grease arrestor: _____

   b Maintenance and cleaning of grease arrestors: _____

📋 **WORKSHEET 2**

Student name: _____

Enrolment year: _____

Class code: _____

Competency name/Number: _____

Task: Review the section 'Domestic treatment plants', then complete the following.

1   Name and describe the two categories of waste water.

a   _____

_____

b   _____

_____

2   Explain the basic difference between aerobic and anaerobic treatment.

_____

_____

_____

_____

3   Refer to Table 13.2. A septic tank is a water-based primary treatment system. What are the options for effluent dispersal?

_____

_____

_____

_____

4   Refer to Table 13.2. An aerated waste-water treatment system is a water-based secondary treatment system. On what types of properties can surface irrigation not be used?

_____

_____

_____

_____

5 Obtain an application to install an on-site effluent treatment/dispersal system and any guidelines from your local (council) authority. In some areas, these will be available on the internet. What is the name of your local (council) authority?

_____

6 The following plan represents a house connected to a septic tank and an on-site dispersal system at 12 Field Road, Springfield. Using the application to install an on-site effluent treatment/dispersal system obtained earlier, complete the application form and determine the holding capacity of the septic tank. *Note:* The holding capacity must be determined from your local authority requirements and regulations. The fall on the land is 1:40. *Note:* If septic tanks are not permitted in your area, select another approved system, such as an aerated waste-water treatment system.

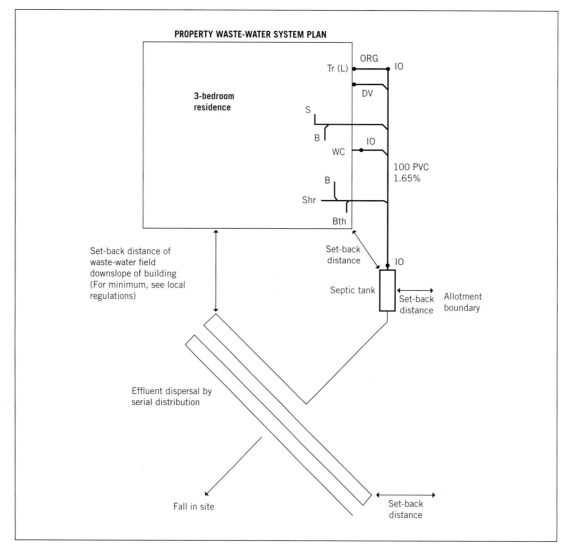

Source: Gary Cook.

a Holding capacity of septic tank/treatment unit: _____

b From your local authority requirements and regulations, what are the minimum buffer/set-back distances between the following: _____

   i Dispersal trench down slope of habitable building: _____

   ii Septic tank and habitable building: _____

   iii Septic tank and allotment boundary: _____

   iv Dispersal trench upslope of allotment boundary: _____

7   In some situations, the material of the septic tank may determine its position on the site. Explain why.

_____

_____

_____

_____

8   Complete the following:

In a septic tank, anaerobic bacteria act on organic matter, breaking it down into
_____, which settles to the bottom of the tank, and _____,
which is formed at the surface.

9   What must be periodically removed from a septic tank?

_____

_____

10 When installing a septic tank, what vertical distance is required between the inlet and outlet?

_____

11 Before installing a septic tank, a base is required. Explain, how a base is prepared.

_____

_____

WORKSHEET 3

Student name: _____

Enrolment year: _____

Class code: _____

Competency name/Number: _____

Task: Review the section 'Effluent disinfection systems', then complete the following.

1 What are two methods used for effluent disinfection?

   a _____

   b _____

2 Name three situations where disinfection of effluent is necessary:

   a _____

   b _____

   c _____

3 One method of effluent disinfection is chlorination. How does chlorination achieve this?

   _____

   _____

   _____

4 Why is an electro-chlorinator not effective when the source of effluent is rainwater?

   _____

   _____

5 Why are high levels of residual chlorine not desirable?

   _____

6 Complete the following:

   After inspecting the site and effluent disinfection system, you will be prepared to maintain the system to ensure that:

   a the system will perform in accordance with _____ and local regulatory requirements.

   b the effluent is thoroughly _____, and the discharge meets the treatment requirements.

   c the effluent can be tested for _____ or other specified measures, and the dosage rates are adjusted to achieve the levels and stability specified by Australian Standards and local regulatory requirements.

7   What is the purpose of taking a grab sample of outflow effluent?

_____

8   What component in UV units must be replaced periodically?

_____

9   When performing maintenance on a UV unit, why is the temperature of effluent measured?

_____

_____

 **WORKSHEET 4**

Student name: _____

Enrolment year: _____

Class code: _____

Competency name/Number: _____

Task: Review the section 'On-site dispersal systems', then complete the following.

1   What are six categories of on-site dispersal systems?

a   _____

b   _____

c   _____

d   _____

e   _____

f   _____

2   What are the two main factors that determine the length of an absorption trench?

a   _____

b   _____

3   Under what circumstances might authorities not permit the use of absorption trenches or transpiration beds?

4   Neatly draw a cross-section of an absorption transpiration bed.

5 Neatly draw an on-site dispersal system on a hillside of a site with a fall of 1:40 from the outlet of a septic tank.

6 List four installation requirements for surface sprinkler irrigation of treated effluent.

a _____

b _____

c _____

d _____

7 List four common requirements for maintenance of aerated waste-water treatment systems (AWTS).

a _____

b _____

c _____

d _____

8 Where surface irrigation is installed, what must not be allowed onto the irrigated area?

_____

9 Complete the following:

Trenches for _____ systems are excavated level.

10 What can cause absorption trenches or transpiration beds to become clogged?

_____

11 Why is it necessary to locate underground services before installing on-site dispersal systems?

_____

12 What type of laboratory is used for testing effluent water?

_____

WORKSHEET 5

Student name: _____

Enrolment year: _____

Class code: _____

Competency name/Number: _____

Task: Review the section 'Clean-up and finalisation of documentation', then complete the following.

1 Why is it important to ensure that access covers are correctly fitted?

_____

_____

2 Research your local (council) authority and sewerage authority requirements for lodging documentation. Circle your responses to the following, and list who is responsible.

Description	Required	Responsibility
a Installation of pre-treatment facilities		
'As-constructed' plans must be lodged	YES / NO	_____
Maintenance records must be lodged	YES / NO	_____
Compliance (or Completion) Certificate must be lodged	YES / NO	_____
b Installation of domestic treatment plants		
'As-constructed' plans must be lodged	YES / NO	_____
Maintenance records must be lodged	YES / NO	_____
Compliance (or Completion) Certificate must be lodged	YES / NO	_____
c Maintenance of effluent disinfection systems		
Maintenance records must be lodged	YES / NO	_____
Compliance (or Completion) Certificate must be lodged	YES / NO	_____
d Installation of on-site dispersal systems		
'As-constructed' plans must be lodged	YES / NO	_____
Maintenance records must be lodged	YES / NO	_____
Compliance (or Completion) Certificate must be lodged	YES / NO	_____

WORKSHEET 6

Student name: _____

Enrolment year: _____

Class code: _____

Competency name/Number: _____

Task: Practical exercises

All exercises in the previous five worksheets must be completed and checked by your trainer/teacher before you attempt the following practical exercises.

The following provides broad guidance on the minimum requirements. Individual trainers and teachers have flexibility to create and combine exercises.

Your trainer/teacher should provide preliminary guidance, conduct an induction, and ask underpinning knowledge questions.

1. Research your local sewerage authority requirements, and install and test two different pre-treatment facilities from a sanitary plumbing or drainage system of a building.

2. Research your local (council) authority requirements, then mark out, plan and install an approved domestic treatment plant.

3. Research your local (council) authority requirements.
   Option 1: Determine the requirements for and install a perforated pipe in an absorption trench, with the trench being at least 3 m in length, running from a distribution pit or outlet of a domestic treatment plant.
   Option 2: Determine the requirements for and install a surface or sub-surface irrigation on-site dispersal system.

4. Given plans and specifications, demonstrate the maintenance procedure for an effluent disinfection system for a domestic treatment plant.

5. Indicate the requirements for storage and handling of any chemicals required, by referring directly to the relevant MSDS.

GLOSSARY

A

Air admittance valves Devices fitted in a sanitary plumbing system which allow air to enter, but prevent escape of air and gases.

Angle of repose The slope at which dumped or excavated soil is naturally stable and does not fall away.

B

Backfill Material used to refill trench.

Bearers Timber laid across a trench, resting on pressure pads on the surface. Lower toms and walers are suspended from upper walers, which in turn are suspended from the bearers.

Boundary trap A trap used to prevent air from the network utility operator sewerage system entering the property drain.

Branch drains Sections of drain that convey or are intended to convey discharges from one or more fixture discharge pipes to the main drain.

C

Capping Timber nailed to toms to help position the tom between the walers.

Certificate of Compliance A certificate that is numbered and in the approved form of the relevant regulatory authority, completed and lodged by the licensed plumber.

Cleat Block of wood nailed to soldiers to locate and support toms.

Closed sheeting Vertical timbers used to fully cover and support a trench wall.

E

Exfiltration The unintended escape of water from any drainage system.

F

Fixture unit A unit of measure that expresses the hydraulic loading imposed by fixtures and appliances, and is used in the sizing of sanitary plumbing and drainage systems.

G

Gibault A fitting used for jointing pipes that incorporates rubber 'o' rings, which are compressed to achieve a seal by bolted slip-on flanges.

Grinder A low-capacity pump for small applications used to shred the sewage into small particles prior to it passing through an impeller.

I

Infiltration The unintended entry of water into any drainage system.

Inlet pit A chamber that provides for the entry of stormwater to a stormwater drain and allows access for inspection and maintenance.

Inspection shaft A shaft connected to a junction in the sanitary drain for the purposes of allowing ready access for inspection and maintenance.

Invert The lowest point in the arc of an internal surface of any pipe or fitting.

L

Lacing Timber used to position and suspend a waler.

Legal point of discharge (LPD) The connection point provided by the network utility operator for the connection of a property stormwater drain to the stormwater system.

M

Macerator A positive displacement pump containing a macerating device to reduce the size of solid matter in waste water.

Main drain The main section of drain on a property sewerage drain which conveys or is intended to convey the discharge of all sanitary fixtures and appliances to the network utility operator sewerage system.

Mutrator A pump with a combined cutter and impeller within the one device.

R

Rising main A pressure pipe connected to the outlet of a pump or pumps through which sewage is transported.

S

Soffit The highest point in the arc of an internal surface of any pipe or fitting.

Soldier Vertical upright timber used for supporting a trench wall.

Spoil pile Excavated material placed beside a trench.

Stormwater pits Chamber constructed or fitted in a stormwater drain for the purpose of providing access for inspection, testing and maintenance.

Submersible pump A pump that is designed to operate while submerged.

Sullage A small centrifugal pump used for raising waste originating from waste fixtures from a lower to a higher level. (Not suitable for handling soil waste from closet pans, etc.)

Surcharge Overflow from a sewer caused by overloading or failure in the network utility operator sewerage system.

Surface pump A mechanical device used for raising fluids from a lower level to a higher level, which is accessible for maintenance at ground level without the need for a ladder or lifting accessories.

T

Timber tongs Tool used for the placing of toms from the surface.

Tom Horizontal timber (strut) used to hold soldiers against a trench wall or to press walers apart in a closed sheeted trench.

Trade waste Water containing waste that is discharged from commercial and industrial premises other than stormwater, sewage or unpolluted water.

Trade waste agreement A written agreement between a network utility operator and business operator for the purpose of defining the acceptable parameters of quality and quantity of waste arising from business, trade or manufacturing processes.

Trench Generally regarded as an excavation that is deeper than 1.5 m, longer than its depth and width, and is for the purpose of laying a pipe or cable.

W

Walers Horizontal member used to keep concrete forms from bulging.

INDEX